Game Theory and Policymaking in Natural Resources and the Environment

Game theory has become one of the main analytical tools for addressing strategic issues in the field of economics and is expanding its influence in other fields of social sciences. With the increased level of extraction of natural resources and pollution of environments, game theory gains its place in the literature and it is seen more and more as a tool for policymakers and not only for theoreticians.

The book is structured in four parts dealing with the management of natural resources, the negotiation aspects of water management, water allocation through pricing and markets, and how conflicts and regulation shape the management of the environment. The first part explores game theory concepts applied to fisheries and grazing lands, which are two important natural resources. In the next two parts, several game theory methodologies are considered in the negotiation approach to water management and approaches to water pricing and markets. The last section looks at environmental protection as the end process of the interplay between conflict and regulation.

This book includes chapters by experts from developing and developed countries that apply game theory to actual issues in natural resources and the environment. As such, the book is extremely useful for graduate students and technical experts interested in the sustainable management of natural resources and the environment. It is also relevant to all game theory and environmental economics students.

Ariel Dinar is a Lead Economist in the Development Research Group of the World Bank and Adjunct Professor at the Johns Hopkins School of Advanced International Studies, Washington, DC, USA.

José Albiac is a Research Fellow at CITA, an agricultural research centre of the Government of Aragón, Zaragoza, Spain.

Joaquín Sánchez-Soriano is an Associate Professor at the Department of Statistics, Mathematics and Computer Science of the University Miguel Hernández, Elche, Spain.

Routledge Explorations in Environmental Economics

Edited by Nick Hanley
University of Stirling, UK

Greenhouse Economics
Value and Ethics
Clive L. Spash

Oil Wealth and the Fate of Tropical Rainforests
Sven Wunder

The Economics of Climate Change
Edited by Anthony D. Owen and Nick Hanley

Alternatives for Environmental Valuation
Edited by Michael Getzner, Clive Spash and Sigrid Stagl

Environmental Sustainability
A Consumption Approach
Raghbendra Jha and K. V. Bhanu Murthy

Cost-Effective Control of Urban Smog
The Significance of the Chicago Cap-and-Trade Approach
Richard F. Kosobud, Houston H. Stokes, Carol D. Tallarico and Brian L. Scott

Ecological Economics and Industrial Ecology
Jakub Kronenberg

Environmental Economics, Experimental Methods
Edited by Todd L. Cherry, Stephan Kroll and Jason F. Shogren

Game Theory and Policymaking in Natural Resources and the Environment
Edited by Ariel Dinar, José Albiac and Joaquín Sánchez-Soriano

Game Theory and Policymaking in Natural Resources and the Environment

Edited by Ariel Dinar, José Albiac, and Joaquín Sánchez-Soriano

Routledge
Taylor & Francis Group
LONDON AND NEW YORK

First published 2008
by Routledge
2 Park Square, Milton Park, Abingdon, Oxon OX14 4RN

Simultaneously published in the USA and Canada
by Routledge
270 Madison Ave, New York, NY 10016

Routledge is an imprint of the Taylor & Francis Group, an informa business

Typeset in Times New Roman by
RefineCatch Limited, Bungay, Suffolk
Printed and bound in Great Britain by
Biddles Digital, King's Lynn

British Library Cataloguing in Publication Data
A catalogue record for this book is available from the British Library

Library of Congress Cataloging in Publication Data
Game theory and policy making in natural resources and the environment / edited by Ariel Dinar, José Albiac and Joaquín Sánchez-Soriano.
p. cm.
Includes bibliographical references and index.
1. Natural resources—Management—Mathematical models. 2. Environmental policy—Mathematical models. 3. Game theory. I. Dinar, Ariel, 1947– II. Albiac, José. III. Sánchez-Soriano, Joaquín.
HC85.G36 2008
333.701′5193—dc22
2007032312

ISBN 10: 0–415–77422–5 (hbk)
ISBN 10: 0–203–93201–3 (ebk)

ISBN 13: 978–0–415–77422–2 (hbk)
ISBN 13: 978–0–203–93201–8 (ebk)

Contents

Illustrations

Figures

Tables

Contributors

Jiro Akita is Professor of Economics at the Graduate School of Economics and Management, Tohoku University. He is currently Vice-Chair of the Department of Economics. His research area is international finance, macroeconomics, and environmental economics. His translated book on environmental economics has become one of the major references in Japan.

José Albiac is Researcher at the Department of Agricultural Economics, Agrifood Research and Technology Center (Government of Aragon, Spain). His main fields of work are water resource management and forest economics.

Stefan Ambec is Research Fellow in Economics at the Grenoble Applied Economics Laboratory, National Institute for Agricultural Research (INRA-GAEL), at the University of Grenoble, France. His main research interests are in environmental economics and natural resources, and industrial organization. His works on river management have been published in the *Journal of Economic Theory* and the *Canadian Journal of Economics*.

Rita De Agostini graduated in Mathematics in 2004 at the University of Eastern Piedmont, with a Masters thesis on water management. Her main interest is in applications of game theory.

Philippe Delacote is PhD Researcher at the European University Institute (Florence, Italy). His field of specialization lies at the frontier of development and environmental economics. His precedent work focused mainly on forest economics, more precisely the use of forest products by poor agricultural households and corruption issues in the context of forest exploitation.

Ariel Dinar is a Lead Economist in the Development Research Group of the World Bank and Adjunct Professor at the Johns Hopkins School of Advanced International Studies, Washington, DC, United States. His work focuses on application of game theory to issues of water and natural resource conflicts at regional and international levels.

Lars Ehlers is Associate Professor in Economics at the Université de Montréal, Canada, a member of the Center for Interuniversity Research in Quantitative Economics (CIREQ), and an elected member of the Council of the Society of Social Choice and Welfare. He is a specialist in matching theory (with applications to entry-level labor markets and public school choice) and he has worked on social choice and axiomatic mechanism design. He has had numerous papers published in the *Journal of Economic Theory* and *Games and Economic Behavior*.

Kyle J. Emerick is a graduate student in the Department of Agricultural and Resource Economics at the University of Arizona. He is also a Graduate Research Assistant with the Climate Assessment for the Southwest project (CLIMAS) of the University of Arizona's Institute for the Study of Planet Earth.

Stefano Farolfi is a CIRAD Senior Researcher in Environmental and Natural Resource Economics. He is Senior Lecturer and Research Fellow at the Centre for Environmental Economics and Policy in Africa (CEEPA), University of Pretoria, South Africa. He is experienced in water management and governance and is currently leading several research projects and is a member of the external advisory board of the Water Research Commission of South Africa.

Franklin M. Fisher is Jane Berkowitz Carlton and Dennis William Carlton Professor of Economics, Emeritus at the Massachusetts Institute of Technology. He is Chair of the Water Economics Project and the senior author of its book, *Liquid Assets*, and of sixteen other books.

Vito Fragnelli is Associate Professor of Game Theory at the University of Eastern Piedmont. He organized the International Meeting on Game Practice and the Environment held in Alessandria in 2002 and edited the proceedings with C. Carraro. He is Scientific Director of the Summer School on Game Theory and Operations Research. He is responsible for or cooperates in several Italian and European research projects. He is author of about twenty papers published in international journals and books.

George B. Frisvold is Professor in the Department of Agricultural and Resource Economics at the University of Arizona. He is also an Investigator with the Climate Assessment for the Southwest project (CLIMAS) of the University of Arizona's Institute for the Study of Planet Earth.

Aurora García-Gallego is Associate Professor and Co-Director at the Laboratori d'Economia Experimental, Universitat Jaume I, Castellón, Spain. She holds a PhD (1994) from the European University Institute, Florence, Italy. She has had articles published in the *Journal of Economic Behavior and Organization*, *International Journal of Industrial Organization*, *Regional Science and Urban Economics*, *International Review of Law and*

Economics, *Environmental and Resource Economics*, and *Ecological Economics*.

Nikolaos Georgantzís is Associate Professor and Director at the Laboratori d'Economia Experimental, Universitat Jaume I, Castellón, Spain. He holds a PhD (1993) from the European University Institute, Florence, Italy. He has had articles published in the *Journal of Economic Behavior and Organization*, *International Journal of Industrial Organization*, *Regional Science and Urban Economics*, *Journal of Economics and Management Strategy*, *Environmental and Resource Economics*, and *Ecological Economics*.

Rachael E. Goodhue is Associate Professor in the Department of Agricultural and Resource Economics at the University of California, Davis. Her research addresses a number of areas, including contracting, industrial organization, agricultural organization, agricultural policy, pesticide use and regulation, and natural resource use policies and property rights.

Annette T. Huber-Lee leads Theme 5: Global and National Food and Water System of the Challenge Program on Water and Food. She is Research Fellow at the International Food Policy Research Institute and Senior Research Fellow at the Stockholm Environment Institute.

Haruo Imai is Professor of Economics at the Institute of Economic Research, Kyoto University. He has served as editor of the *Economic Studies Quarterly* (now called the *Japanese Economic Review*). His research field is in game theory, mathematical economics, microeconomics, and their application in industrial economics and environmental policy. He is the leader of several research projects.

Praveen Kujal is Associate Professor at the Universidad Carlos III de Madrid, Spain. He holds a PhD (1994) from the University of Arizona, United States. His published work includes articles in the *Economic Journal*, *Journal of International Economics*, *Journal of Economic Behavior and Organization*, *International Journal of Industrial Organization*, *Research Policy*, *Economics Letters*, and *Information Economics and Policy*.

Xiaokai Li, of the Institute of Water Sciences, Beijing Normal University, is a water professional and manager of international water resource development investments with wide exposure to transboundary water issues.

Xueyu Lin, of the Institute of Water Sciences, Beijing Normal University, is a renowned academician in applied hydrology and groundwater studies and management.

Marko Lindroos holds a PhD in Economics. He is a Lecturer at the Department of Economics and Management, University of Helsinki. His main research field is game theory and bioeconomic modeling.

Nancy McCarthy is Research Fellow at the International Food Policy Research Institute (IFPRI) and Levy Fellow at George Mason University School of Law. Her research emphasizes property rights and collective management. She joined IFPRI in 1996 to research property rights and risk in livestock management as a joint postdoctoral fellow with IFPRI and the International Livestock Research Institute.

Gordon R. Munro is Professor Emeritus with the Department of Economics and the Fisheries Centre, University of British Columbia, and Visiting Professor with the Centre for the Economics and Management of Aquatic Resources, University of Portsmouth. He has been involved, for over thirty years, in research on fisheries management problems, particularly those arising under the New Law of the Sea. He has published widely on both theoretical and policy issues. Along with research, he has undertaken consulting work for Asia–Pacific Economic Cooperation, the Food and Agriculture Organization of the United Nations, the Organisation for Economic Co-operation and Development, the United Nations Development Programme, and the Department of Fisheries and Oceans Canada.

Hidenori Niizawa is Professor of Environmental Economics in the School of Economics, University of Hyogo. His research area is environmental policy, emissions trading, the Kyoto Protocol, and post-Kyoto Protocol. He is chief editor of the Annual Report of the Society of Environmental Economics and Policy Studies, and is the project leader of some research funds of the Kyoto Protocol.

Fioravante Patrone is Professor of Game Theory at the Faculty of Engineering, University of Genoa, Italy. He is author of more than fifty publications in mathematical analysis and game theory. He has been the promoter of the Game Practice meetings that started in 1998 and of the international summer schools, Game Theory + . He has served as Director of the Interuniversity Centre for Game Theory and Applications for many years.

Pedro Pintassilgo holds a PhD in Natural Resource Economics. He is currently an Assistant Professor at the Faculty of Economics, University of Algarve. His main research field is game theory applied to high seas fisheries.

Gordon C. Rausser is Robert Gordon Sproul Distinguished Professor in the Department of Agricultural and Resource Economics, University of California, Berkeley. He is a leading scholar on the forces driving the dramatic changes under way in agriculture, such as new technologies, globalization, population change, and economic growth. He has developed several highly respected research programs in such areas as public policy for food and agricultural sectors, futures and options markets, political economics of policy reform, environmental and natural resource analysis, and applied econometrics.

Kate Rowntree is Professor of Geography at Rhodes University in Eastern Cape Province of South Africa. She has been engaged in research in water resource management since 1990 and has published on Kenyan, South African, and European water management issues. Since 1997 she and her students have been actively engaged in a project in the Kat Valley in Eastern Cape, supporting the participation by all stakeholders in the development of a catchment management plan. This project has embraced the spirit and intent of South Africa's national Water Act of 1998, which promotes integrated water resource management through stakeholder participation.

Joaquín Sánchez-Soriano is Professor of Statistics and Operations Research at the Faculty of Engineering and Director of the Operations Research Center at University Miguel Hernández of Elche. His current research interests include game theory and its applications, in particular to radio resource management, transportation problems, and market design, and operations research applications to telecommunications problems and water resource management.

Haifeng Shi, of the Department of Planning, Ministry of Water Resources, Beijing, has extensive experience in water resource planning and management policies in China.

Leo K. Simon has research interests ranging from theoretical game theory to simulation modeling to applied political economy. His most recent theoretical work introduces a new class of incomplete information games, called aggregation games. His current work in political economy compares the agrienvironmental policy formation processes in Europe and the United States, and the role played by the World Trade Organization in these two regions.

Sophie Thoyer is a Senior Lecturer at the Department of Agricultural and Resource Economics at the Higher National School of Agronomy (ENSAM), Montpellier. Her main research work focuses on the political economy of agricultural and environmental reforms, with a particular focus on the modeling of negotiation processes. She also works on the comparison between negotiation instruments and market-like instruments, such as auctions, to allocate agrienvironmental contracts to farmers.

Paloma Zapata-Lillo is Associate Professor of Mathematics at the Universidad Nacional Autónoma de México. Her areas of interest are the theories of repeated games and evolutionary games. She has written the books *Fundamentación Matemática de los Algoritmos Simpliciales* (Mathematical Foundation of the Simplicial Algorithms) and *Economía, Política y Otros Juegos: Una Introducción a los Juegos no Cooperativos* (Economics, Politics, and Other Games: An Introduction to Noncooperative Games), the latter to be published soon. A third book, *Los Juegos del Comportamiento Social* (Social Behavior Games) is in preparation.

Technical Editor

John Dawson is a freelance writer and editor based in Nairobi, Kenya. He edits, upgrades, writes, and rewrites documents for a number of international organizations, including the World Bank and the World Health Organization, and attends conferences as a report writer for the United Nations. His two published books, *Quiz Setting Made Easy* and *East Africa Alive*, reflect his interest in mind puzzles and wildlife.

Acknowledgments

This book is one of the products of the Sixth Meeting on Game Theory and Practice Dedicated to Development, Natural Resources and the Environment, held in July 2006 in Zaragoza, Spain. The meeting, one of a series of biennial meetings on game theory and practice commencing in 1998, was co-funded by the Government of Aragon, Spain, and the Spanish Ministry of Education and Science, and was hosted by the Mediterranean Agronomic Institute of Zaragoza (IAMZ-CIHEAM).

Many people and organizations were involved in the process leading to the publication of this book and it is our honor to mention them here.

We would like to acknowledge the efforts of our friends in the Organizing Committee of the meeting, Fioravante Patrone, Rashid Sumaila, and Dunixi Gabiña. We are indebted to the support from the members of the meeting's Scientific Committee, Serdar Güner, Carlo Carraro, Marc Kilgour, Fioravante Patrone, Michael Maschler, Stef Tijs, Rashid Sumaila, Ignacio Garcia-Jurado, H. Peyton Young, Henk Folmer, Gian-Italo Bischi, Vito Fragnelli, Leon A. Petrosjan, and David W.K. Yeung.

The 100 participants of the meeting and the 65 who presented papers benefited greatly from the professional organization by IAMZ-CIHEAM, led by the Institute's deputy director, Dunixi Gabiña, and director, Luis Esteruelas, and supported by the Center for Agrofood Research and Technology (CITA), Government of Aragon.

Special thanks are due to the reviewers of the various chapters: Dirk Engelmann, Victor Galaz, Serdar Güner, Brian H. Hurd, Kieran Kelleher, Pham Do Kim Hang, Van W. Kolpin, Takanoby Kosugi, Raul P. Lejano, Carmen Marchiori, Stefano Moretti, Elinor Ostrom, Wu Xun, and David Yeung, who provided us with guidance and feedback on needed modifications and revisions.

The technical editing of the chapters deserves special mention, mainly because it was not an obvious job. The final product, which all of us are proud of, was meticulously edited in a very careful process led by John Dawson.

Ariel Dinar, José Albiac, and Joaquín Sánchez-Soriano
June 2007

Abbreviations and acronyms

ATO	ambito territoriale ottimale (optimal territorial area)
CDM	clean development mechanism
CDM-EB	clean development mechanism Executive Board
CER	certified emission reduction
CQ	common quota
CUT	Comité de Unidad de Tepoztlán (Unified Tepoztlán Committee)
ERU	emission reduction unit
EU-ETS	European Union Emissions Trading Scheme
FAO	Food and Agriculture Organization of the United Nations
ICCAT	International Commission for the Conservation of Atlantic Tuna
IQ	independent quota
NEIO	new empirical industrial organization
OECD	Organisation for Economic Co-operation and Development
RFMO	regional fisheries management organization
SAGE	schéma d'aménagement et de gestion des eaux
SDAGE	schéma directeur d'aménagement et de gestion des eaux
UNFCCC	United Nations Framework Convention on Climate Change
WAS	water allocation system
WCPFC	Western and Central Pacific Fisheries Commission
WUA	Water Users Association

1 Game theory

A useful approach for policy evaluation in natural resources and the environment

José Albiac, Joaquín Sánchez-Soriano, and Ariel Dinar

This chapter sets the tone for the book by posing the intriguing question of whether or not game theory has developed sufficiently, and the conflicts and problems associated with sharing of natural resources and environmental amenities have worsened enough, that game theory may assist in evaluating various policies aimed to improve their management. The chapter uses several examples from the cases discussed in the book as well as some examples not included in the book to convince the reader that there are a number of opportunities to apply game theory to policy issues in natural resources and the environment. The chapter also highlights the problems that still exist and the aspects that have to be addressed in the interpretation of the results and in the application of the various game theory approaches.

1.1 Background

Since the middle of the twentieth century, principles, concepts, and methodologies originating in the theory of games have been successfully applied to such diverse fields as economics, politics, evolutionary biology, computer science, statistics, mathematics, accounting, social psychology, law, epistemology, and ethics, providing analytical, insightful ideas and explanations to various important problems in each of these fields. Particularly, the significant role of game theory in economics and social sciences has been recognized by the award of the Nobel Prize for Economics to game theorists on two occasions, in 1994 and 2005.

Generally speaking, game theory could potentially be useful in any context where there are two or more agents facing conflict of interest, with the final result of these interactions depending on their (strategic) behavior. Thus, with globalization and openness of societies, and with the increased level of extraction of natural resources and pollution of environments, game theory gains its place in the literature and is increasingly seen as a tool not only for theoreticians but also of actual benefit to policymakers.

Since the industrial revolution, technological advances have triggered a massive increase in production, wealth, and population to levels that appear to be unsustainable. This extraordinary growth in human activities is pressuring natural resources and leading to extensive environmental damage, which in turn threatens the proper functioning of ecosystems and the well-being of many societies that rely on their services.

As population increases and availability of natural resources remains constant or decreases, the potential for conflict over management, extraction, and allocation of basic natural resources such as land, water, and fisheries becomes more likely, and the resultant increase in negative environmental externalities—arising, for example, from the stress put on fisheries, forests, and water—affects individuals, groups, and territories. One important implication is that strategic behavior by individuals and groups becomes more essential if they are to maintain their livelihoods and continue to survive.

What makes water resources a natural candidate for cooperative game theory applications?

"Indeed this is a question that intrigues many scholars involved in this field. First, water-related conflicts involve usually a small number of stakeholders (players) that are interrelated to each other. Therefore, there is a greater scope for strategic behavior among players in water-related conflicts. Second, the level of externalities associated with water utilization is a big incentive to cooperate. Externalities include (a) the zero-sum (or constant-sum) outcomes of unilateral use of the resource (e.g., if party B uses more of the resource in the aquifer, less is left to party A, and also the cost of pumping party A faces is much more substantial due to the depth to the water table), and (b) the negative impact of water quality degradation that party A—the upstream—imposes on party B—the downstream (in the case of a river). Third, the great economies of scale associated with water infrastructure make it more attractive to build bigger rather than smaller water projects, hence providing incentives to joint considerations. Fourth, water projects are in many cases multi-objective ones, leading to inclusion interests. Therefore, a major issue of water projects investment and management is the allocation of the cost and benefits among the various stated project objectives (e.g., sub-sectors, groups of beneficiaries). And fifth, many water problems are transboundary in nature, leading to inter-jurisdictional, interregional, or international conflicts. Since the players and the problems will last, it is likely that cooperation may be attractive for part or all the players." (See additional discussion in Chapters 5, 6, 7, and 10.)

(Dinar et al. 2007)

The compelling reason for the application of game theory to environmental and natural resource problems is that these problems stem from interdependence among agents, through their interrelated actions and strategies. Not only are the outcomes of decisions by agents interrelated, but also individual decisions are often taken without knowledge of the decisions of other agents (see box above for the case of water resources).

The public good aspects of natural resources at local or global scales, and the externalities associated with them, make their management challenging, as there are incentives to free riding. Sustainable management calls for control mechanisms designed to induce collective action and cooperation among stakeholders.

There are many examples of lack of enforcement by authorities or absence of any authority; for example, in the cases of carbon emissions, air quality, water resource quantity and quality (especially for groundwater), and loss of biodiversity and natural capital. And when enforcement is in place, there are also problems of asymmetric information between the regulatory agency and the agents using the resource.

Game theory demonstrates that under noncooperative solutions, each individual agent maximizes its own benefit taking into account that other agents also maximize their individual benefits. Noncooperation is driven by the structure of incentives, theoretical dimensions of which include the prisoner's dilemma game, the so-called tragedy of the commons, and free riding (Axelrod 1984; Hardin 1968).

Cooperative solutions could result from binding agreements with built-in penalties that are enforced by the agents themselves, called self-enforcing agreements. In such a setting, a characteristic function is defined that computes for each coalition of players the total benefits that all members of the coalition can attain by themselves. The best-known noncooperative game solution is the Nash equilibrium of the game, while the full cooperation game solution maximizes the coalition payoff, in which case the agents have to find a reasonable distribution of the additional benefits obtained from cooperation. These points are demonstrated in Figure 1.1, using economic concepts that are applied to a problem of pollution abatement.

Figure 1.1 depicts a situation of pollution abatement where linear marginal benefit and cost functions are assumed. MB_i are marginal benefits and MC_i are marginal costs from pollution abatement by each player i, and MB are total marginal benefits from abatement. Under A^0, the noncooperative state, there is no effort by players on pollution abatement. The noncooperative solution A^{NC} is the Nash equilibrium where players equalize individual marginal benefits MB_i with individual marginal costs MC_i. The level of abatement in the full cooperative solution A^C maximizes welfare and applies the condition for efficient provision of public goods $MB = \Sigma_i MB_i = MC_i$. The specification of the marginal benefit function requires knowledge of biophysical processes and pollution damages to ecosystems. When this information is not available, the optimum level of abatement A^C is not known. In

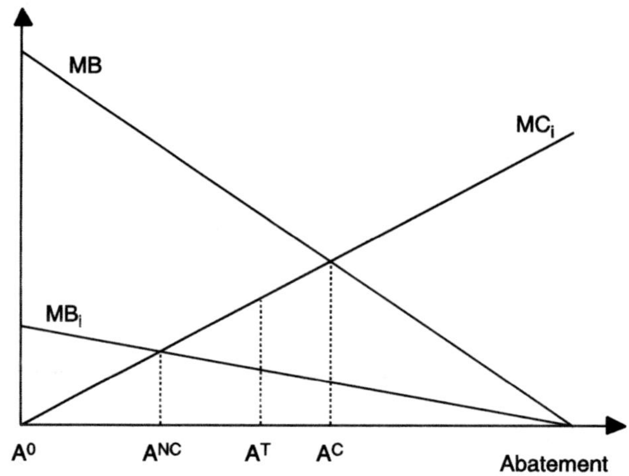

Figure 1.1 Pollution abatement under noncooperative and cooperative solutions.

such a case the alternative is to establish an abatement threshold A^T, where cooperation implies minimizing total abatement costs across players to reach the threshold.[1]

Full cooperation is not usually an equilibrium because of possible free riding by some players, and also because some players may end up being worse off due to the nature of the cooperative solution and their relatively good status quo situation. To make sure that all players improve upon the individual Nash equilibrium solution, gains could be redistributed through side payments. This is the usual outcome of partial cooperation in real-world situations, which involve both cooperating and free-riding players, and where negotiations lead to agreements with built-in incentives that reward or penalize individuals joining or disrupting the agreement.

Several mechanisms have been suggested to redistribute gains among players. In practice, allocation mechanisms are frequently based on equity rules, which are a type of social norm. Percentage reductions in emissions have been applied in the Montreal and Kyoto Protocols, and percentage allocation of resources is embodied in many water resource agreements to reduce water extractions or to share river flows. Other allocation mechanisms include the Shapley value, Nash bargaining, and cost-sharing rules. These will not be explained here.

Cost-sharing rules could be implemented using taxes, tradable permits and lump sum fines or subsidies. In the case of pollution, side payments may contradict the polluter pays principle as pollution abatement may require compensation to polluters, and also anticipated side payments may induce abatement efforts below noncooperation.

1.2 Focus of the book

Given this background, readers who are not specialist economists or game theorists are invited to read the book to gain insight into the usefulness of applying game theory to various real-life issues. The reader of the various chapters of this book will find examples of game theory applications to familiar problems of natural resource and environment management and their policy implications. Among the fields considered are pollution, fishery management, land management, water management (including negotiation and pricing), and environmental management (including conflict and regulation). These applications are distributed over the next fourteen chapters, which were originally presented as papers at a conference dedicated to game theory and practice in natural resources and the environment in Zaragoza, Spain, in 2006.[2] The meeting in Zaragoza was the sixth in a series of biennial meetings on game theory and practice commencing in 1998.[3]

1.3 Components of the book

The book is structured in four parts dealing with the management of natural resources, the negotiation aspects of water management, water allocation through pricing and markets, and how conflicts and regulation shape the management of the environment. The first part explores game theory concepts applied to fisheries and grazing lands, which are two important natural resources. In the next two parts, several game theory methodologies are considered in the negotiation approach to water management and approaches to water pricing and markets. The last part looks at environmental protection as the end process of the interplay between conflict and regulation.

The first part includes three chapters addressing international fisheries and grazing rights. Chapter 2, by Munro, makes an assessment of the relevance of game theory to the management of internationally shared fisheries, presenting the game theory concepts and demonstrating their essential role for a sound understanding of the policy issues and the design of workable measures. Chapter 3, by Goodhue and McCarthy, analyzes pastoralist systems in Africa where several groups with different property rights share the resource. The authors compare the different property rights arrangements (traditional, private, common), indicating the conditions for the strengthening of traditional or private property rights. Chapter 4, by Pintassilgo and Lindroos, analyzes coalition formation in straddling stock fisheries by applying game theory to a fishery represented by the Gordon–Schaefer model and the North Atlantic bluefin tuna fishery. Results show that cooperation agreements break down because of free riding by third-party countries, and enforcement against noncooperation is the key issue in protecting straddling stock fisheries.

The second part deals with negotiation in water management. Chapter 5, by Dinar, Farolfi, Patrone, and Rowntree, compares alternative negotiation (role playing) and mediation (cooperative game) mechanisms in solving water allocation problems in the Kat basin in South Africa. Both negotiation and cooperative game approaches result in similar benefit shares to players from various scenario arrangements, providing complementary valuable guidance for future water allocation decisions. Chapter 6, by Ambec and Ehlers, investigates the river-sharing problem among riparian users. Acceptable outcomes from cooperation require coalition stability and fairness, and they can be implemented through negotiation rules or decentralized water markets. Chapter 7, by Goodhue, Rausser, Simon, and Thoyer, explores water allocation among stakeholders by emphasizing the importance of the structure of the negotiation process on eventual outcomes. The Rausser–Simon bargaining model is used to illustrate these questions in the Adour basin in France, and to reveal basic relationships between negotiation structure and bargaining power. Chapter 8, by Frisvold and Emerick, examines net gains from large rural–urban water transfers, considering such issues as bilateral monopoly, asymmetric capacity and bargaining power, and negative externalities. Results indicate that irrigation technology subsidies may have unintended effects that hinder water conservation, and that while water transfers benefit contracting parties they may also entail negative environmental and third-party externalities.

The third part looks at water pricing and markets as instruments for water management. Chapter 9, by Fisher and Huber-Lee, looks into the resolution of water conflicts by using the water allocation system tool. The empirical results presented show gains from cooperation between Israel, Jordan, and Palestine. The tool can be used also to provide guidelines for infrastructure planning and allocation among sectors, and to solve conflicts in other regions of the world. Chapter 10, by García-Gallego, Georgantzís, and Kujal, investigates the level of centralization and the use of markets to allocate water. Several public and private market structures are simulated in an experimental setting, and results show that centralized public management generates more efficient outcomes in terms of welfare and water quality. Chapter 11, by De Agostini and Fragnelli, explores the design of urban tariff systems in Italy, which comply with principles of fairness and reduction of water wastage. Several water pricing mechanisms are analyzed, taking into account household characteristics and cost-sharing rules, and the implications for water planning by suppliers. Chapter 12, by Li, Shi, and Lin, presents a water allocation model for the Yellow River basin. Different management regimes, such as unregulated withdrawal, quotas, and water markets, are examined by using the Nash–Harsanyi negotiation approach. Large welfare gains are possible by implementing water markets or by enforcing the present flimsy quota regime, although implementation and transaction costs could be quite large.

The fourth part considers the impact of conflict and regulation in shaping environmental management. Chapter 13, by Delacote, analyzes the impact of consumer boycotts using game theory, looking at the war of attrition between consumers and the firm under pressure. The potential for success by environmental boycotts is low because the opportunity cost to consumers seems to be greater than the damage sustained by targeted firms. Chapter 14, by Zapata-Lillo, uses evolutionary game theory to assess the emergence of social awareness for environmental amenities. The analysis shows the conditions and eventual incentives that would strengthen the process of spontaneous collective actions, in order to preserve natural resources and the environment. This knowledge is a key ingredient for the design of sound environmental policies. Chapter 15, by Imai, Akita, and Niizawa, investigates the performance of alternative baseline-setting methods for the clean development mechanism introduced by the Kyoto Protocol. The results demonstrate that the choice of the baseline is important for project evaluation under the clean development mechanism, and may affect the output scale and emissions of firms undertaking such projects.

1.4 Policy messages from the book

If there was a simple way to summarize the possible contribution of game theory to natural resources and the environment, it would be to say that game theory could provide guidance for arrangements that increase the stability of policies aimed at improving management and allocations of the resources among users. The following chapters demonstrate, using many applications of game theory, that it is indeed along these lines that game theory can make tangible contributions.

Game theory not only suggests stable solutions to resource allocation problems but also points to the social loss and volatility of existing arrangements among various agents. The value of awareness raising is not less important than that of solution crafting. As such, the chapters by Munro and by Pintassilgo and Lindroos address global considerations of an international open pool resource—fisheries. International fisheries disputes have been on the agenda of many countries and development agencies attempting to find new and stable regimes. But because of strategic interaction between, and among, those States sharing fishery resources, it is all but impossible to analyze the economics of the management of the resources other than through game theory, which may suggest the necessary conditions for stability of the new regime. Closer examination reveals that the success of the management of straddling fish stocks under existing international agreements is at serious risk unless such agreements contain complementary measures that empower them to combat noncooperative behavior by member States.

Remaining still in the international arena, additional perspectives by Ambec and Ehlers in the case of international water highlight the importance of utility transfers from downstream to upstream countries in order to achieve efficiency and stability of international agreements. The possible transfer arrangements are many (for example water markets, fiscal transfers, sharing rules for benefits and costs, and trade of water for other commodities). The importance of partial coalitions in regional water arrangements is further discussed in the case of more than three countries, where a grand coalition solution might not be manageable. In such a case partial agreements with only few main countries in the basin may provide a reasonably stable agreement.

Providing further support to the importance of partial coalitions, Fisher and Huber-Lee conclude, from an analysis of various possible coalitions in the Jordan River basin, that regional cooperation can benefit the parties in several ways (for example, a joint project for the construction of an effluent treatment plant in Gaza is of potential benefit to both Israel and Palestine). The chapter also highlights an extremely important point for sustainable cooperation as conditions change—development of a flexible tool that allows adjustment of water allocations, with all parties benefiting from the adjustments.

Climate change, an issue of global importance and impact, has been addressed by the Kyoto Protocol. The clean development mechanism (CDM) is among the mechanisms developed under the Kyoto Protocol to reduce levels of greenhouse gases in the atmosphere. Imai, Akita, and Niizawa argue that the CDM baseline level may lead to important outcomes and analyze how alternative CDM baseline-setting methods (for example ex ante, ex post) perform when applied to a CDM project undertaken by a profit-maximizing firm operating in an imperfectly competitive industry. They conclude that it is difficult to draw policy implications from their analysis because of the different policy objectives of the policymakers in each of the nodes comprising the CDM market (host countries, investors, international organizations), and the different outcomes of the ex ante and ex post methods. For example, a firm's marginal cost is lower under the ex post baseline; the ex post baseline results in more pollution credits; price per unit of carbon is lower, and thus consumers are better off, under the ex post baseline; and the net increase in emission is larger under the ex post baseline (though this depends also on the type of market—monopoly or oligopoly).

Sharing resource rights among users within the resource boundaries or transferring resources between resource boundaries are extremely important and timely issues on the agenda of policymakers. Several chapters address these issues and provide relevant policy insights.

Frisvold and Emerick focus on water transfers between sectors or basins. By demonstrating how a game-theoretic analysis of rural–urban water transfers can examine both the gains from trade and the issues of bilateral monopoly, asymmetric capacity, and bargaining power that are likely to affect any agreement, the authors expand the set of policy variables that are relevant for such analysis. First, it is clear from the game theory analysis that, at least in the case of agriculture, pre-existing distortions in agricultural output and input markets should be carefully addressed. Second, it is essential that the analysis includes all necessary physical parameters; otherwise the allocation plan might lead to an unsustainable agreement. The case of transfers in the

US–Mexico border region demonstrates the negative third-party environmental effects that need to be addressed through the adoption of monitoring and management practices, with implications for efficiency and costs in the context of the regional agreement.

In a similar setting, the analysis by Li, Shi, and Lin of the Yellow River allocation in China provides messages that could be of value to policymakers elsewhere. Several options, including unregulated allocation, prior agreed water quotas by subbasins and user types, and clearly defined water rights, cover many of the cases where water allocation is being practiced. The best-performing option, namely clearly defined water rights, maximizes social benefits through water conservation and efficient use. However, there is a risk of increase in actual water use resulting from a water market without proper regulation, leading to a need for minimum ecological flows to be enforced by the regulator. This point is also discussed by Frisvold and Emerick, and in the case of the Kat basin in South Africa in the chapter by Dinar et al.

Additionally, Goodhue and McCarthy (in Chapter 3) focus on grazing systems that are subject to competition among various herders. As in the case of the Yellow River allocation, there are conditions under which it may be desirable to define property rights under an alternative grazing system to increase social welfare. However, in the case of grazing there are indigenous institutions that provide much more stability to the game. It may be the case that it is preferable to strengthen the traditional system, perhaps by reaffirming rights to forage and water and eliminating any expropriation of these resources by specific user groups.

Two chapters introduce negotiation games in the water sector. Negotiation is a form of game where the players follow a set of rules for achieving their interests. Negotiations allow the players to include multi-issues in their objective function. The questions always asked are whether or not negotiations can produce a stable agreement and whether or not they can address all multiple sets of the impact of relevant issues on the agenda of the players.

Goodhue et al. (in Chapter 7) focus on water transfer games, where the value of access to the water source (quota or right) can only be assessed within the context of the composition of the group of negotiators and their preferences. There are broader lessons from the chapter regarding the design of water allocation negotiation processes. When defining an interest group and identifying stakeholders to participate in a specific negotiation, policymakers should seek to define key objectives and distinguishing characteristics of specific stakeholders, perhaps through a prenegotiation process whereby the participants share information. Another important factor affecting the game outcome is the stochastic nature of the physical system. Under such conditions, the major natural situations have to be kept separated in order to allow a meaningful outcome.

Questions may be asked as to whether or not the negotiation process is too lengthy or too costly. Dinar et al. (in Chapter 5) compare cooperative game theory and role-playing game solutions related to water allocations and payoff distributions. They suggest that such comparison is useful for policy purposes as it allows assessment of the nature of the assumptions and, if needed, a revisitation of them. A particularly important policy implication could be derived for the role local representative groups may undertake with regards to fairness and environmental sustainability in decentralization of management of river basins.

Another aspect of water allocation and efficient use has been addressed via market games and pricing games. The water market mechanism has been criticized for leading to unfair outcomes because of some agents' greater economic or political power over other less strategic or weaker players. García-Gallego, Georgantzís, and Kujal, applying experimental economics to a decentralized water system, argue that in a system where agents can learn from past actions, the decisionmakers' incentives dominate possible efficiency losses because of an increased complexity of the underlying system managed separately by each type of agent. The policy implication is that where the market mechanism is vulnerable to private agents' anticompetitive strategies when a market mechanism is to be implemented to allocate water resources, competition among decentralized private owners should be safeguarded and promoted by the regulator.

De Agostini and Fragnelli argue that a declarative tariff system can lead to water savings by introducing penalties for overuse and prizes for underused quota. An important point that could be relevant for policymakers that introduce pricing is the inequitable impact that a pricing system can have on small users and large users, if the latter are small in number. Addressing the fairness issue through various game theory allocation approaches (for example bankruptcy, Shapley value, and Owen value) provides several rules for computing the tariff for each user and keeping the water provider's budget covered.

When dealing with strategic behavior, the environment and natural resources are issues that often generate organized responses from groups that are affected by policies or unilateral behavior.

Delacote explores the conditions under which a consumer boycott upon environmental considerations may be successful. One policy implication for consumer groups is that the ability to affect the behavior of producers depends heavily on the share represented by the boycotting group in total demand. But large consumers usually have high boycotting costs and thus are less likely to participate, leading to less effective or failed impact. Environmental impact-improving policies may include awareness building, informing, and educating, all of which can induce a decrease in overall polluting consumption, which would in turn reduce environmental degradation, and increase the share of population more likely to participate in the boycotting. The game theory model suggests that while environmental policies and consumer boycotts may not be good substitutes, they could be effective complements.

Considering the impact of environmental degradation on indigenous peoples in Mexico, Zapata-Lillo, using an evolutionary game approach, shows that sufficiently large human conglomerates may act both by forcing themselves to change their own environment-damaging activities, and by preventing those of the large corporations and State powers. Evolutionary games may be useful in comparing policies and evaluating how some may endanger the existence of many communities important to the protection of ecology and the environment.

1.5 Conclusion

This book deals with several issues related to natural resources and the environment, including use of such resources as fisheries, climate change, water transfer and management, land allocation and management, and citizens' responses to environmental disputes with governments and corporations. The different game-theoretic approaches developed and applied to the case studies in the individual chapters provide meaningful messages that can be of use to parties involved in each of the cases and to policymakers in general.

The aim of this book, then, is to demonstrate to policymakers, in an understandable form, the potential applications of game theory to issues of direct relevance to them. As Munro says in Chapter 2, after observing that game theory concepts are, as yet, poorly understood by policymakers: "What this requires of economists is that they become effective expositors, taking the results of their game theory analysis and expressing these results in a form that can be readily understood and appreciated by the practitioners."

Notes

1. See Perman et al. 2003 and Hanley and Folmer 1998 for additional details.
2. The Sixth Meeting on Game Theory and Practice Dedicated to Development, Natural Resources and the Environment was funded by the government of Aragon and the Spanish Ministry of Education and Science, and was hosted by the Mediterranean Agronomic Institute of Zaragoza (CIHEAM).
3. Genoa, Italy (1998), Valencia, Spain (2000), Hilvareenbeek, the Netherlands (2002), and Elche, Spain (2004) were general in nature; another meeting focusing on game practice and the environment was held in Alessandria, Italy (2002).

References

Axelrod, R. (1984) *The Evolution of Cooperation*, New York: Basic Books.

Dinar, A., Dinar, S., McCaffrey, S., and McKinney, D. (2007) *Bridges over Water: Understanding Transboundary Water Conflicts Negotiation and Cooperation*, New Jersey: World Scientific.

Hanley, N. and Folmer, H. (1998) *Game Theory and the Environment*, Cheltenham: Edward Elgar.

Hardin, G. (1968) "The tragedy of the commons", *Science*, 162: 1243–8.

Perman, R., Ma, Y., McGilvray, J., and Common, M. (2003) *Natural Resource and Environmental Economics*, Harlow: Pearson Education.

2 Game theory and the development of resource management policy

The case of international fisheries

Gordon R. Munro[1]

This chapter is concerned with the relevance, if any, of game theory to a major resource management issue, namely the management of internationally shared fishery resources. It is argued that the economics of the management of such resources cannot be understood other than through the lens of game theory. Several elementary game theory concepts are discussed that are of utmost policy relevance, but which are, as of yet, poorly understood by most policymakers. In addition, the chapter discusses a key policy problem in the management of shared fishery resources that demands a game-theoretic analysis. The required analysis, however, has yet to be developed.

2.1 Introduction

This chapter considers the impact, actual and potential, of game theory upon policymakers concerned with a major fishery resource management issue. The issue is that of the management of what the Food and Agriculture Organization of the United Nations (FAO) terms "internationally shared fish stocks" (FAO 2002), which can be defined as fish stocks, not confined to a single national jurisdiction, that are exploited by two or more States (FAO 2002; 2003: 7.1.3).

While there are freshwater shared fish stocks, the management issue will be examined solely within the context of marine capture fisheries.[2] These fisheries, worldwide, have annual harvests in the order of 80 million tonnes, with a first-sale value of approximately US$70 billion, and are estimated to provide employment (direct and indirect) for as many as 200 million persons (Garcia and Newton 1997).[3] It is estimated that, of the world capture fishery harvests, up to one third can be accounted for by internationally shared fish stocks (Munro et al. 2004).

The issue of internationally shared fish stock management arose out of the Third United Nations Conference on the Law of the Sea, 1973–82. This conference, which was to have a revolutionary impact on the management of world marine capture fisheries,[4] brought forth the 1982 United Nations Convention on the Law of the Sea (1982 Convention, hereafter).

The 1982 Convention achieved the status of international treaty law in 1994, and has come to serve as the bedrock source of the "rules of the game" for the management of world marine capture fishery resources (United Nations 1982).

Under the 1982 Convention, coastal States are given the right to establish 200-nautical-mile exclusive economic zones off their coasts. To all intents and purposes, coastal States have property rights to the fishery resources encompassed by the exclusive economic zones (McRae and Munro 1989). It was estimated, at the time of the Third Conference on the Law of the Sea, that the exclusive economic zones, if spread throughout the world, would encompass fishery resources accounting for 90 percent of the world's marine capture fishery harvests (Alexander and Hodgson 1975).

Capture fishery resources are, with few exceptions, mobile. Consequently, it was recognized during the 1973–82 Conference that the typical coastal State would find that it was sharing some of the fishery resources of its exclusive economic zone with neighboring coastal States, or with so-called distant water fishing States, operating in the remaining high seas adjacent to the exclusive economic zone.

Since the close of that conference, it has come to be recognized that the management of internationally shared fishery resources is one of the most significant resource management issues to have arisen under the exclusive economic zone regime. The FAO categorizes these internationally shared fish stocks as follows:

- A Transboundary fish stocks: fishery resources that are to be found in two or more neighboring exclusive economic zones
- B Straddling fish stocks (broadly defined): fish stocks that are to be found both within the exclusive economic zone and the adjacent high seas[5]
- C Discrete high seas fish stocks: fish stocks that are confined to the remaining high seas (FAO 2003, 7.1.3; Munro et al. 2004)[6]

Categories A and B, it must be emphasized, are not mutually exclusive. There are many examples of transboundary fish stocks that also cross the exclusive economic zone boundary into the adjacent high seas.

There are several elementary game theory concepts that have significant, if not profound, policy consequences in the real world of fisheries management. These concepts are, at the time of writing, only poorly understood by most policy practitioners, although there are hopeful signs that this is beginning to change. There is, however, an emerging issue in the management of shared fishery resources that has yet to be addressed adequately by game theorists.

The discussion commences with two background sections designed to provide an outline of the relevant resource management issues. The two sections also incorporate a brief review of the development of the economic analysis designed to address these issues. The background sections are then followed by specific applications of game theory concepts to the policy issue at hand.

2.2 Review of the management of internationally shared fishery resources: stage one—transboundary fish stocks

The background survey begins with a discussion of the relatively straightforward case of transboundary fish stock management. This is followed by an examination of the more complex case of straddling fish stocks.

Economists view capture fishery resources, as they do all natural resources, as a form of natural capital, assets that are capable of yielding a stream of economic returns (broadly defined) through time. Since fishery resources are capable of growth (like forests, but unlike minerals), these resources—natural capital—can be managed on a sustained basis, essentially by skimming off the growth through harvesting. This also means that the resources can provide economic benefits to society indefinitely. It means further that one can, within limits, engage in positive investment in the natural capital by harvesting less than the growth.

World capture fishery resources have, however, been seen historically as the quintessential common pool resource, in that, in the past at least, it was deemed to be too costly to put in place effective property rights to the resource. The common pool aspects of the resource have been seen to lead to resource overexploitation and economic waste.

Economists have traditionally analyzed management of capture fishery resources confined to the waters of a single State by contrasting management of a fishery by an all-powerful social manager—the ideal, the target—with a pure, open-access, common pool fishery, in which the resource is exploited by a very large number of fishers, and in which there is a complete absence of resource property rights or government regulations. The models developed are referred to as bioeconomic models, because they represent an explicit fusion of marine biology and economics.

In any event, the all-powerful social manager is seen as managing the portfolio of natural capital assets, in the form of the capture fishery resources, in such a manner as to maximize the economic returns from the resources—loosely referred to as resource rent—to society through time. In the contrasting case of an open-access, common pool fishery, no fisher has any incentive to invest in the resources. No rational investor will incur the cost of investment (forgoing harvests today), unless a future return from the investment, exceeding the cost of the investment, could be expected. No fisher, in these circumstances, could expect such a return on a resource investment. Any fisher who refrains from harvesting is likely to do no more than increase the harvests of its competitor. The fishers will, in fact, have an incentive to engage in extensive resource disinvestment. The economic theory of fisheries management demonstrates that, in these circumstances, the fishery resources will be driven down to the point that the economic rent forthcoming from the fishery will have been eliminated. One of the pioneers of modern fisheries economics, H. Scott Gordon, characterized the resultant economic rent-free equilibrium as "bionomic equilibrium" (Gordon 1954).

There is no question that, at bionomic equilibrium, there will, from society's point of view, have been excessive disinvestment in the fisheries' natural capital (overexploitation) (Clark 1990; Bjørndal et al. 2000). Bionomic equilibrium is to be seen as a benchmark of resource management undesirability.

While the overexploitation of particular local fisheries has been a matter of concern for many centuries, the overexploitation of the great ocean fishery resources was not a concern until the first half of the twentieth century, because these resources were seen as being inexhaustible. Perhaps pure open access would lead to the loss of some economic benefits, but at least the resources were safe from dangerous overexploitation, as seen from a biological standpoint. Attempting to regulate such fisheries was hardly worth the effort, or so it was argued.

This point of view found its clearest expression in the doctrine of the freedom of the seas, as propounded by the seventeenth-century Dutch jurist, Hugo Grotius, in his volume *Mare Liberum* ("The Free Sea"). Under this doctrine, the oceans are classed either as the territorial sea of coastal States or (the remainder) as the high seas. The territorial sea is a narrow strip of water, by tradition no wider than 3 nautical miles.[7] The resources in the high seas are deemed *res communis*, the property of all, and thus open to exploitation by all.

The doctrine of the freedom of the seas, as it pertained to fisheries, rested upon two fundamental premises:

- The high seas fishery resources are inexhaustible.
- Coastal States are unable to control effectively resource exploitation activities beyond their territorial seas (Orrego Vicuña 1999).

These premises were defensible in Grotius' day. Given the then existing state of fishing technology, the high seas fishery resources were, to all intents and purposes, safe from (biological) overexploitation. It was too costly (not to say dangerous) to exploit them extensively (Orrego Vicuña 1999).

Since the seventeenth century, and in particular since the late nineteenth century, both of these premises have become increasingly untenable. This decline in tenability has, in turn, led to a steady erosion of the freedom of the seas, as it pertains to fisheries. The erosion is not complete, however. A residue persists that continues to exacerbate the difficulty of managing international fishery resources.

Rapid advances in fishing technology, for example the shift from sail to steam, reduced harvesting costs and thereby increased the vulnerability of ocean fishery resources. By the end of World War II it was apparent that the great ocean fishery resources were anything but inexhaustible. The erosion of the freedom of the seas doctrine appeared first in the form of international conventions designed to put restrictions on fishing activities in certain segments of the high seas. The International Commission for the Northwest

Atlantic Fisheries, 1949–77, which attempted to impose some management rules over Atlantic high seas fisheries off North America, from Greenland to the Carolinas in the United States, provides one example.

Following the end of World War II, several coastal States attempted, unilaterally, to extend their jurisdiction over seabed resources beyond their territorial seas. In order to prevent a chaotic extension of coastal State marine jurisdiction, the United Nations convened a series of Conferences on the Law of the Sea. The First and Second Conferences did little to address fisheries issues. The Third Conference (1973–82), as has been seen, revolutionized marine capture fisheries management, and led, through the establishment of the exclusive economic zone regime, to a massive erosion of the freedom of the seas, as it related to fisheries. With only 10 percent of capture fishery harvests being accounted for by fishery resources in the remaining high seas, the freedom of the seas seemed, as far as fisheries were concerned, to be all but irrelevant in 1982.

As noted, negotiators in the Third United Nations Conference on the Law of the Sea, and outside observers, were quick to discern that the coming exclusive economic zone regime would carry with it the problem of management of shared fishery resources. Since fishery resources lying in the remaining high seas were seen to be of minor importance, it seemed obvious that the only shared fishery resources worthy of serious consideration were category A fish stocks—transboundary stocks occurring in neighboring exclusive economic zones. Thus, it is no surprise that the development of the economics of shared fish stock management commenced with a focus solely on transboundary stocks.

When economists first began to write on this issue in the mid-1970s, the articles forthcoming were generally not clearly argued and as a result lacked policy value. There was a simple reason for this. The articles failed to recognize the fact that there will, except in unusual circumstances, be a strategic interaction between, and among, States sharing a fishery resource. To take an elementary example, consider two coastal States, A and B, sharing a transboundary fishery resource. The harvesting activities of A will (except under unusual circumstances) have an impact upon the harvesting opportunities of B, and vice versa—hence the strategic interaction.

Research into the economics of shared fish stock management made no real progress until this strategic interaction was recognized explicitly. This meant that economists studying this issue had to draw upon the theory of strategic interaction, more commonly known as the theory of games. Economists cannot analyze the economics of the management of internationally shared fishery resources, with the hope of providing useful insights to policymakers, other than through the lens of game theory.

The lesson has been learned. Economists' models of shared fish stock management are now blends of their bioeconomic models, used to analyze the economics of the management of fishery resources confined to the exclusive economic zone of a single State, and game theory.

With this in mind, economists, approaching the issue of the management of transboundary fish stocks, have to address two questions. These are:

- What are the consequences of coastal States sharing such a resource managing the resource noncooperatively?
- What conditions must be met if a cooperative fisheries management arrangement is to be stable over the long run?

Needless to say, if the answer to the first question is that the negative consequences of noncooperative resource management are negligible, then the second question becomes of no interest.

With regards to the first question, let it first be noted that if neighboring coastal States sharing a transboundary fish stock attempt to manage the resource noncooperatively they are not necessarily in violation of the 1982 Convention, which specifies that coastal States "shall seek . . . to agree upon the measures necessary to coordinate and ensure the conservation and development of such stocks" (United Nations 1982: Article 63(1)). Importantly, however, the coastal States are not required to reach an agreement. If the States negotiate in good faith, but are unable to reach an agreement, then each coastal State is to manage its segment of the resource independently in accordance with the other provisions of the 1982 Convention (United Nations 1982; Munro et al. 2004). Hence, noncooperative management of the transboundary fish stock is countenanced under the 1982 Convention.

The first question is addressed, appropriately, by drawing upon the theory of noncooperative games, with the model of Nash (1951) being the most popular among economists. The question was first examined in 1980 in two articles appearing almost simultaneously, by Clark (1980) and Levhari and Mirman (1980). Both come to essentially the same conclusion, namely that one can anticipate a prisoner's dilemma type of outcome, in which the players (the coastal States) will be driven to adopt policies that will lead to overexploitation of the resource.[8] Clark goes so far as to argue that if the coastal States are symmetrical (identical in all relevant respects) the outcome will be comparable to the bionomic equilibrium to be found in open-access fisheries confined to a single exclusive economic zone (Clark 1980).[9]

The basic nature of the prisoner's dilemma outcome, in a fisheries context, can be illustrated as follows. Consider a transboundary fishery resource shared by two coastal States, A and B, and suppose further that there is no significant resource management cooperation between the two. A and B manage their respective segments of the resource on their own, in accordance with the provisions of the 1982 Convention. If A were to undertake to restrict harvests in order to invest in the resource, the benefits from this action would not be enjoyed by A alone, but would rather be shared with B. What assurance would A have that B would also undertake to conserve the resource? Since there is no cooperation, the answer is none. It is only too possible that B would be content to be a free rider taking advantage of A's resource investment efforts. In these circumstances, it is likely that A will conclude that the return on its resource investment would be less than the cost, and that its best course of action ("strategy") would be to do nothing. B could be expected to come to the same conclusion.

Worse, A has to allow for the possibility that B might deliberately deplete the resource. If A seriously believes this, then it could decide that its best strategy is to strike first. Once again, B could follow the same line of reasoning.

Thus one can conclude that a failure on the part of neighboring coastal States to cooperate can have severe consequences. To those versed in game theory, the conclusion is apparent. However, it will be argued that, in the world of policymakers, while all policymakers acknowledge that cooperation is advantageous, there is not full recognition of the possible consequences of noncooperation.

In analyzing cooperative resource management arrangements, economists naturally draw upon the theory of cooperative games, with the model of Nash (1953) once again being the most popular. The number involved in a typical transboundary fishery arrangement is small, so that considerable progress can be made with simple two-player models (Munro 1979).

In the cooperative management of transboundary fish stocks, a key issue is that of allocation among cooperating coastal States. It is not the only issue, however. The cooperative management arrangement must, in addition, deal with the issue of the optimal management strategy through time. There is no guarantee that the players will have identical management goals. A third, and critical, issue is that of implementation and enforcement of the cooperative arrangement.

The simple two-player cooperative game models bring to light two fundamental conditions that must be met if the cooperative resource management arrangement is to be stable. They also bring forth one aspect of cooperative management that can serve to increase the likelihood of stability being achieved. The first condition is straightforward, and easily described. The solution to the cooperative game—the cooperative management agreement—must be collectively rational, in the sense that there does not exist another agreement that could make one player better off without harming the others.[10]

The second condition is obvious, but is often ignored in practice, as shall be illustrated at a later point. It is that the solution must be individually rational, in the sense that each and every player has to be assured of receiving a return, a payoff, from the cooperative arrangement at least as great as it would receive under noncooperation. This assurance has to last throughout the life of the arrangement. In game-theoretic terms, these minimum payoffs are referred to as the threat point payoffs, and are normally assumed to be those arising from the solution to a noncooperative game.

This second fundamental condition is closely linked to the third issue with which a cooperative fisheries management arrangement must deal, namely implementation and enforcement. Assume that the allocations among the players, the cooperating coastal States, are seen by all to be fair, but suppose further that at least one player has no confidence in the enforcement mechanisms. That player, supposing that there will be no effective checks on cheating, may quickly conclude that its actual payoff under the cooperative arrangement will be less than what it would receive under noncooperation. Acting rationally, the player could be expected to refuse to enter into the cooperative arrangement.

The aspect of cooperative games that can increase the likelihood of a stable cooperative resource management arrangement arising involves side payments, which are transfers that can take many forms. A cooperative transboundary fish stock game without side payments can be defined as one in which the economic returns to coastal State A from the cooperative arrangement will be determined solely by the harvest taken by coastal State A's fleet within that State's exclusive economic zone. What holds true for coastal State A holds true for all other coastal States in the cooperative arrangement.

There are at least two advantages of side payments. First, the existence of side payments shifts the allocation focus from the sharing of the harvest to the sharing of the economic returns from the fishery. The two are not necessarily the same. Consider the case of coastal States sharing the resource having different management goals. As Munro (1987) remarked, this difference in management goals more often than not reflects a difference in the valuation that the States place on the resource. In order to maximize the economic returns from the resource, it is necessary to allow the management preferences of the State placing the greatest value on the resource to hold sway. In order, in turn, for this to happen, that State will have to be prepared to compensate its partner, or partners, which involves the use of side payments.[11]

A current real-world example involves different groups of users of a resource, rather than States per se. The Atlantic salmon, which extends from the Atlantic coast of North America to Western Europe, has suffered a precipitous decline as a consequence of ineffective cooperative management. Among the users of the resource are commercial fishers of several States, anglers in the rivers of North America and Europe, and conservation groups, which place a value on the continued existence of the salmon. The anglers and conservationists place a substantially greater value on the salmon than do the commercial fishers. Through a private sector initiative, the North Atlantic Salmon Fund was established in 1989. The fund is supported by anglers and conservation groups, but also obtains assistance from governments. Through its fund-raising activities, the North Atlantic Salmon Fund has been able to buy up quotas issued to groups of commercial Atlantic salmon fishers in several States, with very beneficial consequences for the resource.[12] The buying up of the quota is a side payment exercise, pure and simple.[13]

The second advantage of side payments is that they increase the scope for bargaining. This helps to ensure (but does not guarantee) that the individual rationality condition will be met for all players.

The early articles on cooperative management of transboundary fish stocks (for example Munro 1979) made several controversial assumptions. One of these assumptions attracted a host of critics, namely the assumption that cooperative management agreements, upon being achieved, are binding through time. These critics were then faced with the task of attempting to show how one establishes stable cooperative management agreements when they are not binding through time. The most successful attempt came from a group of economists based in Helsinki, the most prominent member of which was Veijo Kaitala (see, for example, Kaitala 1985).

The main threat to stability in a nonbinding cooperative resource management agreement is that of cheating. Kaitala demonstrates that the problem can be addressed with credible mutual threat schemes (Kaitala 1985).

The Helsinki group did much more than address the problem of possible cheating, however. It brought to light a problem that has a high degree of relevance in the real world of policymaking. The problem arises from the fact that shifting underlying conditions through time can have an impact on the relative bargaining power of the players.

The problem was addressed by Kaitala and Pohjola (1988), who consider a simple two-player game in which the cooperative agreement between them is nonbinding. Cooperation begins after the resource has been heavily exploited. The two players differ, but only in terms of costs of fishing. Side payments are given full sway, with the result that the low-cost player effectively buys out its high-cost partner. A credible mutual threat scheme is in place. The high-cost player's bargaining power is low at the beginning of the arrangement, but will change as the resource is rebuilt under cooperative management. Essentially, the high-cost player's threat point payoff will increase as the resource rebuilds. Kaitala and Pohjola demonstrate that, even with a mutual threat scheme in place, it could pay the high-cost player to breach the agreement after the stock is rebuilt. To prevent this from occurring, the low-cost player must, when putting a side payments scheme into effect at the beginning of the cooperative resource management program, recognize that its partner's bargaining power will increase over time (Kaitala and Pohjola 1988).

The Kaitala–Pohjola model is deterministic, in the sense that the future can be predicted accurately. Hence, the low-cost player can plan accordingly. As shall be pointed out, experience from the real world shows that shifts affecting relative bargaining power can often be no more than anticipated. They cannot be predicted with any accuracy, in terms of either timing or direction of effect. Real-world experience also demonstrates that such shifts can readily destabilize cooperative resource management agreements that are ostensibly binding in nature.

2.3 Review of the management of internationally shared fishery resources: stage two—straddling fish stocks

The anticipation, throughout the Third United Nations Conference on the Law of the Sea, that fishery resources in the remaining high seas beyond the exclusive economic zones would be of minor importance, proved to be dramatically wrong. While discrete high seas stocks have yet to achieve great importance, there was, following the close of the conference, extensive exploitation of the high seas segments of straddling stocks, which undermined coastal State attempts to manage those stocks found within exclusive economic zones.[14]

The high seas segments of straddling fish stocks, which are subject to exploitation by both coastal States and distant water fishing States, are covered in Part VII: High Seas of the 1982 Convention (see Articles 87 and 116–120, in particular). The articles related to fisheries in Part VII are very unclear regarding the rights, duties, and obligations of coastal States on the one hand, and those of distant water fishing States on the other, with respect to the high seas segments of straddling fish stocks (Munro 2000).

This lack of clarity arose, presumably, for two reasons. The first was the apparent unimportance of fishery resources in the remaining high seas, which has already been noted. The second was that the drafters of the 1982 Convention had a difficult balancing act to perform. The principle of the freedom of the high seas, including freedom to fish, is set forth in Article 87 of the Convention (United Nations 1982). This freedom had to be balanced against the interests of coastal States in managing the exclusive economic zone segments of the straddling stocks (United Nations 1982: Article 116; Munro et al. 2004).

The lack of clarity of the fisheries articles of Part VII of the 1982 Convention made it difficult to effect cooperative management of the resources. The experience of the 1980s, following the conclusion of the Third United Nations Conference on the Law of the Sea, and the early 1990s, demonstrated that the economist's model of noncooperative management of transboundary stocks applied, without modification, to straddling stocks. Case after case of overexploitation of such resources emerged.

An example is provided by Alaska pollock, the species that has historically yielded the largest harvest in the North Pacific. Large concentrations are to be found in the Bering Sea. A significant part of the resource straddles the American zone and a high seas enclave between the American and Russian zones, referred to as the "Doughnut Hole". The management of the straddling stock was noncooperative. The pollock resources in the Doughnut Hole were not just overexploited but, in the words of the FAO, plundered (FAO 1994). As Munro et al. (2004: 45) remark, "the overexploitation of straddling/highly migratory fish stocks worldwide . . . bears powerful testimony to the predictive powers of the economic analysis of the noncooperative management of such resources".

The growing concern over the state of world straddling fish stocks led the United Nations to convene an international conference to address the issue—the United Nations Conference on Straddling Fish Stocks and Highly Migratory Fish Stocks (1993–95)—which adopted in 1995 what is commonly referred to as the United Nations Fish Stocks Agreement.[15] The agreement, which achieved the status of international treaty law in late 2001, is not meant to replace any part of the 1982 Convention, but is rather designed to buttress the Convention (Bjørndal and Munro 2003).[16]

Under the terms of the United Nations Fish Stocks Agreement, straddling stocks (broadly defined) are to be managed on a region-by-region basis through regional fisheries management organizations (RFMOs). The RFMOs are to have as members both coastal States and distant water fishing States. Examples are provided by the Northwest Atlantic Fisheries Organization, the Northeast Atlantic Fisheries Commission, and the Western Central Pacific Fisheries Convention.

The question then becomes, to what extent do the economic game theory models developed for transboundary fish stocks have to be modified when dealing with straddling fish stocks? One part of this question has already been answered. The model of noncooperative management of transboundary stocks can be applied without modification to straddling stocks (Sumaila 1999).

When one turns to the cooperative management of the resources, the answer is quite different. The economic game theory model of cooperative management of transboundary stocks requires substantial modification when the issue of cooperative management of straddling stocks is confronted. First, one can anticipate that the number of players in the typical straddling stock game will be large. In the analysis of transboundary fish stock management, considerable progress can be made using two-player models. Two-player models are simply inadequate for straddling stocks. Economists are compelled to employ models in which the number of players exceeds two, often by a wide margin. This, in turn, means that they have to allow for the possibility that players will form themselves into subcoalitions. The coalition of all the players together in a fisheries game is referred to as the grand coalition.

With subcoalitions possible, it is no longer sufficient to be concerned about the individual rationality condition being satisfied. For the solution to the cooperative game to be stable through time, the solution must also be such that no subcoalition believes that it would be better off on its own, playing competitively against the remaining members of the grand coalition.

Second, in contrast to transboundary stock management, the number and nature of the players cannot be expected to be constant through time. Some members of the RFMO are distant water fishing States, the fleets of which are very mobile. An original, or "charter", member of an RFMO may withdraw. More importantly, a distant water fishing State, hitherto not a member of the RFMO, may apply for membership. The United Nations Fish Stocks Agreement makes it clear that the charter members of an RFMO cannot bar prospective new members outright (United Nations 1995: Articles 8, 10, and 11; Munro et al. 2004). This gives rise to the so-called new member problem (Kaitala and Munro 1993).

The third difference falls under the heading of free riding, which can be defined as enjoyment of the fruits of cooperation by nonparticipants in the cooperative management arrangement. Let it be conceded that the boundary between free riding and noncompliance may be fuzzy.

Be that as it may, while Munro et al. (2004) maintain that free riding is theoretically possible in the management of transboundary fish stocks, they are hard pressed to come up with any real-world examples. Free riding, by contrast, has historically been a very serious problem in the management of straddling stocks.

The difference arises largely because of the distinction between illegal and unregulated fishing. According to the FAO, illegal fishing involves fishing by one State, or entity, in the exclusive economic zone of another State without the latter's permission, or willful noncompliance with the management provisions of an RFMO by a member of the RFMO (FAO 2001: 3.1). Thus, if a nonmember free rides by fishing without permission in the exclusive economic zone of a coastal State member of the RFMO, the nonmember is engaging in poaching, and the affected coastal State can take vigorous action.

On the other hand, if vessels flying the flag of a nonmember of an RFMO fish in the high seas portion of the area governed by the RFMO in a manner inconsistent with the management provisions of the RFMO, such vessels are deemed to be engaging in unregulated fishing (FAO 2001: 3.3.1). Unregulated fishing is a much vaguer concept than illegal fishing and reflects the influence of the lingering freedom of the seas doctrine. While unregulated fishing is deemed to be morally reprehensible, it has, in the past, been unclear what RFMO members can to do curb such activities. If nonmembers can engage in such unregulated fishing with impunity, then the disincentive for nonmembers to free ride by fishing in the high seas under RFMO governance will be weak indeed.

Measures are now being taken to deal with the problem of unregulated fishing. As shall be related, the United Nations, through the FAO, and the High Seas Task Force based at the Organisation for Economic Co-operation and Development (OECD), are actively attempting to address the problem (FAO 2001; OECD 2006).

Applied game theorists, using what is known as a coalitional bargaining approach, have addressed the free-riding problem in straddling stock management (Pintassilgo 2003; Pintassilgo and Lindroos 2008). The fundamental concept of stand-alone stability is introduced. The grand coalition, that is an RFMO, is stand-alone stable if "no player is interested in leaving the cooperative agreement to adopt a free-rider behavior" (Pintassilgo 2003: 183).

Pintassilgo (2003) applies this coalitional bargaining analysis to the case of the bluefin tuna fishery of the Eastern North Atlantic and Mediterranean, which is currently under the management of an RFMO in the form of the International Commission for the Conservation of Atlantic Tuna. He argues convincingly that, if there are no effective curbs on unregulated fishing, the grand coalition of the players in the Eastern North Atlantic and Mediterranean bluefin tuna fishery game is not stand-alone stable. In other words, the RFMO can be expected to collapse. If unregulated fishing is in fact effectively curbed, the prospects for the RFMO are much brighter.

Next to be considered is the so-called new member problem. Kaitala and Munro (1997) argue that the new member problem can lead to a subtle form of free riding that has nothing to do with unregulated fishing. Consider the case of a group of charter members of an RFMO undertaking to rebuild a hitherto overexploited straddling stock. They engage in a resource investment program over time. As they are about to enjoy the fruits of their investment, they are approached by a prospective new member, which agrees to abide by the management regime of the RFMO, but which demands a pro rata share of the total allowable catch, and thus net economic returns from the fishery, free of charge. If the new member's demands are acceded to, the new member will effectively be free riding by enjoying a share of the return on the resource investment while having borne none of the cost of the investment (Kaitala and Munro 1997; Munro et al. 2004). Kaitala and Munro demonstrate that anticipation of such new member free riding could lead to charter RFMO members calculating that their expected cooperative payoffs would fall below their threat point payoffs. In other words, the individual/subcoalition rationality condition would not be satisfied, with the consequence that the RFMO would be stillborn.

Kaitala and Munro argue that a solution to the problem could lie in granting the charter members collective property rights to resources encompassed by the RFMO, and then allowing new members to buy their way in, much like new entrants to a domestic individual transferable quotas fishery (Kaitala and Munro 1997). While Kaitala and Munro do not say so explicitly, if the charter members are granted effective collective property rights to the high seas fishery resources encompassed by the RFMO, unregulated fishing will be transformed into illegal fishing, and the high seas portion of the RFMO will become high seas in name only.

If the aforementioned effective property rights are not created, the charter members can try to turn poachers into gamekeepers by persuading free riders to join the club, that is, to become new members. Then, however, a dilemma arises. If prospective new members are granted too generous a share of the returns from the fishery, the RFMO could be undermined by the implicit free riding. If the prospective new members believe that they are being offered too little, they will return to their explicit free-riding ways.[17] The clear implication of the Pintassilgo and Lindroos analysis (Pintassilgo 2003; Pintassilgo and Lindroos 2008) is that there will be cases in which there is no way out, and the dilemma will remain unresolved (see also Munro 2006).

With the review now complete, it becomes appropriate to turn to the question of the relevance of this game-theoretic analysis to the real world of policymaking in internationally shared fishery resources. The implication arising from the review is that the game-theoretic analysis is highly relevant. The question is now discussed more fully.

2.4 Relevance of game theory analysis to real-world policymaking in internationally shared fishery resources

An assertion made in an earlier part of the chapter now bears repeating, namely that it is all but impossible to analyze effectively the economics of internationally shared fish stock management other than through the lens of the theory of strategic interaction—game theory. It can also be added that game theory analysis and concepts, applied to the management of such shared stocks, have long since moved beyond the realm of academia, and appear in publications forthcoming from such international bodies as the World Bank, the OECD, and the FAO.[18]

First to be examined in the ensuing sections are three interrelated concepts from elementary game theory that have a high degree of relevance to real-world fisheries management but which are imperfectly understood by policy-makers. They are (a) the individual/subcoalition rationality condition; (b) the prisoner's dilemma; and (c) side payments. This will be followed by a discussion of two nonelementary issues, namely (d) the analysis of the problem of unregulated fishing in the emerging RFMO regime; and (e) the question of the dynamic resilience of cooperative fisheries management arrangements. It will be argued that the dynamic resilience question has yet to be fully explored by game theorists.

2.4.1 Individual (subcoalition) rationality condition

The 2002 Norway–FAO Expert Consultation on the Management of Shared Fish Stocks, to which reference was made at an earlier point, established several working groups, including Working Group A on resolving allocation issues. In its report to the Expert Consultation plenary, Working Group A stated that it intended to list general features of the allocation issue that it felt were not sufficiently understood in the realm of policy. One of these was:

> the basic requirement for stable long term cooperation: it has to be recognized that each and every participant in a cooperative arrangement must anticipate receiving long term benefits from the cooperative arrangement that are at least equal to the long term benefits, which it would receive, if it refused to cooperate. This fact, which should be obvious, is often ignored in practice.
>
> (FAO 2002: 8)

An example of this fact being insufficiently acknowledged is provided by the Treaty between the government of Canada and the government of the United States of America Concerning Pacific Salmon, 1985, governing the management of Pacific salmon from northern California to the Gulf of Alaska (Figure 2.1).[19] The Canada–United States Pacific salmon fisheries

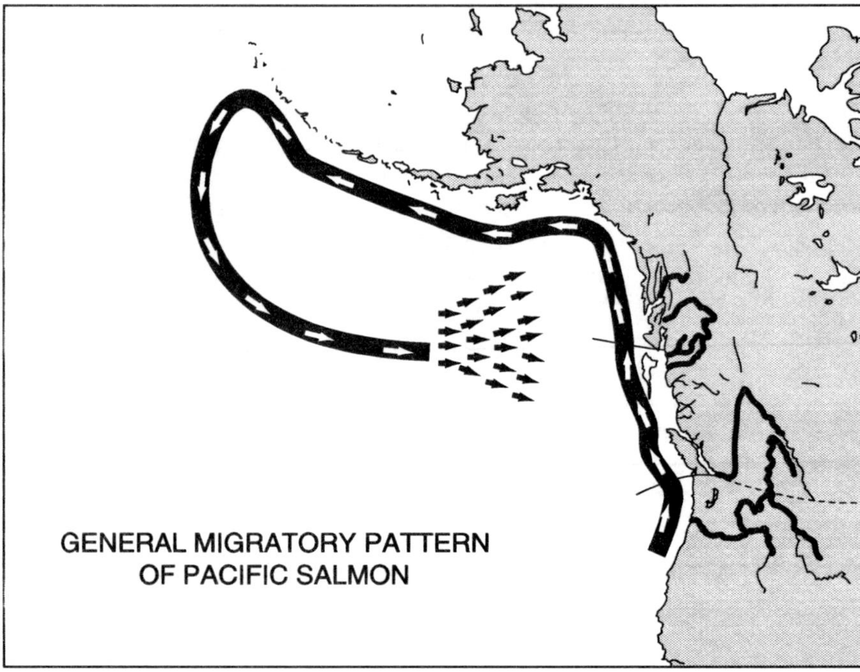

Figure 2.1 General migratory pattern of Pacific salmon.
Source: Miller and Munro 2004.

game is reasonably complex. While Canada can be regarded as a single player, the United States is really a coalition consisting of Alaska, the combined states of Washington and Oregon, the so-called treaty Indian tribes of Washington and Oregon, and the United States federal government. Bargaining, historically, has proceeded in two stages. There is a subgame played among the players in the United States coalition, followed by a game between the coalition and Canada.

Having said that, both the United States and Canada are wealthy developed States. Both take pride in the quality of their fisheries research and management. At the time that the treaty was being negotiated, Pacific salmon was the most important resource for the fishing industries of the states of Oregon, Washington, and Alaska, and the province of British Columbia.

After many years of painful negotiations, a treaty was concluded between the governments of Canada and the United States in 1983. The treaty was blocked in the United States Senate by the senior senator from Alaska. The senator, by taking this action, was neither irrational nor malicious. A historical review shows that Alaska was worse off under the 1983 version of the treaty than it would have been in the absence of a treaty (Munro and Stokes 1989).

Expressing outrage at the blocking of the treaty, Canada reverted to competitive behavior through deliberate overfishing, as the theory would predict

(Munro and Stokes 1989). The states of Washington and Oregon and the treaty tribes, though not Alaska, suffered from Canada's reversion to competitive behavior.[20] The strategy was effective. Washington and Oregon and the treaty tribes were compelled to renegotiate with Alaska. Treaty negotiations recommenced and the treaty was ratified in 1985. The solution to the cooperative game was only temporary, however. By the early 1990s, environmental shifts had brought about a situation in which Alaska once again found itself having nothing to gain from the treaty (Huppert 1995).

In the early 1990s, the subcoalitions reconfigured, with Canada entering into a de facto subcoalition with Washington and Oregon and the treaty tribes. Alaska played competitively against this subcoalition. The cooperative game crumbled. While the treaty, in a strictly legal sense, remained in place, it entered into a state of paralysis that lasted for six years. The prisoner's dilemma played itself out, with damaging consequences for the resources (Miller et al. 2001).

With the consequences of noncooperation becoming increasingly evident, a series of negotiations took place that resulted in a Canada–United States agreement being signed in 1999 to patch up the treaty, which appeared to satisfy the Alaskan individual rationality condition. How stable the solution to the new cooperative game will prove to be over time remains to be seen (United States Department of State 1999; Miller et al. 2001).

2.4.2 Prisoner's dilemma

The 1982 Convention does, as has been pointed out, allow for the possibility that States or other entities sharing a fishery resource will not be able to achieve an agreement and will, as a result, manage their segments of the resource independently, without the benefit of cooperation. While some States sharing a resource appear to have at least some degree of awareness of the consequences of noncooperation, in other cases States seem only dimly aware of the potentially damaging consequences of noncooperation.

An example of States being caught by surprise by the consequences of ineffective cooperation is provided by the case of the South Tasman Rise trawl fishery (Figure 2.2), the details of which are to be found in Munro et al. 2004. The resource is an orange roughy stock that straddles the boundary of the Australian fishing zone (that is, its exclusive economic zone) and the adjacent high seas. Orange roughy is usually exploited during its spawning phase, when it is concentrated. Orange roughy is highly valued, and is extremely slow growing. It is thus very vulnerable to overfishing.

Australia became aware of the resource in 1997, and saw New Zealand as the only other State likely to attempt to harvest the resource. The Australian authorities approached their New Zealand counterparts with the objective of establishing a cooperative management arrangement. Australia and New Zealand are both developed, and have a common cultural background, since both are former British colonies. Each country has an exemplary

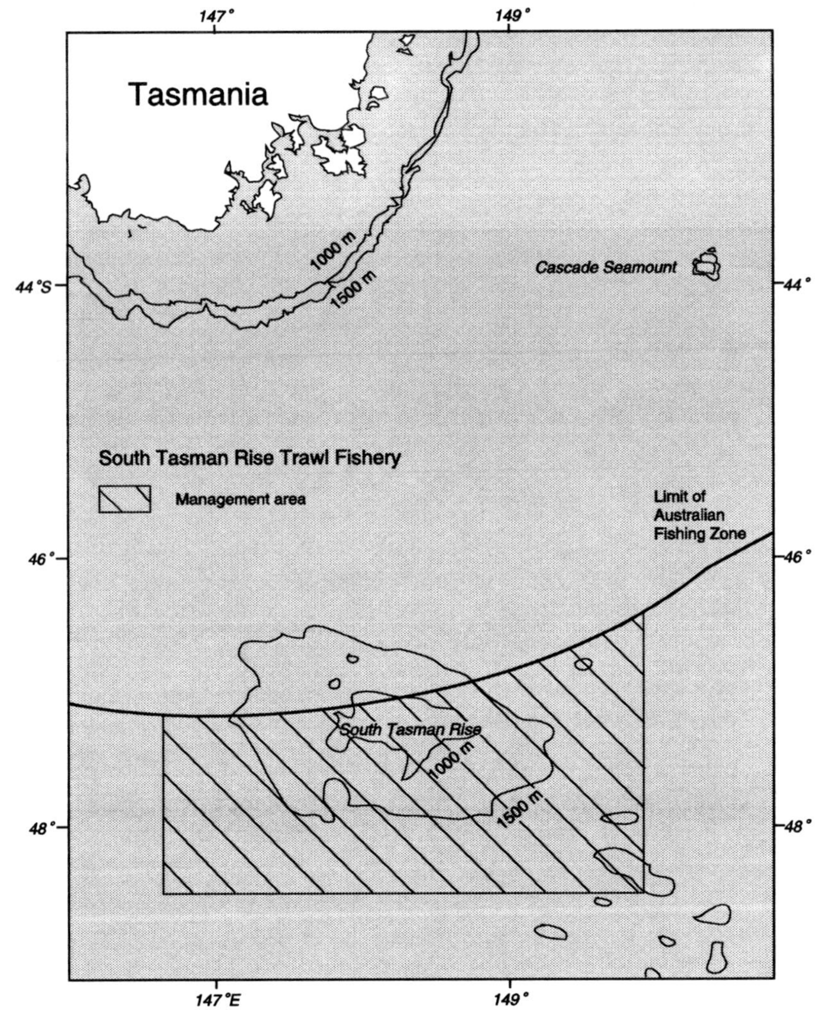

Figure 2.2 South Tasman Rise trawl fishery.
Source: Munro et al. 2004.

record of domestic fisheries management. Establishing an effective cooperative management arrangement would seem straightforward.

An agreement was put into place in late December 1997. The implementation was, however, faulty. The first fishing season under the agreement was to commence on 1 March 1998. Allowable activities of the two fleets during January and February 1998 were ill defined, thus creating an unanticipated window of noncooperation. Given that the two fleets probably regarded one another with some suspicion during this period, and that the resource was highly valued and vulnerable, the outcome was not difficult to predict, applying the principles of game theory.

The Australians acted first. The Australian fleet took much of the total allowable catch, set by the December 1997 agreement, before the March start date. The New Zealand fleet fished after the start date regardless. The next year the Australian fleet promised to behave, but the New Zealand fleet greatly exceeded its assigned share of the total allowable catch—no surprise, given that it had been stung in the previous year. The situation was further aggravated by unexpected free riding by a third party, in the form of South African flagged vessels (Munro et al. 2004).

In 2000, Australia and New Zealand signed a well thought-out, airtight, cooperative management agreement, which, among other things, dealt with third-party free riding. Effectively, there had been a set of repeated prisoner's dilemma games leading ultimately to cooperation. The cooperation had, however, come too late. The resource had been heavily overfished between 1998 and 2000. Given the extremely slow-growing nature of the resource, it will be a very long time before the resource recovers, if in fact it ever does (Munro et al. 2004).[21]

In another situation, Angola, Namibia, and South Africa share fishery resources, particularly hake, in the Benguela Current region (Figure 2.3). Sumaila et al. (2005), undertaking a study for a United Nations-sponsored project in the region, looked at the advisability of these States entering into a full-scale cooperative management arrangement, with particular emphasis on hake. With regards to the fishery resources examined in the study, they could find no convincing evidence of the prisoner's dilemma at work (at least that could be ascribed to the sharing of stocks). There was no obvious overexploitation of the resources. Thus, tacit cooperation appeared to prevail. The study did, nonetheless, recommend that the coastal States proceed to full cooperation for several reasons, an important one being that tacit cooperation could turn out to be a weak reed; since there has been uncertainty and confusion about the true extent of the sharing of the fishery resources among the three coastal States, the apparent tacit cooperation was perhaps based upon the fishing industries' temporary ignorance of the impact of their fishing activities upon one another (Sumaila et al. 2005).

2.4.3 Side payments

Side payments, which involve transfers in some form, serve, it will be recalled, both to shift the focus from sharing the harvests to sharing the economic benefits from the fishery, and to increase the scope for bargaining. Until recently the value of side payments has been understood only vaguely by policymakers. There are, however, some encouraging signs that this may be changing.[22]

An example of the message getting through, as it were, is provided by the Canada–United States Pacific Salmon Treaty, to which reference has already been made. The fish are produced in rivers, streams, and lakes. After spawning the fish go downriver to the ocean, where they may live for several years.

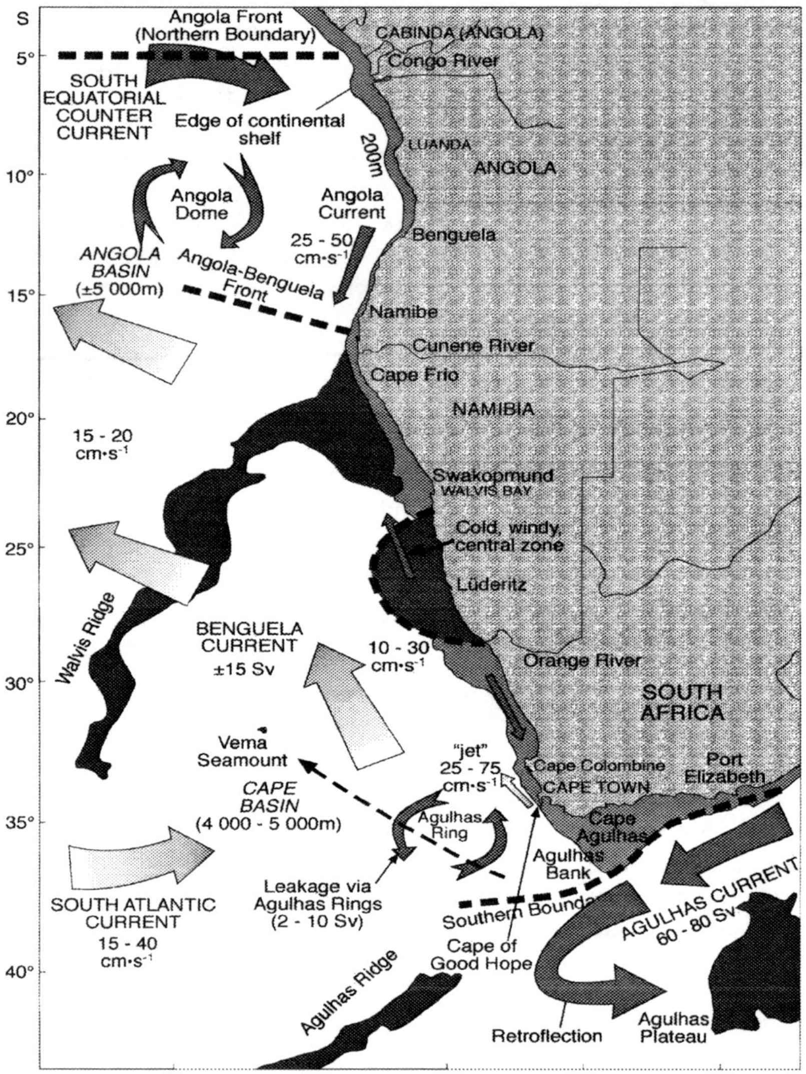

Figure 2.3 Benguela Current large marine ecosystem.
Source: Sumaila et al. 2005.

They are harvested as they are about to return to their freshwater habitat, where the survivors spawn and die. The resource is shared by virtue of the fact that Canadian fishers inevitably intercept and catch some American-produced salmon, and that American fishers inevitably intercept Canadian-produced salmon.

The original treaty made no provision for side payments between Canada and the American subcoalition. American interceptions were to be balanced by Canadian interceptions. Over time, it became evident that this "fish-for-fish" rule seriously narrowed the scope for bargaining. The 1999 agreement designed to repair the treaty, to which reference was made earlier, does contain provisions for side payments, although they are certainly not labeled as such. The side payments are modest, but what is important is that the precedent has been set.

A second example is provided by the Pacific Island States of the Western and Central Pacific. This region has the largest stock of tropical tuna resources in the world. The resources are both transboundary and straddling in nature.

The Pacific Island States undertook to manage their resources cooperatively in 1979, using the emerging United Nations Law of the Sea Convention as a framework. The tuna resources are not spread evenly throughout the region, tending to concentrate around the equator. Two subcoalitions emerged: the "haves", States close to the equator, such as Papua New Guinea; and the "have-nots", States more distant from the equator, such as Fiji. There is clear evidence that there have been side payments from the haves to the have-nots, with the objective of enhancing cooperation, although the term transfers, let alone side payments, has never been used (Munro 1990; Munro et al. 2004).[23]

The Pacific Island States, along with neighboring Indonesia and the Philippines and distant water fishing States operating in the region, for example the United States and Japan, have recently established what is undoubtedly the largest RFMO in the world, the Western and Central Pacific Fisheries Commission (WCPFC) (Figure 2.4). The subcoalition that is the Pacific Island States has a coordinating body in the form of the Pacific Islands Forum Fisheries Agency. A recent publication on the issue of allocations within the WCPFC, coauthored by a former deputy director of the Forum Fishing Agency, states the following:

> A number of economists and other parties have indicated that "game theory" offers prospects for examining the nature of cooperative and non-cooperative approaches to allocation. In particular, it has been suggested that the best way forward is likely to go beyond simply allocating rights (e.g., shares of a total allowable catch of particular species or species group or the equivalent) to national fleets. A more sophisticated approach, involving "side payments" or "negotiation facilitators" . . . may be required.
>
> (Willock and Cartwright 2006: 5)

The need for this "more sophisticated" approach is demonstrated by an issue that concerns the WCPFC at the time of writing. The issue provides an example of the potential for side payments to resolve differences between, and among, players regarding resource management objectives and plans.

The WCPFC governs the management of four tuna species: skipjack, albacore, bigeye, and yellowfin. Concern has been expressed about the

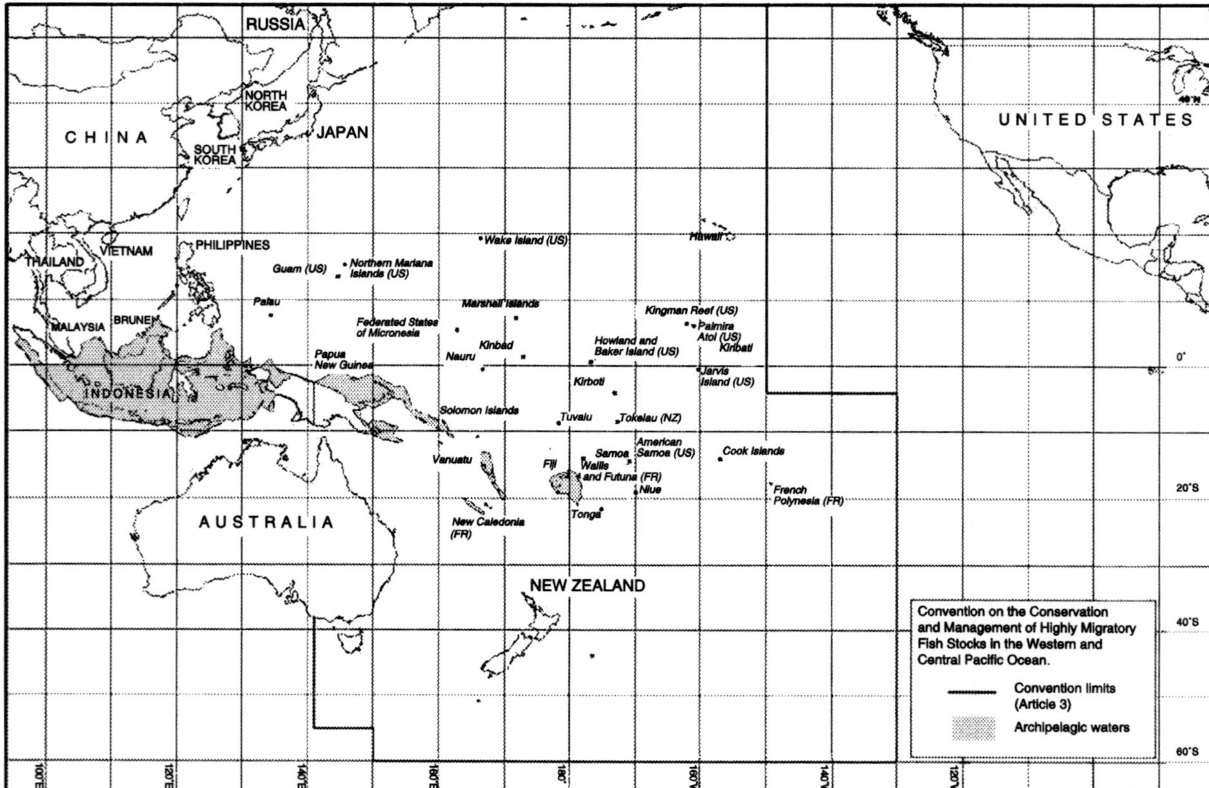

Figure 2.4 Western and Central Pacific Ocean: Convention limits.
Source: Munro et al. 2004.

current status of two of these species—yellowfin and bigeye (Reid 2006). The Scientific Committee of the WCPFC has recommended, as one possible solution, a 15 percent reduction in fishing effort across the board for all tuna species. The extensive intermingling of the tuna species (FAO 2005) almost certainly underlies this across-the-board recommendation. A senior economist employed by the Forum Fishing Agency has undertaken an assessment of the economic consequences of the Scientific Committee's across-the-board proposal (Reid 2006).

There is good reason to believe that the proposal would increase the global economic benefits from the fishery to the WCPFC grand coalition. The one difficulty is that the targeting of the four species is not uniform throughout the region. Reid concludes in his assessment that the vessels of distant water fishing States operating on the high seas portion of the WCPFC would benefit substantially from the proposed resource investment. The Pacific Island States, as a group, would actually suffer an economic loss. Needless to say, if no further action is taken, the Pacific Island States subcoalition can be expected to block the proposal—the subcoalition rationality condition once again.

What then, asks Reid, is to be done? Noting the support for side payments (negotiation facilitators) in the Report of the Norway–FAO Expert Consultation on the Management of Shared Fish Stocks (FAO 2002), he concludes:

> To overcome the difficulties inherent in obtaining agreement on management measures, members of the WCPFC will need to give serious consideration to the possibility of the use of "negotiation facilitators," or "side payments," in order to ensure that the costs and benefits of any such management measures are borne equitably between members.
>
> (Reid 2006: 7)

This particular case is relatively easy, in that it is one of evident winners compensating evident losers. Nonetheless, if the advice were to be followed, an important precedent would have been set.

A further comment is in order. Those economists who originally wrote about the value of side payments (for example Munro 1979, 1987) discussed them in the context of two-player games involving the management of a single species. In the RFMO regime, one has to deal with far more than two players. Furthermore, many cooperative fisheries arrangements are concerned not with just one species but several, as in the case of the WCPFC. If one thinks of a fishery resource as a natural capital asset, these arrangements are thus called upon to manage cooperatively a portfolio of such assets. FAO economist Rolf Willmann has asserted that as the number of players increases and the resource portfolio broadens, the scope for the use of side payments can be expected to increase exponentially (Rolf Willmann, personal communication).[24]

2.4.4 Unregulated fishing

The new international Law of the Sea, as it pertains to fisheries, is not rigid and static; rather it is in a process of evolution through State practice. Game theory analysis can contribute by helping to direct the evolution along appropriate lines.

The problem of unregulated fishing provides an example. The FAO has put forward a plan of action to curb what it refers to as illegal, unreported, and unregulated fishing (FAO 2001). The efforts of the FAO are being supplemented by the OECD (2006). It is of utmost importance that economists, undertaking game-theoretic analysis of the economic management of shared fish stocks,[25] support the efforts of the FAO and the OECD by transmitting to policymakers the results of their analysis, demonstrating the consequences of unchecked unregulated fishing.

There is evidence that the economist's message is beginning to gain some acceptance within the community of international legal experts. Rayfuse (2006), at the Sharing the Fish Conference 2006 in Fremantle, Australia, raised the question of whether a new legal paradigm was called for in international fisheries. She argued that the freedom of the seas, as it pertained to fisheries, had seen its day, and explicitly raised the question of whether "RFMOs [should be] vested with 'property rights' in high seas fisheries" (Rayfuse 2006: slide 8). She acknowledged that her questions would be highly controversial.[26] What is important is that they are now up for open discussion in the international legal community.

2.4.5 Resilience

The final issue pertains to the dynamic resilience of cooperative fisheries management regimes through time. As has been mentioned, the importance of shifting threat point payoffs over time was raised by Kaitala and Pohjola as long ago as 1988. It was also pointed out, however, that the Kaitala and Pohjola model is limited by the fact that it is deterministic (that is, it assumes that the future is known fully).

For an example of the problem of resilience, the Canada–United States Pacific Salmon Treaty once again provides a source. When the treaty was ratified in 1985, there was more or less balance between Canadian and American interceptions. Furthermore, the Alaskans had been placated. If environmental and economic conditions had remained unchanged through time, the treaty would have remained stable.

Environmental conditions did not remain unchanged. What was not realized at the time the treaty was ratified was that a climate regime shift was under way. The regime shift was to have a strongly negative impact on Pacific salmon stocks off southern British Columbia and the states of Washington and Oregon. The regime shift was also to have a very positive impact upon the Pacific salmon stocks off northern British Columbia and Alaska (Miller et al. 2001; Miller and Munro 2004).

By the early 1990s, the effects of the climate regime shift had become evident. The interception balance was upset and, more importantly, the Alaskan individual rationality condition was no longer satisfied. The treaty, as noted before, seized up.

There is no evidence that the Alaskans cheated. The treaty was, and is, effectively legally binding. What the experience revealed is that the disaffected player does not have to resort to cheating to destabilize the cooperative game. The player can easily create difficulties so that the cooperative process becomes unworkable (Miller and Munro 2004). The Canada–United States Pacific Salmon Treaty has been repaired for the time being. As noted, it remains to be seen whether or not the repaired version survives.

Another example, recently resolved at the time of writing, involves the Norwegian spring-spawning herring stock in the North Atlantic. This stock, which historically has been one of the largest fish stocks in the North Atlantic, migrates, when it is healthy, between Norway and Iceland. The resource crashed in the early 1970s. The remnants of the resource were confined to Norwegian waters and were subject to a harvest moratorium. By the mid-1990s the resource had recovered and recommenced its migratory pattern.

The resource, which is both transboundary and straddling, is now subject to exploitation by Russia, the Faeroe Islands, and the European Union, as well as Norway and Iceland. In time, a cooperative management arrangement was established under the framework of the 1995 United Nations Fish Stocks Agreement. The arrangement came complete with what were close to side payments, and was put forward as a model of cooperative management of a straddling stock (Munro 2000).

The harvest allocations are based, more or less, on zonal attachments of the resource, determined by the quantity of stock and time spent in each player's zone during the resource's migration. Norway and Russia can be seen as constituting a subcoalition (T. Bjørndal, personal communication). The Norway–Russia subcoalition, due either to unexpected shifts in resource migratory patterns or to faulty earlier biological research, claimed in 1993 that its share fell far short of what its zonal attachment dictates. It began acting as if its subcoalition rationality condition was not being satisfied (Bjørndal and Munro 2005). The subcoalition played competitively against the rest.

There was not a complete breakdown in the cooperative resource management arrangement, which could have led to the resource crashing once again. Nonetheless, the prisoner's dilemma has started to make its presence felt. The players began unilaterally raising their quotas. Furthermore, the players had previously entered into side deals to allow one another to take part of their quotas in each other's zones. This was done to increase the size of the global economic pie (Bjørndal and Munro 2001). Norway announced that all players, save Russia, were to be barred from its zone, thereby diminishing the global economic pie (Bjørndal and Munro 2005).

In late 2006 to early 2007, the cooperative game was restored. The players, having peered into the abyss, pulled back (Terje Lobach, personal communication).

Thus, both the Canada–United States Pacific Salmon Treaty and the Norwegian spring-spawning herring cooperative arrangements have, in their histories, displayed a lack of dynamic resilience in the face of uncertainty. Working Group A of the Norway–FAO Expert Consultation on the Management of Shared Fish Stocks completed its list of general features of the allocations issue that are insufficiently understood by emphasizing the importance of cooperative resource management arrangements having sufficient resilience (FAO 2002: 8).

Miller and Munro (2004) argue that shocks, particularly environmental ones, can be anticipated, but they cannot be predicted. While the need for ensuring flexibility is plain, they claim that they can obtain no clear guidance on this question from the existing literature on the application of game theory to resource management issues (Miller and Munro 2004). Such guidance is still lacking.

2.5 Conclusion and policy implications

This chapter has been concerned with the relevance, if any, of game theory analysis to a specific resource management issue, namely the management of internationally shared fishery resources. The immediate response to the question is that it is all but impossible to analyze the economics of the management of these resources other than through the lens of game theory. Strategic interaction between, and among, those States sharing the resources lies at the heart of the resource management problem.

It has been argued that there are several basic, indeed elementary, game theory concepts that are of direct and immediate relevance to policymakers. It was also argued that these concepts are, as yet, poorly understood by policymakers. What this requires of economists is that they become effective expositors, taking the results of their game theory analysis and expressing these results in a form that can be readily understood and appreciated by the practitioners.

This is an urgent matter. The implementation of the RFMO regime, described in this chapter, is an ambitious undertaking. Game theory has a great deal to say about the necessary conditions for the stability of the new regime. If the regime proves, in the end, to be unstable and collapses, the consequences for world capture fishery resources, many of which are already threatened, will be severe.

Notes

1. The author wishes to express his gratitude for the generous support provided by the Sea Around Us project of the Fisheries Centre, University of British

Columbia, which is, in turn, sponsored by the Pew Charitable Trust of Philadelphia, United States. The author would also like to express his appreciation for the useful and thoughtful comments provided by an anonymous reviewer.

2. Capture fisheries involve the harvesting of fish in the wild, as opposed to aquaculture, popularly referred to as fish farming.
3. World capture fishery harvests have been more or less stationary since the mid-1990s. Hence, one can continue to accept the Garcia and Newton estimate.
4. The negotiations on fisheries issues were more or less completed by 1975.
5. The United Nations makes a distinction between highly migratory fish stocks and straddling fish stocks (see for example United Nations 1995). Highly migratory fish stocks, tuna primarily, are, because of their highly migratory nature, to be found both within the exclusive economic zone and the adjacent high seas. Straddling fish stocks are all other fish stocks (except anadromous and catadromous stocks) to be found both within the exclusive economic zone and the adjacent high seas. The distinction was made largely for political reasons, and can be defended on neither biological nor economic grounds (Munro et al. 2004). Hence, the two are merged into what shall be termed straddling fish stocks broadly defined.
6. There is not uniform agreement on the categorization of these fish stocks. While there is no disagreement on the definitions of straddling and discrete high seas stocks, what might be termed a second school of thought prefers to use the term "transboundary stocks" as the generic term and to use the term "shared stocks" to denote those fish stocks crossing the exclusive economic zone boundary of one coastal State into the exclusive economic zone(s) of one, or more, other coastal State(s). See Van Houtte 2003.
7. It now extends to 12 nautical miles (United Nations 1982: Article 3).
8. The term "prisoner's dilemma" is derived from a theoretical game that explores the possible payoffs that may result when two prisoners, arrested for the same crime but kept separate, adopt various bargaining strategies intended to maximize their own benefit. See http://plato.stanford.edu/entrics/prisoner dilemma/.
9. With the usual proviso that there can be exceptions in unusual circumstances.
10. Another way of expressing this condition, using some economists' jargon, is to say that the solution to the cooperative game must be Pareto optimal.
11. This has come to be known as the compensation principle (Munro 1987).
12. See North Atlantic Salmon Fund, www.nasfonline.org.
13. Surely, however, Atlantic salmon must be considered to be a straddling stock. The answer is no. Salmon, in both the Atlantic and the Pacific, constitute a special case. Under Article 66 of the 1982 Convention, directed high seas fishing of salmon is made illegal (United Nations 1982: Article 66; Burke 1994).
14. The reason for this surprising development lies mainly in the fact that distant water fishing State fleets were driven out of the newly formed exclusive economic zones. This is discussed in detail in Munro 2000.
15. The full title is the Agreement for the Implementation of the Provisions of the United Nations Convention on the Law of the Sea of 10 December 1982 Relating to the Conservation and Management of Straddling Fish Stocks and Highly Migratory Fish Stocks (United Nations 1995).
16. That is, to correct the weaknesses of Part VII of the 1982 Convention.
17. It is true that, under the terms of the 1995 Agreement, States are not to allow their vessels to exploit fishery resources under RFMO jurisdiction, unless the State is a member of the RFMO or has committed itself to abide by the management regulations laid down by the RFMO (United Nations 1995). There are two difficulties with this, however. Until the RFMO regime becomes declared part of customary international law, these requirements are binding only upon States that are party to the 1995 Agreement. At this point, many fishing States are not

(High Seas Task Force 2006). Second, there is the problem of enforcement, given that those who flout the 1995 Agreement in this manner are deemed to be engaging in unregulated, as opposed to illegal, fishing (see Munro et al. 2004 for details).

18. For examples see Agüero and Gonzalez 1996; Munro et al. 2004; OECD 1997.
19. See Pacific Salmon Commission website: www.psc.org/pubs/treaty.pdf.
20. Canada was generally on very good terms with the states of Washington and Oregon, and the treaty tribes of Washington and Oregon. The Canadian strategy can be described as that of punishing one's friends.
21. There is a moral to this story. The moral is not that the two coastal States should be censured. The moral is, rather, that, if coastal States such as Australia and New Zealand can fall into the prisoner's dilemma hole, then no States coming together to manage a shared fishery resource cooperatively can consider themselves to be safe. The predictive power of the economist's model of noncooperative management of shared fishery resources proves to be brutally strong.
22. Part of the problem may lie in the terminology, a not uncommon problem in game theory. To many, the term side payment sounds suspiciously like "bribe". The 2002 Norway–FAO Expert Consultation discussed side payments, and recommended their use. Many participants at the Expert Consultation, however, preferred to substitute the less provocative term "negotiation facilitators" (FAO 2002).
23. One example is provided by a treaty that the Pacific Islands signed with the United States in 1987. Under the treaty the United States fleet has multilateral harvesting rights. American license fees were to be paid into a single fund that the Pacific Islands were to allocate among themselves. The Pacific Islands agreed that 75 percent should be paid out on the basis of American harvests within their respective exclusive economic zones. The remaining 25 percent was to be divided equally, without reference to American harvests. Effectively, this meant that the have-nots would receive more than could be justified by American harvests in their exclusive economic zones. Hence, the have-nots received implicit side payments from the haves (Munro 1990).
24. Willmann notes that no game theorist has yet to confirm or repudiate the assertion. He is confident that some game theorist will rise to the challenge (Rolf Willmann, personal communication).
25. In particular, game theory analysis involving the coalitional bargaining approach.
26. Rather than criticizing her, Professor Rayfuse's legal colleagues at the conference congratulated her for raising questions that very much needed to be asked.

References

Agüero, M. and Gonzalez, E. (1996) "Managing transboundary stocks of small pelagic fisheries: problems and options", World Bank Discussion Paper 329: Fisheries Series, World Bank, Washington, DC.

Alexander, L.M. and Hodgson, R.D. (1975) "The impact of the 200-mile economic zone on the Law of the Sea", *San Diego Law Review*, 12: 569–99.

Bjørndal, T. and Munro, G. (2001) "The management of high seas fisheries and the implementation of the U.N. Agreement of 1995: problems and prospects", in R. Johnston and A. Shriver (eds) *Proceedings of the Tenth Biennial Conference of the International Institute of Fisheries Economics and Trade*, International Institute of Fisheries Economics and Trade, Corvallis.

Bjørndal, T. and Munro, G. (2003) "The management of high seas fisheries resources and the implementation of the UN Fish Stocks Agreement", in T. Tietenberg and H. Folmer (eds) *The International Yearbook of Environmental and Resource Economics 2003/2004*, Cheltenham: Edward Elgar.

Bjørndal, T. and Munro, G. (2005) "Shared fish stocks and high seas issues", in A. Weintraub, T. Bjørndal, R. Epstein, and C. Romero (eds) *Management of Natural Resources: A Handbook of Operations Research Models, Algorithms, and Implementations*, International Series in Operations Research and Management Science, Amsterdam: Kluwer Academic Publishers.

Bjørndal, T., Kaitala, V., Lindroos, M., and Munro, G. (2000) "The management of high seas fisheries", *Annals of Operations Research*, 94: 183–96.

Burke, W.T. (1994) *The New International Law of Fisheries: UNCLOS 1982 and Beyond*, Oxford: Clarendon Press.

Clark, C.W. (1980) "Restricted access to a common property resource: a game-theoretic analysis", in P. Liu (ed.) *Dynamic Optimization and Mathematical Economics*, New York: Plenum Press.

Clark, C.W. (1990) *Mathematical Bioeconomics: The Optimal Management of Renewable Resources*, 2nd edn, New York: Wiley Interscience.

FAO (Food and Agriculture Organization of the United Nations) (1994) "World review of highly migratory species and straddling stocks", FAO Fisheries Technical Paper 337, Rome.

FAO (Food and Agriculture Organization of the United Nations) (2001) *International Plan of Action to Prevent, Deter and Eliminate Illegal, Unreported and Unregulated Fishing*, FAO, Rome.

FAO (Food and Agriculture Organization of the United Nations) (2002) *Report of the Norway-FAO Expert Consultation on the Management of Shared Fish Stocks, Bergen, Norway, 7–10 October 2002*, FAO Fisheries Report 695, Rome.

FAO (Food and Agriculture Organization of the United Nations) (2003) *Code of Conduct for Responsible Fisheries*, FAO, Rome.

FAO (Food and Agriculture Organization of the United Nations) (2005) "Review of the state of world marine fishery resources", FAO Fisheries Technical Paper 457, Rome.

Garcia, S. and Newton, C. (1997) "Current situation, trends and prospects in world capture fisheries", in E. Pikitch, D. Huppert, and M. Sissenwine (eds) *Global Trends in Fisheries Management*, American Fisheries Symposium 20, American Fisheries Society, Bethesda.

Gordon, H.S. (1954) "The economic theory of a common property resource: the fishery", *Journal of Political Economy*, 62: 124–42.

High Seas Task Force (2006) *Closing the Net: Stopping Illegal Fishing on the High Seas*, Governments of Australia, Canada, Chile, Namibia, New Zealand, and the United Kingdom, World Wide Fund for Nature (WWF), World Conservation Union (IUCN), and the Earth Institute at Columbia University. Available online: www.high-seas.org (accessed February 20, 2007).

Huppert, D.D. (1995) *Why the Pacific Salmon Treaty Failed to End the Salmon Wars*, SMA 94–1, School of Marine Affairs, University of Washington, Seattle.

Kaitala, V.T. (1985) *Game Theory Models of Dynamic Bargaining and Contracting in Fisheries Management*, Institute of Mathematics, Helsinki University of Technology, Helsinki.

Kaitala, V.T. and Munro, G. (1993) "The management of high seas fisheries", *Marine Resource Economics*, 8: 313–29.

Kaitala, V.T. and Munro, G. (1997) "The conservation and management of high seas fishery resources under the New Law of the Sea", *Natural Resource Modeling*, 10: 87–108.

Kaitala, V.T. and Pohjola, M. (1988) "Optimal recovery of a shared resource stock: a differential game with efficient memory equilibrium", *Natural Resource Modeling*, 3: 91–119.

Levhari, D. and Mirman, L.J. (1980) "The great fish war: an example using a dynamic Courant-Nash solution", *Bell Journal of Economics*, 11: 649–61.

McRae, D. and Munro, G. (1989) "Coastal State 'rights' within the 200-mile exclusive economic zone", in P. Neher, R. Arnason, and N. Mollet (eds) *Rights Based Fishing*, Dordrecht: Kluwer.

Miller, K. and Munro, G. (2004) "Climate and cooperation: a new perspective on the management of shared fish stocks", *Marine Resource Economics*, 19: 367–93.

Miller, K., Munro, G., McDorman, T., McKelvey, R., and Tydemers, P. (2001) "The 1999 Pacific Salmon Agreement: a sustainable solution?" Canadian–American Public Policy Occasional Paper 47, Canadian–American Center, University of Maine, Orono.

Munro, G. (1979) "The optimal management of transboundary renewable resources", *Canadian Journal of Economics*, 12(3): 355–76.

Munro, G. (1987) "The management of shared fishery resources under extended jurisdiction", *Marine Resource Economics*, 3: 271–96.

Munro, G. (1990) "Extended jurisdiction and the management of Pacific highly migratory species", *Ocean Development and International Law*, 21: 289–307.

Munro, G. (2000) "The UN Fish Stocks Agreement of 1995: history and problems of implementation", *Marine Resource Economics*, 15: 265–80.

Munro, G. (2006) "International allocation issues and the high seas: an economist's perspective", paper presented at the Sharing the Fish Conference 2006, Fremantle, Australia, February 26–March 2, 2006.

Munro, G. and Stokes, R. (1989) "The Canada–United States Pacific Salmon Treaty", in D. McRae and G. Munro (eds) *Canadian Oceans Policy: National Strategies and the New Law of the Sea*, Vancouver: University of British Columbia Press.

Munro, G., Van Houtte, A., and Willmann, R. (2004) "The conservation and management of shared fish stocks: legal and economic aspects", FAO Fisheries Technical Paper 465, Rome.

Nash, J.F. (1951) "Noncooperative games", *Annals of Mathematics*, 54: 289–95.

Nash, J.F. (1953) "Two-person cooperative games", *Econometrica*, 21: 128–40.

OECD (Organisation for Economic Co-operation and Development) (1997) *Towards Sustainable Fisheries: Economic Aspects of the Management of Living Marine Resources*, Paris: OECD.

OECD (Organisation for Economic Co-operation and Development) (2006) *High Seas Task Force*. Available online: www.high-seas.org (accessed April 15, 2007).

Orrego Vicuña, F. (1999) *The Changing International Law of High Seas Fisheries*, Cambridge: Cambridge University Press.

Pintassilgo, P. (2003) "A coalition approach to the management of high seas fisheries in the presence of externalities", *Natural Resource Modeling*, 16: 175–97.

Pintassilgo, P. and Lindroos, M. (2008) "Application of partition function games to the management of straddling fish stocks", this volume, Chapter 4.

Rayfuse, R. (2006) "Regional allocation issues or zen and the art of pie cutting", presentation to the Sharing the Fish Conference 2006, Fremantle, Australia,

26 February–2 March 2006. Available online: www.fishallocation.com/papers/index.html (accessed February 21, 2007).

Reid, C. (2006) "Economic implications of an implicit allocation of bigeye harvest rights through an across the board reduction in effort levels in the Western and Central Pacific tuna fishery", paper presented at the Sharing the Fish Conference 2006, Fremantle, Australia, 26 February–2 March 2006. Available online: www.fishallocation.com/papers/index.html (accessed February 21, 2007).

Sumaila, U.R. (1999) "A review of game-theoretic models of fishing", *Marine Policy*, 23: 1–10.

Sumaila, U.R., Munro, G., and Keith, H. (2005) *Benguela Current Commission (BCC) Economic Study*, prepared for the United Nations Office for Projects Services (UNOPS) and Benguela Current Large Marine Ecosystem (BCLME) Programme. Available online: www.earthmind.net/marine/docs/bcc-economic-study.pdf (accessed February 19, 2007).

United Nations (1982) *United Nations Convention on the Law of the Sea*, United Nations Doc. A/Conf.62/122.

United Nations (1995) *United Nations Conference on Straddling Fish Stocks and Highly Migratory Fish Stocks: Agreement for the Implementation of the United Nations Convention on the Law of the Sea of 10 December 1982 Relating to the Conservation and Management of Straddling Fish Stocks and Highly Migratory Fish Stocks*, United Nations Doc. A/Conf./164/37.

United States Department of State (1999) *Diplomatic Note No. 0225 from Canada to the United States; Reply; Attached Agreement, June 30.*

Van Houtte, A. (2003) "Legal aspects in the management of shared fish stocks: a review", in *Papers Presented at the Norway–FAO Expert Consultation on the Management of Shared Fish Stocks, Bergen, Norway, 7–10 October 2002*, FAO Fisheries Technical Paper 422, Rome.

Willock, A. and Cartwright, I. (2006) *Conservation Implications of Allocation under the Western and Central Pacific Fisheries Commission*, WWF Australia and TRAFFIC Oceania, Sydney.

3 Traditional grazing rights in sub-Saharan Africa and the role of policy

Rachael E. Goodhue and Nancy McCarthy

Traditional sub-Saharan pastoralist systems are characterized by mobility, which allows pastoralists to respond to the "patchy" nature of rainfall and pasture. In many cases, multiple groups can use an area and have different types of property rights. This chapter develops a model that incorporates the key features of traditional property rights using fuzzy sets, and examines when the traditional system will be privately and socially preferable to private property and common property systems. The chapter concludes by discussing the implications for policies regarding land management institutions, including when it may be desirable to strengthen the traditional system.

3.1 Introduction

Pastoralist systems play an important economic role in sub-Saharan Africa. According to Dixon et al. (2001), pastoralist systems account for 7 percent of the agricultural region including sub-Saharan Africa, North Africa, and the Middle East. Thornton et al. (2002) estimate that approximately 10 percent of the poor in sub-Saharan Africa are pastoralists. Given the importance of pastoralism, efforts by governments and international donors to alleviate poverty and encourage economic growth must include policies designed to improve the welfare of pastoralists.

The traditional land management institutions underlying pastoralist systems in the region have been subject to a number of significant stressors over the last few decades: increasing populations; private expropriation of communal grazing lands for private use in grazing or agriculture; government-sponsored privatization, including some induced by land titling programs; localized rangeland degradation; increased, increasingly violent, conflicts between tribes; and severe droughts (Oba 1992; Swallow 1994; Bruce and Mearns 2002; McPeak 2003; World Bank et al. 2003; Cotula et al. 2004; Toulmin 2005; Hammel 2001; Hundie 2006). In many areas, these factors have led to the weakening or disappearance of traditional institutions. As a result, policymakers considering interventions designed to improve pastoralists' welfare must consider land management institutions explicitly. Should traditional institutions be strengthened or supported? Are they still viable, in a given situation? Would pastoralists be made better off if alternative land management institutions were introduced, or strengthened if they have replaced traditional ones?

Policymakers' interest in land management institutions corresponds to a growing consensus by governments, nongovernmental organizations, and researchers that many previous property rights reform initiatives have failed (Graham 1988; Sserunkuuma and Olson 2001; Bruce and Mearns 2002; Cotula et al. 2004). Early reform efforts tended to center on attempts to privatize common use resources by assigning them to individuals or small groups, such as Kenya's group ranches, as part of efforts to transform pastoralist systems into commercial ranching systems (Fratkin 2001). Often, such policies led to increased sedentarization among pastoralist groups, as did efforts to "modernize" pastoralists' economic systems by introducing improved livestock production methods and increasing the land devoted to agriculture (Pantuliano 2002). This in turn led to greater privatization of land and water resources. More recently, there have been recommendations to clarify resource use; that is, to limit the use of clearly defined water and grazing resources to specific, clearly defined groups of users, and exclude others (Hammel 2001; Haro et al. 2005).

Each type of policy prescription is linked to a familiar economic paradigm. Privatization is linked to the economic efficiency of private ownership, relative to common resource exploitation. Access clarification is an outgrowth of the common property literature, which often models resource use as a noncooperative game. Studies such as those of Ostrom (1990) and Thompson and Wilson (1994) have concluded that the inefficiency of common use can be mitigated in cases where the common property regime is a well-defined user group using a clearly defined resource. Such conditions encourage users to cooperate in resource management, which increases social returns. Access clarification is viewed also as a means of correcting an open-access situation, where a resource is available to all users, when this regime results in overuse of the resource or deterioration of the resource base. Bruce and Mearns (2002: 16) conclude that while the link between policy and paradigm is "attractive in theory, there has been little success with such 'fixed-boundary' common-property approaches".

In part, this failure may be due to policymakers' misinterpretation of the conclusions of the common property literature regarding well-defined boundaries.[1] While seminal common property studies such as those of Netting (1972, 1976, 1981) studied cases of successful common grazing land management with fixed resource boundaries, fixed user definitions, and specific written, formal rules regarding resource access and use that applied the same guidelines to all eligible users, the common property literature also encompasses studies of the successful management of fugitive resources, such as fish populations. The intention of the term "well-defined" as a requirement for successful common property management is simply to indicate that it must be possible to exclude potential users who do not have rights to utilize the resource (Ostrom 2002).

Another factor that may contribute to the failure of both types of policy prescriptions is a tendency among some analysts and some policymakers to ignore or underestimate the importance of the relationship between a society's culture and its institutions, such as its system of property rights (Ensminger 1996). An institution evolves within a specific cultural context, and its performance is dependent on that context. Imposing that institution within a different cultural context ("institutional monocropping") may lead to a very different outcome (Evans 2004).

Observers have identified the imprecision, or fuzziness, of access rights as one of the strengths of pastoralist systems in sub-Saharan Africa and elsewhere (Lane and Moorhead 1994; Scoones 1994; Toulmin 1994). The objectives of this chapter are to explore the key features of traditional pastoralist systems, incorporate these features into a noncooperative game-theoretic model of resource use, use the model to compare the performance of the traditional system to noncooperative common property and private property systems, and identify implications for policy.

The remainder of the chapter is organized as follows. Section 3.2 discusses important features of traditional pastoralist systems, and section 3.3 discusses measures of their performance. Section 3.4 provides background on fuzzy set theory and its use in game theory. Section 3.5 presents the noncooperative fuzzy game that models the key features of traditional pastoralist systems. Section 3.6 presents results comparing the performance of the traditional pastoralist system modeled as a noncooperative fuzzy game to the performance of alternative property rights regimes, including private property modeled as a profit maximization problem, and common property modeled as a noncooperative game. Section 3.7 concludes with a discussion of the policy implications of the analysis in the context of drought and global climate change, and the broader implications for government policies regarding property rights and pastoralist systems.

3.2 Nature of traditional pastoralist systems

Sub-Saharan Africa's pastoralist systems are characterized by three key features: mobility, shared use of forage and water resources, and the imprecise nature of the institutional rules governing common resource use and the underlying property rights. These institutional features are responses to the nature of the ecological system pastoralists face. Average rainfall is low in sub-Saharan Africa, and its variability is high. As a result, forage is "patchy", so that productivity can vary substantially within the range available to a given pastoralist or pastoralist group. Mobility allows pastoralists to access areas with relatively high rainfall realizations, rather than settling themselves in a single location. Shared use allows multiple pastoralists or pastoralist groups to include a specific area in their resource portfolio. The property rights regime enables the management of grazing areas to reflect resource limitations without excluding lower-priority users in order to secure rights for higher-priority users.

Rights to forage and water are defined within an extremely complex system. The boundaries of grazing areas are not well defined, except perhaps at the broadest, tribal level, nor is the set of users with rights to a given area (McPeak and Barrett 2001; Haro et al. 2005). Often, relatively productive areas have a primary user or manager group, and a large number of secondary users, whose use is managed by the primary user group to a greater or lesser extent (Cossins and Upton 1987; Niamir 1991; Oba 1992). Within the imprecise nature of the definition of a user group, different users can have different degrees of rights, and a number of factors can influence the ability of a given pastoralist or pastoralist group to utilize a specific resource (Cossins and Upton 1987; Coppock 1994). Kinship, networks, and other relationships, such as age groups, can all affect a group's use of a specific resource (Haro et al. 2005). Rights also vary depending on the users' livestock. In most cases, home herds with lactating and young animals receive higher priority when resources are limited. Satellite herds are much more mobile, so they have low priority for grazing near the base camp or a given water resource (Cossins and Upton 1987; Coppock 1994). Finally, access may depend on conditions in other grazing areas to which a group also has rights (Hogg 1990; Solomon et al. forthcoming). If a group's core areas are suffering from drought, then it is more likely to have a greater degree of access sanctioned by other users than would be the case if its home range was in good condition. Increased access to other areas in times of drought is often part of a mutual insurance agreement, with the arrangement reciprocated by the other user group when weather conditions are reversed (Bartels et al. 1990). In sum, the forage available to a given group at a given time is very difficult to define precisely.

3.3 Performance of traditional pastoralist systems: existing measures

Given its complexity and its ill-defined boundaries, the property rights system underlying traditional sub-Saharan pastoralist systems does not closely resemble any of the standard economic models of property rights regimes: open access, noncooperative common property, or private property. The question then arises: is the traditional system then less efficient than alternative systems? Some analysts argue that the definition of property rights is an essential part of what makes a traditional pastoralist system responsive to its risky, patchy environment.

> Overlapping claims to resources, shifting assertions of rights and continuous contestation and negotiation of access rules dominate tenurial arrangements in uncertain environments. The solution is not to impose particular tenure types on a variable setting; whether these are uniquely

> communal or private they are unlikely to work. Instead, the need for flexible tenure arrangement must be recognized. . . . Customary tenure systems operate shared, overlapping forms of tenure rights in such settings as maintaining strict boundaries is usually untenable.
>
> (Scoones 1994: 27)

This assessment runs counter to the standard economic analyses of the relationships between property rights and economic efficiency. Conventionally, well-defined resource boundaries and user groups are predicted to lead to more efficient resource allocation. As discussed in section 3.2, the examined systems do not possess these characteristics, except perhaps at extremely aggregated levels, such as tribes. In order to determine whether or not it may be worthwhile to strengthen traditional land management institutions, the performance of these institutions must be compared to those of alternative systems. In the case of pastoralist systems, one can begin with two physical measures: the stocking rate, or the number of animals per unit of land; and the energy produced by the system.

3.3.1 Stocking rate

The stocking rate is sometimes used by economists as a measure of the efficiency of an institutional system in maximizing profits. In the conventional static models of open access and noncooperative common property, users will stock more animals than would maximize the profits of the group of users as a whole, because each user ignores the effect of its animals on the grazing available for other users' animals. This externality leads to a suboptimal outcome. Under open access, users continue to increase the stocking rate until profits are driven to zero, while under noncooperative common property users still realize some profits. If a single user controlled the pasture, or if all users cooperated when choosing the stocking rate, then the profit-maximizing stocking rate would be achieved. Intuitively, this externality translates easily into a dynamic economic model when grazing today reduces currently available forage and affects the ability of the pasture to regenerate and provide additional forage in the next period.

3.3.2 Carrying capacity and energy production

Range scientists have a measure of the stocking rate that represents the dynamic *physical* optimum: carrying capacity. Loosely speaking, an area's carrying capacity is the number of livestock per unit area that can be sustained over time. If higher stocking rates are used, then the area will degrade, making it less productive for future grazing. Carrying capacity was originally developed for use in areas with higher average rainfall and lower rainfall variability than sub-Saharan Africa. Given these differences, scientists have debated its relevance for the systems modeled here. In a related debate, some range scientists have argued that weather shocks, rather than livestock populations, determine range conditions, which suggests that pastoralists are not using their resources at an unsustainable rate. Empirical studies have tended to show that degradation of grazing areas is not a serious concern, with the exception of localized degradation near towns and important water resources (McPeak 2003).

Cossins and Upton (1987) take a different approach to evaluating the performance of a traditional sub-Saharan pastoralist system. For the case of the Borana pastoral system of southern Ethiopia, they document the physical productivity of the area, the water management system, management of cattle and other livestock, features of the household and settlements, and market participation. They then estimate the net energy, animal liveweight, and protein offtake produced by the system. The result compares favorably with the net energy produced by other livestock-based systems with similar average rainfall, including average performers among Kenyan ranches, and Australian cattle stations in three different districts. Given the rainfall, they conclude that it would be difficult to enhance the performance of the pastoralist system. In a closely related article, they examine possible policy interventions designed to improve the system's efficiency (Cossins and Upton 1988). They conclude that improving access to markets has little scope for increasing efficiency, and that increasing dry season water supplies in some areas may lead to overstocking, actually reducing productivity. They suggest that supplemental feeding for calves may result in small productivity improvements.

3.3.3 Other approaches

In another comparison in their 1987 article, Cossins and Upton find that although the pastoralist they study produces less food than an agropastoralist in the nearby Ethiopian highlands, the pastoralist produces more cash output. Hence, they conclude that the pastoralist will be better off, provided there is sufficient access to markets for livestock and food. Van den Brink et al. (1995) undertake a more general evaluation of the performance of a traditional pastoralist system compared to an agricultural system. Rather than a detailed empirical study of a specific area and people, they construct a dynamic programming model, and discuss its relevance to climatic and social conditions in the West African Sahel. They compare the relative performance of a pastoralist system to an agricultural system as a function of average rainfall and rainfall variability. They find that in areas where rainfall is sufficiently low and variable the pastoralist system is more productive than the agricultural system. Their analysis does not consider the possibility of noncooperative interactions between multiple pastoralists; implicitly, they consider a single profit-maximizing user, or a group of perfectly cooperative users.[2]

3.3.4 Summary of performance measures

These evaluations suggest that pastoralist systems can be productive adaptations to specific ecosystems. However, these evaluations do not indicate whether or not the performance of these systems would be improved if a different property rights regime replaced the complex traditional system. Cossins and Upton (1987, 1988) take property rights as given. Van den Brink et al. (1995) focus on mobility, although they discuss management institutions in the Sahel and their effects on resource use, and distinguish them from an open-access situation.

In order to address this omission in the existing literature, and compare the performance of the traditional system to alternative property rights regimes, fuzzy set theory is used to model the traditional property regime as a noncooperative game, and compare its performance to a noncooperative common property regime and a private property regime. The latter two regimes, conventional economic property rights paradigms, have been associated with failed policy prescriptions for sub-Saharan pastoralist systems; this analysis explores whether the omission of critical features of these systems was a contributing factor. Because these conventional paradigms rely on standard set theory, they are ill suited to modeling complex property rights systems with imprecise definitions of resources and user groups. The model used in this analysis utilizes fuzzy set theory, which generalizes the concept of set membership in order to model complex systems.

3.4 Fuzzy set theory

Economic applications of fuzzy sets in game theory can be grouped into two categories. The first set of applications uses fuzzy logic to model players' strategies in a game (Bogataj and Usenik 2005; Aristidou and Sarangi 2006). Those in the second category use fuzzy sets to describe aspects of an economic system modeled as a noncooperative game, such as firms' information sets and consumer preferences (Dompere 1995; Greenhut et al. 1995; Mansur 1995; Goodhue 1998). This analysis falls into this category, termed "restricted fuzzy games", because players have crisply defined strategies and beliefs over their opponents' actions and only parameters affecting their decisions are fuzzy. The remainder of this section introduces fuzzy set theory, which is used to model traditional property rights in a noncooperative game regarding pastoralists' grazing decisions in section 3.5. As discussed in section 3.2, property rights under traditional systems are contingent and incomplete. Fuzzy sets are capable of modeling these characteristics in a direct, analytically tractable way.

Fuzzy set theory was introduced as a means of modeling complex systems (Zadeh 1965). When a system is very complex, it becomes very complicated to model precisely. Fuzzy set theory allows for imprecision in the way specific objects or events are categorized, in order to reduce the computational costliness of modeling the system. It does so by relaxing set boundaries, so that a specific object or event can be a partial (fuzzy) member of both a set and its complement.

A simple illustration of the value of fuzzy sets in the context of the analysis in this chapter regards rainfall and drought. Does an annual rainfall that is 80 percent of average annual rainfall constitute a drought year for forage and water resources? Using conventional set theory, either that rainfall level is in the set "rainfall defining a drought year", or it is not. The relationship between annual rainfall and drought-limited resources is not so clear-cut. The allocation of the rainfall between the short and long rainy seasons matters, as does the previous year's rainfall. Even temperatures may affect the outcome. One possibility for the analyst is to introduce rules relating all of these factors to each other and resource levels into his or her model, and solve the resulting complex system. Another is to utilize fuzzy sets to model the relationship between annual rainfall and drought. The extent to which a given level of annual rainfall is associated with a drought year is defined by that level's "degree of membership" in the set "rainfall defining a drought year".

In addition to reducing complexity, another advantage of fuzzy set theory is that it is well suited to incorporating verbal variables into a model. For example, fuzzy sets are used to represent uncertain outcomes, using information obtained from experts. It can be difficult or impossible to elicit specific probability density functions from individuals. It can be easier to elicit rough groupings of outcomes into categories such as "highly possible", "possible", and "not very possible".

3.5 Noncooperative fuzzy game-theoretic model of grazing decisions under traditional property rights

This section develops and solves a noncooperative game between two pastoralists utilizing two pastures, when traditional property rights are modeled using fuzzy sets. "Fuzzy rights", or the fuzzy property regime, are implemented by using absolute and relative rainfall shocks as imprecise indicators of a pastoralist group's access to a given area.

3.5.1 Players and strategies

Consider two profit-maximizing pastoralists (or pastoralist groups), A and B. Each pastoralist has a herd of fixed size a and b, respectively. Normalizing the price of output to 1 and ignoring inputs other than forage, pastoralist i's profits are equal to i's animals' total weight gain (or any other measure of animal productivity, such as a combination of weight gain and milk production). Each pastoralist chooses the share of a season spent on each of two pastures, 1 and 2, in order to maximize weight gain. Each pastoralist takes the other's pasturing decision as given when making this decision, so the two pastoralists are playing a noncooperative Nash game over their use of the available pasture resources. In equilibrium, neither pastoralist will wish to alter its pasturing decision.

3.5.2 Production

The two pastures, 1 and 2, are identical in every way except that their rainfall realizations are independently and identically distributed. On a pasture, weight gain is a linear-quadratic function of the number of animals (McCarthy 1996). It increases linearly with the total forage αr_j, which is a function of rainfall r_j. The effect of an increased number of animals on a pasture when both pastoralists use it for some share s of the season reduces available forage relative to total forage by $-\frac{\beta(a+b)}{1-s}$. Both pastoralists observe the rainfall realizations for both pastures, where $0 < r_j \leq 1$. This assumption can be interpreted as representing network communications regarding the quantity and quality of forage available in various areas.

3.5.3 Fuzzy property rights

In order to implement the noncooperative game-theoretic model in the context of the traditional property rights system, consider the case where the pastoralists' rights to a given pasture are asymmetric. Each pastoralist i has a home pasture, to which i has relatively strong rights: 1 is A's home pasture, and 2 is B's home pasture. Each pastoralist has weaker rights regarding the other pastoralist's home pasture. These weaker rights represent the share of the season that A can graze pasture 2, or that B can graze pasture 1. To represent the difference in rights between the two pastures, define the fuzzy set $Access_i$ as "pasture to which pastoralist i has access", and represent i's right to use a pasture as the degree of membership of that pasture in the fuzzy set. Each pastoralist i's home pasture is a full member of $Access_i$, so its degree of membership in $Access_i$ is 1.

The membership of the nonhome pasture in $Access_i$ depends on absolute rainfall realizations and relative rainfall realizations on the two pastures. The lower the rainfall realization in i's home pasture, the greater i's fuzzy right to the other pasture, and the lower the rainfall realization in i's home pasture relative to the rainfall realization in the other pasture, the greater i's fuzzy right to the other pasture. For analytical convenience, rainfall is normalized to lie on the closed interval [0,1]. Low absolute rainfall and low relative rainfall are represented by two fuzzy sets. Fuzzy set notation takes the following form: *set* = {*degree of membership of item j in set/item j, for all j*}. In this case, the items in the two rainfall-related fuzzy sets are pastures 1 and 2. The fuzzy set corresponding to low absolute rainfall is the statement "Pasture j is in a drought year", or

$$drought = \{max(0,1\text{-}\gamma r_j)/\ pasture\ j,\ j = \{1,2\}\},$$

where $\gamma > 1$, so that some rainfall realizations are not considered to be members of the *drought* fuzzy set at all. The fuzzy set corresponding to low relative rainfall is the statement "Pasture j is in more of a drought year than pasture $\neq$ j is", or

$$worse = \{max(0,\ r_2\text{-}r_1)/pasture\ 1,\ max(0,\ r_1\text{-}r_2)/pasture\ 2\}.$$

This fuzzy set defines the pasture that receives less rainfall as being worse off. The degree of membership indicates how much worse off the pasture with less rainfall is.

The membership of *i*'s nonhome pasture in the fuzzy set defining *i*'s access to the two pastures, "pastures to which *i* has access", or $Access_i$, is a fuzzy function of the pasture's absolute and relative rainfall. Consider pastoralist A. Because pasture 1 is A's home pasture, it has a degree of membership of 1 in $Access_A$. Pasture 2 has a degree of membership that is no greater than 1, based on the minimum of its membership in the *drought* fuzzy set, the *worse* fuzzy set, and the common property stocking rate, X_{A2}. In other words, the more forage 2 has, and the more forage it has *relative* to 1 (even if it has a small absolute amount of forage), the greater the access A has to it. Notice that if 1's rainfall realization is greater than 2's then A has no access rights to pasture 2. The noncooperative common property stocking rate is included simply to recognize that A will never use pasture 2 more than the profit-maximizing share of the season under a conventional noncooperative common property regime, because it will never be profitable to do so. For pastoralist A, $Access_A$ takes the following form:

$$Access_A = \{1/pasture\ 1,\ min(drought, worse, X_{A2})/pasture\ 2\}.$$

For notational convenience, pasture 2's membership in $Access_A$ is labeled P_{A2}, and pasture 1's membership in $Access_B$ is labeled P_{B1}.

3.5.4 Equilibrium of the fuzzy game

Given these specifications, the pastoralists will choose the share of the season to spend on each pasture. Consider the case where $r_1 < r_2$, so that A has rights to pasture 2. A chooses whether to remain on pasture 1 for the entire season or to move to pasture 2 for some share of the season. If A remains on pasture 1, A's returns are $ar_1 - \beta a$. If A moves to pasture 2 for some share s of the season, A's returns are

$$s\left(ar_2 - \frac{\beta(a+b)}{1-s}\right) + (1-s)(ar_1 - \beta a).$$

For a given s, A will move to pasture 2 if

$$ar_2 - \frac{\beta(a+b)}{1-s} > ar_1 - \beta a.$$

Similarly, if $r_1 > r_2$, then for a given s B will move to pasture 1 if

$$ar_1 - \frac{\beta(a+b)}{1-s} > ar_2 - \beta b.$$

Comparing the two conditions, it is apparent that at most one pastoralist will move in the equilibrium outcome associated with a given pair of rainfall realizations. Furthermore, given that pastoralist i chooses to move, i will move to the other pasture for the maximum amount of time permissible under i's fuzzy right.

Figure 3.1 illustrates when each pastoralist will move under the fuzzy property system for a specific set of parameter values: γ, the parameter defining the *drought* fuzzy set, is 2, the forage productivity parameter a is 12, the externality parameter β is 0.5, and herd sizes a and b are normalized to 1. Rainfall realizations for pasture 1 are listed on the vertical axis, and rainfall realizations for pasture 2 are listed on the horizontal axis. Mobility is depicted

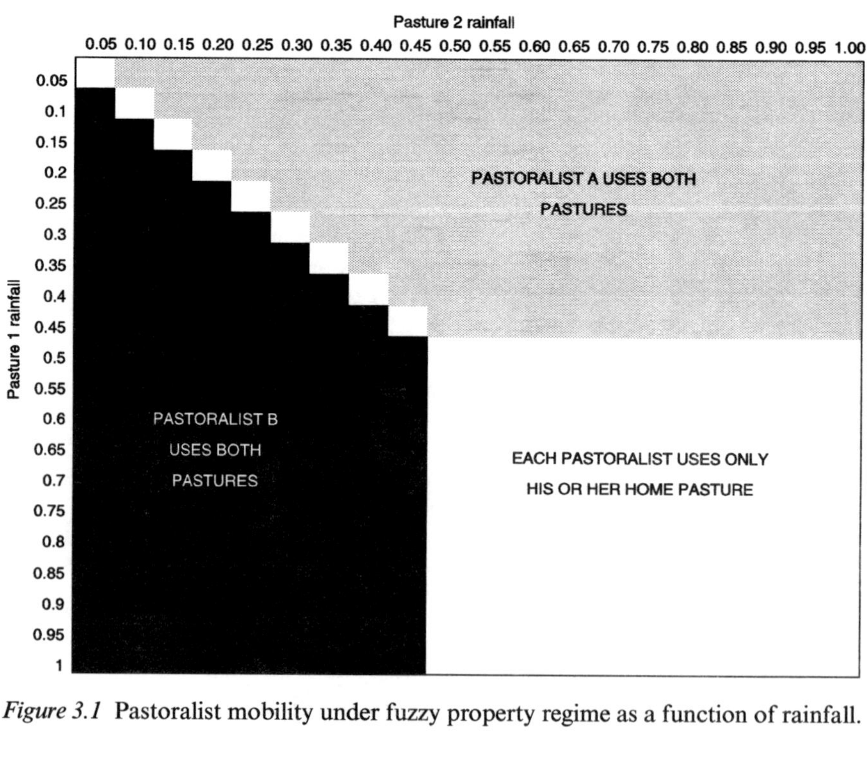

Figure 3.1 Pastoralist mobility under fuzzy property regime as a function of rainfall.

as a function of rainfall realization pairs, with different outcomes represented by the different areas on the graph. Each cell corresponds to a rainfall realization pair, and the areas reflect the pairs for which each pastoralist moves to the other pastoralist's home pasture for part of the season, and for which neither pastoralist moves. The black-shaded area indicates rainfall realization pairs for which pastoralist B will first graze pasture 2 and then move to pasture 1, and the crosshatched area indicates rainfall realization pairs for which pastoralist A will first graze pasture 1 and then move to pasture 2. These areas indicate pairs of rainfall realizations for which one pastoralist has access to the other pastoralist's home pasture. Notice that when the rainfall realizations for the two pastures are the same, neither pastoralist will move. In such cases, the membership of both pastures in the *worse* fuzzy set is zero, so that access to the other pastoralist's home pasture is zero for both pastoralists. Notice that if both pastures receive rainfall realizations that are high enough in both absolute and relative terms, then neither pastoralist will move. In these cases, the membership of both pastures in the *drought* fuzzy set is zero, so that access to the other pastoralist's home pasture is zero for both pastoralists.

3.6 Evaluating the performance of the fuzzy property regime

This section evaluates the effect of the fuzzy property regime on the welfare of an individual pastoralist, and on the social welfare of the two pastoralists. In order to compare the performance of the fuzzy property regime to the performance of common and private property regimes, one must consider when mobility occurs, and the share of the season the two pastoralists share a single pasture. This requires the specification of a rainfall distribution, which is assumed to be uniformly distributed on the interval [0,1].

3.6.1 Comparing fuzzy property and private property

Under a private property regime, the pastoralists remain on their home pastures regardless of rainfall realizations. For a given pair of rainfall realizations, pastoralist i's profits are higher under private property than fuzzy property if the pair of rainfall realizations is such that the other pastoralist has the right to share i's home pasture for part of the season. Conversely, i's profits are higher under fuzzy property than private property if the pair of rainfall realizations is such that i would have the right to migrate and share the other pastoralist's home pasture for part of the season under the fuzzy property regime.

Using the same parameters as Figure 3.1, Figure 3.2 compares A's profits from the two property rights regimes for the same set of rainfall realization pairs. The crosshatched area indicates pairs of rainfall realizations for which A's returns are higher under the fuzzy property regime than under the private property regime. This area corresponds to the crosshatched area in Figure 3.1 that indicated pairs of rainfall realizations for which A would move to B's home pasture for part of the season. The black area indicates pairs of rainfall

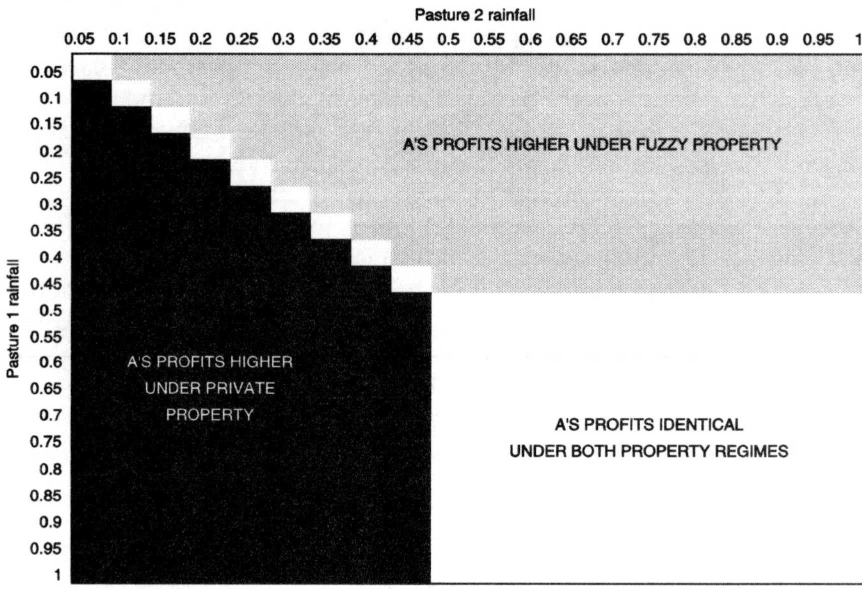

Figure 3.2 Comparison of A's profits under fuzzy property and private property regimes as a function of rainfall.

realizations for which A's returns are higher under the private property regime than under the fuzzy property regime. This area corresponds to the black area in Figure 3.1 that indicated pairs of rainfall realizations for which B would move to A's home pasture for part of the season. When neither pastoralist moves for part of the season, for the pairs of rainfall realizations in the white area in Figure 3.1, profits are identical under the two property regimes, indicated by the white area in Figure 3.2, because the pastoralists' returns are obtained from grazing only their respective home pastures for the entire season.

Figure 3.3 compares social welfare under the fuzzy property regime and the private property regime, using the same parameters as Figures 3.1 and 3.2. As indicated by the crosshatched areas in Figure 3.3, for most of the rainfall realization pairs where A is worse off under the fuzzy property regime, the gains to B are larger than A's losses, and vice versa, resulting in a net improvement in social welfare. When that is not the case, as indicated by the black area in Figure 3.3, rainfalls are relatively similar and low. In such instances, the cost of sharing limited resources is greater than the insurance it provides, from a social perspective.

While Figures 3.2 and 3.3 plot which regime a pastoralist or society prefers for each pair of rainfall realizations, the expected welfare under each system must be computed in order to determine which regime is preferable ex ante. A pastoralist, and society as a whole, will prefer the fuzzy property regime ex ante if the insurance it provides through mobility outweighs the externality of

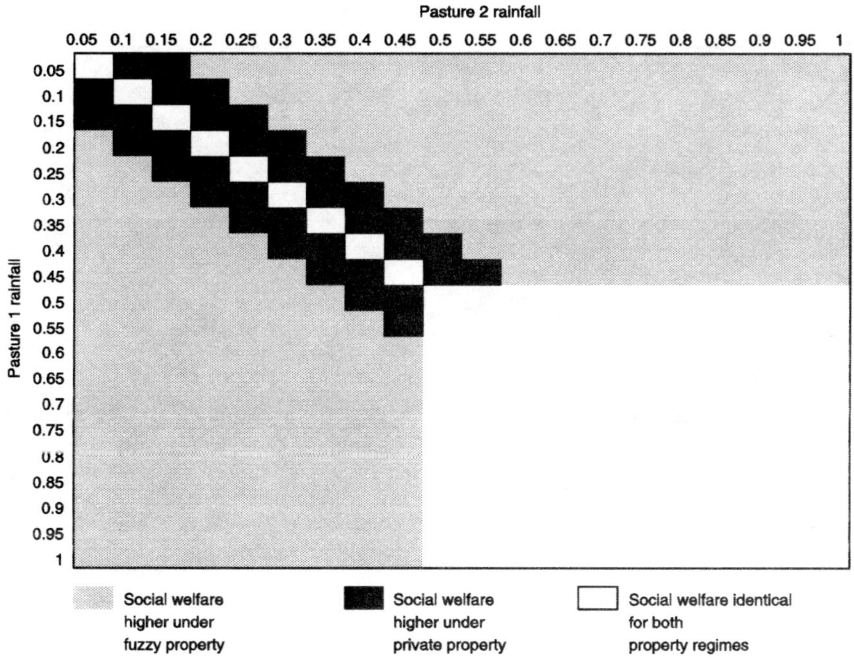

Figure 3.3 Comparison of social welfare under fuzzy property and private property regimes as a function of rainfall.

having both pastoralists share the pasture with more forage for part of the season. This trade-off is a function of parameter values and the distribution of rainfall realizations. For given values of other parameters, including herd sizes a and b, externality parameter β, and drought parameter γ, provided a is sufficiently large each pastoralist's welfare and thus social welfare will be higher under the fuzzy property regime. Essentially, if there is very little forage, even if conditions are good, there is not much benefit to sharing it.

3.6.2 Comparing fuzzy property and common property

Under a common property regime, each pastoralist has complete rights to both pastures. In fuzzy set terminology, the fuzzy set $Access_i$ becomes $Access_i$ = {1/pasture 1, 1/pasture 2}. Both pastoralists are mobile under the common property regime, as they are under the fuzzy property regime. However, because access is not a function of absolute, as well as relative, rainfall, under common property one pastoralist will move unless the rainfall realizations are identical. Thus, for a given set of parameter values mobility occurs under more pairs of rainfall realizations under the common property regime than under the fuzzy property regime, as illustrated in Figure 3.4.

As in Figure 3.1, mobility is depicted as a function of rainfall realization

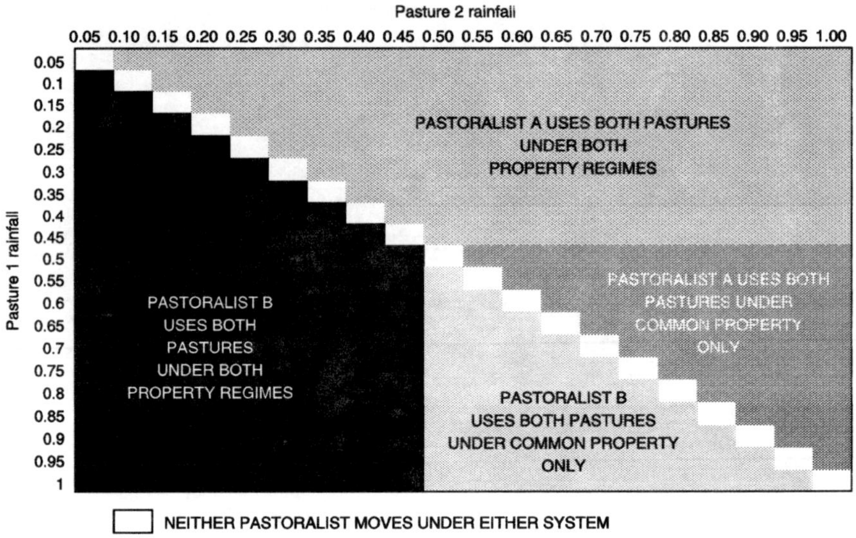

Figure 3.4 Differences in mobility between fuzzy property and common property regimes as a function of rainfall.

pairs by the different areas in Figure 3.4. Here, however, the figure compares mobility under two property rights regimes, so there are more areas to consider. Three areas represent pairs of rainfall realizations for which mobility is the same for the two regimes. The white cells indicate pairs of rainfall realizations for which neither pastoralist will move under either regime. The black and crosshatched areas in Figure 3.4 indicate rainfall realization pairs for which one pastoralist moves to the other's home pasture for part of the season under both regimes. The black-shaded area indicates rainfall realization pairs for which pastoralist B will first graze pasture 2 and then move to pasture 1, and the crosshatched area indicates rainfall realization pairs for which pastoralist A will first graze pasture 1 and then move to pasture 2 under both common property and fuzzy property.

The grey and dotted areas illustrate differences in mobility between the two regimes. Under the fuzzy property regime, if both pastures have sufficiently high rainfall realizations, neither pasture is considered to be a member of the *drought* fuzzy set, so neither pastoralist has access rights to the other's home pasture. However, under common property one pastoralist will always benefit from moving to the other's home pasture unless rainfall is the same for both pastures. The grey area includes rainfall realization pairs for which A will move to B's home pasture for some share of the season under the common property regime, but will remain on pasture 1 under the fuzzy property regime. The dotted area includes rainfall realization pairs for which B will move to A's home pasture for some share of the season under the common property regime, but will remain on pasture 2 under the fuzzy property regime.

It is important to keep in mind that Figure 3.4 only compares mobility under the fuzzy property and common property regimes. It does not report differences in the share of the season the pastoralist who moves spends on the other pastoralist's home pasture, differences in profits, or differences in social welfare between the two regimes. Whether the fuzzy property regime will result in higher social welfare than the common property regime depends on whether the value of being able to exclude the other pastoralist from one's home area under some pairs of rainfall realizations outweighs the insurance provided by mobility under other pairs, which depends on parameter values. When the value of exclusion dominates, the fuzzy property regime is preferred, because when both pastures have rainfall realizations that are not members of the *drought* fuzzy set, neither pastoralist will have access to both pastures, even if the rainfall realizations are not identical. In contrast, whenever there is any difference at all in the rainfall realizations, under the common property regime one pastoralist will spend part of the season on the other pastoralist's home pasture, regardless of how high the rainfall realizations are.

3.7 Conclusion and policy implications

Governments and international aid donors have been actively involved in efforts to improve pastoralists' welfare through various policy interventions. Some of these efforts involve reforming the underlying system of property rights. This chapter has defined the traditional rights system analytically using fuzzy sets in order to reflect how it differs from the private property and noncooperative common property regimes generally used in economic analyses. Under some conditions, fuzzy property rights are preferred to private property rights, and to common property rights.

Due to the stresses on traditional institutions, they have weakened in many areas. In some instances they may no longer be viable, even if efforts are made to strengthen them. This section discusses the policy implications of the analysis in two contexts: its predictions regarding the potential impacts of droughts associated with global climate change and possible policy responses, and its contribution to explaining the outcomes of previous policy initiatives.

3.7.1 Fuzzy property rights, climate change, and drought

Global climate change is considered one of the most important barriers to worldwide poverty eradication (World Bank et al. 2003). Although scientific opinion is not unanimous on this point, increased frequency and duration of droughts in sub-Saharan Africa is an expected consequence of global warming, and may, to some extent, already be occurring. Fratkin (2001) reports increases in droughts in Africa in the second half of the twentieth century in the Sahel and the Horn of Africa. The Turkana people in northwestern Kenya assess that their region has been in a prolonged drought since 1999, following an extended drought from 1992 to 1995. Prior to that, the last prolonged drought was in 1979 through 1980, and before that in 1970 (Oxfam 2006).

The fuzzy property regime provides insight into the likely effects of increased generalized and localized droughts associated with climate change. Under the fuzzy property regime, pastoralist i's right to the other pastoralist's home pasture increases with i's home pasture's degree of membership in the *drought* fuzzy set, given the rainfall realization on the other pasture. Thus, under normal conditions of patchy rainfall, mobility allows pastoralists to respond to localized drought conditions. However, under generalized drought conditions, pastoralists' home pasture rainfall realizations are similar, as well as low, which restricts mobility within customarily used areas. This corresponds to the observed behavior of the drought-stricken Turkana, who have followed two strategies: congregating in towns and relief camps, near food aid and water supplies; and moving herds far outside their traditional grazing areas in an attempt to find pasture (Oxfam 2006). The widespread nature of the drought eliminated the value of mutual insurance, which is the primary benefit provided by traditional property rights. Widespread droughts over long periods of time are likely to reduce the viability of traditional institutions, since mutual insurance is not applicable in such cases.

Increased frequency and severity of localized droughts has different implications. Within the fuzzy property regime, this corresponds to increased frequency of mobility, since a greater share of rainfall realizations will be members of the *drought* and *worse* fuzzy sets. The pastoralist with the lower rainfall realization will spend a greater share of the season, on average, on the other pastoralist's home pasture. This will place greater pressure on the areas with relatively good supplies of water and forage in a given year, as was observed in the Ngorongoro Conservation Area in northern Tanzania during the 1997 drought (Galvin et al. 2004). This will reduce returns to the host pastoralist. While this analysis focused on homogeneous pastures, in cases where some pastoralists have naturally more productive pastures, increased use by others may not be offset by the benefits of mutual insurance. In such cases, the profitability of the traditional system may decline, and private property or common property may become most profitable for pastoralists in more productive areas.

3.7.2 Fuzzy property rights and the implications of access clarification policies

The fuzzy property regime provides insight into some of the unintended consequences of policies intended to clarify resource boundaries and user groups. Consider a common outcome of land reform efforts intended to improve resource use through privatizing land rights in favor of individuals or relatively small groups. Pastoralist groups who had home pastures in relatively productive areas would be more likely to benefit from privatization, and hence would be more likely to participate in land titling efforts and land enclosure. Observers have noted that the richer and better-connected individuals within a region have tended to benefit disproportionately from land privatization efforts, such as the group ranches and later privatizations in Kenya's Maasailand (Graham 1988; Thompson and Homewood 2002; Mwangi 2005), and the Widou Thiengoli borehole controlled grazing scheme in Senegal (Cotula et al. 2004).

Apart from any correlation of wealth or connections with the quality of a group's home pasture, the fuzzy property model suggests an additional implication of this outcome, which is consistent with empirical observations in these two cases. Less successful participants in privatization efforts in their home area are likely to do one of three things: abandon pastoralism entirely, either voluntarily or involuntarily; increase reliance on agriculture and other income-generating activities; or become more mobile and attempt to utilize more intensively other grazing areas to which they have traditionally had access. In the latter case, if they are successful then greater stress will be placed on less productive grazing areas, even in years where these areas have relatively high rainfall, and the enclosure efforts will have eliminated or greatly reduced the insurance that the more productive area would have provided in other years. Clearly, this outcome places additional stress on the traditional system in these less productive areas.

Information obtained in interviews with Maasai elders suggested such a pattern emerging in Kenya's Narok District. In recent years, Maasai in more productive areas have leased land to wheat farmers, eliminating access to forage for a portion of the dry season, until after the wheat harvest. This trend has occurred in conjunction with the subdivision of group ranches in these areas into private parcels. Privatization allows individuals to benefit directly from land leasing, while simultaneously restricting the scope of livestock movement for the group as a whole. Consequently, Maasai based in other, less productive areas not suitable for wheat production lost access to an area they could traditionally graze. In the year the interviews were conducted, the interviewees had seen an influx of herders from other areas into the settlement areas they controlled, because they had relatively more pasture. The elders stated that they would deny permission to graze to Maasai who had leased their own land for wheat.

Consistent with the statements made by the interviewees, the analysis suggests that it may be welfare improving for the pastoralists in the less productive areas to eliminate the traditional system and exclude secondary users entirely, or at least to modify the traditional system in a way that would allow them to exclude pastoralists who can no longer reciprocate with access to areas for which they are the primary managers. However, the poorer pastoralists from the productive area who were effectively excluded from the benefits of the privatization process are not considered in this welfare analysis. Balancing their welfare with other considerations is a difficult social and economic trade-off, well beyond the scope of this chapter.

3.7.3 Conclusion

The results of the analysis in this chapter suggest that policy efforts dealing with property rights need to take a holistic view regarding the sustainability of the entire system. If there is no longer a large enough, or diverse enough, area to physically support the fuzzy property regime's traditional mobility patterns, including the lack of mobility when there is insufficient forage in an area to support both its primary users and other users, then it may be desirable to make active efforts to define property rights under an alternative system. Creating rights of exclusion may, in this case, increase social welfare. On the other hand, it may be the case that it is preferable to strengthen the traditional system, perhaps by reaffirming rights to forage and water and eliminating any expropriation of these resources by specific user groups. Empirical studies have found evidence of partial cooperation among pastoralists in the absence of strict formal rules, which suggests that traditional fuzzy rights persist successfully in some areas (McCarthy et al. 2003; McCarthy and Vanderlinden 2004; Benin and Pender 2006).

Determining which of these cases is likely to hold is a difficult problem, and almost certainly cannot be done without extensive stakeholder participation, as in the case reported in Haro et al. 2005. An advantage of using the fuzzy set theory approach to model traditional property rights is that it is well suited to model linguistic variables. This approach could be used as the basis for a negotiation support system for participants in negotiations regarding property rights reform. Doing so would add value because information about the rights underlying possible location choices could be integrated into the model, not just information on the realized location choices. Rather than simply using a history of resource use in specific years and seasons, which are dependent on the pattern of rainfall realizations, as a basis for defining the rights of groups in different areas, fuzzy set theory could combine this information with stakeholders' verbal characterization of the rights underlying these movement choices, and the determinants of these rights.

Notes

1. The authors thank an anonymous referee for identifying and emphasizing this point.
2. Netting (1972, 1976, 1981) reaches similar conclusions in his study of the southern Swiss village of Törbel. Lower-elevation agricultural lands were owned privately, while higher-elevation forests and grazing lands with lower productivity were owned in common. Interestingly, both the common property and private property regimes addressed the patchiness and variability of agricultural and forage production. The common property nature of the grazing land combined with the management of all of the village's cows by a single group of herders allowed all to benefit equally from the available forage, even when it was patchy. Within the different agricultural microclimates, households owned multiple, geographically separated plots.

References

Aristidou, M. and Sarangi, S. (2006) "Games in fuzzy environments", *Southern Economic Journal*, 72: 645–59.

Bartels, G.B., Norton, B.E., and Perrier, G.K. (1990) "The applicability of the carrying capacity concept in Africa", Network Paper Set 30, Paper 30e, Pastoral Development Network, Overseas Development Institute. Available online: www.odi.org.uk/pdn/papers/paper30e.html (accessed February 22, 2007).

Benin, S. and Pender, J. (2006) "Collective action in community management of grazing lands: the case of the highlands of northern Ethiopia", *Environment and Development Economics*, 11: 127–49.

Bogataj, M. and Usenik, J. (2005) "Fuzzy approach to the spatial games in the total market area", *International Journal of Production Economics*, 93–4: 493–503.

Bruce, J.W. and Mearns, R. (2002) "Natural resource management and land policy in developing countries: lessons learned and new challenges for the World Bank", Issue Paper 115, Drylands Program, International Institute for Environment and Development.

Coppock, D.L. (1994) *The Borana Plateau of Southern Ethiopia: Synthesis of Pastoral Research Development and Change 1980–1991*, Systems Study 5, International Livestock Centre for Africa.

Cossins, N.J. and Upton, M. (1987) "The Borana pastoral system of southern Ethiopia", *Agricultural Systems*, 25: 199–218.

Cossins, N.J. and Upton, M. (1988) "Options for improvement of the Borana pastoral system", *Agricultural Systems*, 27: 251–78.

Cotula, L., Toulmin, C., and Hesse, C. (2004) *Land Tenure and Administration in Africa: Lessons of Experience and Emerging Issues*, International Institute for Environment and Development.

Dixon, J. and Gulliver, A., with Gibbon, D. (2001) *Farming Systems and Poverty: Improving Farmers' Livelihoods in a Changing World*, Rome and Washington, DC: FAO and World Bank.

Dompere, K. (1995) "The theory of social costs and costing for cost-benefit analysis in a fuzzy decision space", *Fuzzy Sets and Systems*, 76: 1–24.

Ensminger, J. (1996) "Culture and property rights", in S. Hanna, C. Folke, and K.-G. Mäler (eds) *Rights to Nature: Ecological, Economic, Cultural, and Political Principles of Institutions for the Environment*, Washington, DC: Island Press.

Evans, P. (2004) "Development as institutional change: the pitfalls of monocropping and the potentials of deliberation", *Studies in Comparative International Development*, 38: 30–52.

Fratkin, E. (2001) "East African pastoralism in transition: Maasai, Boran, and Rendille cases", *African Studies Review*, 44: 1–25.

Galvin, K.A., Thornton, P.K., Boone, R.B., and Sunderland, J. (2004) "Climate variability and impacts on East African livestock herders: the Maasai of Ngorongoro Conservation Area, Tanzania", *African Journal of Range and Forest Science*, 21: 183–9.

Goodhue, R.E. (1998) "Sustaining collusion via a fuzzy trigger", *Review of Industrial Organization*, 13: 333–45.

Graham, O. (1988) "Enclosure of the East African rangelands: recent trends and their impact", Network Paper Set 25, Paper 25a, Pastoral Development Network,

Overseas Development Institute. Available online: www.odi.org.uk/pdn/papers/paper25a.html (accessed February 22, 2007).

Greenhut, J., Greenhut, M.L., and Mansur, Y. (1995) "Oligopoly and behavioral uncertainty: an application of fuzzy set theory", *Review of Industrial Organization*, 10: 269–88.

Hammel, R. (2001) "Securing land for herders in Niger", Drylands Issue Paper E102, International Institute for Environment and Development.

Haro, G.O., Doyo, G.J., and McPeak, J.G. (2005) "Linkages between community, environmental, and conflict management: experiences from northern Kenya", *World Development*, 33: 285–99.

Hogg, R. (1990) "An institutional approach to pastoral development: an example from Ethiopia", Network Paper Set 30, Paper 30d, Pastoral Development Network, Overseas Development Institute. Available online: www.odi.org.uk/pdn/papers/paper30d.html (accessed February 22, 2007).

Hundie, B. (2006) "Property rights among Afar pastoralists of northeastern Ethiopia: forms, changes and conflicts", unpublished paper, Humboldt University of Berlin, presented at Pastoralism and Poverty Reduction in East Africa: A Policy Research Conference, Nairobi, June.

Lane, C. and Moorhead, R. (1994) "New directions in rangeland and resource tenure and policy", in I. Scoones (ed.) *Living with Uncertainty: New Directions in Pastoralist Development in Africa*, International Institute for Environment and Development, Intermediate Technology Publications, London.

Mansur, Y. (1995) *Fuzzy Sets and Economics: Applications of Fuzzy Mathematics to Non-Cooperative Oligopoly*, London: Edward Elgar Co.

McCarthy, N. (1996) "Common property and cooperation in rural Mexico", PhD dissertation, University of California, Berkeley.

McCarthy, N. and Vanderlinden, J.-P. (2004) "Resource management under climatic risk: a case study from Niger", *Journal of Development Studies*, 40: 120–42.

McCarthy, N., Kamara, A.B., and Kirk, M. (2003) "Co-operation in risky environments: evidence from southern Ethiopia", *Journal of African Economies*, 12: 236–70.

McPeak, J.G. (2003) "Analyzing and addressing localized degradation in the commons", *Land Economics*, 79: 515–36.

McPeak, J.G. and Barrett, C.B. (2001) "Differential risk exposure and stochastic poverty traps among East African pastoralists", *American Journal of Agricultural Economics*, 83: 674–9.

Mwangi, E. (2005) "The transformation of property rights in Kenya's Maasailand: triggers and motivations", CAPRi Working Paper 35, CGIAR Systemwide Program on Collective Action and Property Rights. Available online: www.capri.cgiar.org/pdf/capriwp35.pdf (accessed February 24, 2007).

Netting, R. (1972) "Of men and meadows: strategies of alpine land use", *Anthropological Quarterly*, 45: 132–44.

Netting, R. (1976) "What alpine meadows have in common: observations on communal tenure in a Swiss village", *Human Ecology*, 4: 135–46.

Netting, R. (1981) *Balancing on an Alp*, Cambridge University Press, Cambridge.

Niamir, M. (1991) "Traditional African range management techniques: implications for rangeland development", Network Paper Set 31, Paper 31d, Pastoral Development Network, Overseas Development Institute. Available online: www.odi.org.uk/pdn/papers/31d.pdf (accessed February 22, 2007).

Oba, G. (1992) “Ecological factors in land use conflicts, land administration and food insecurity in Turkana, Kenya”, Network Paper Set 33, Paper 33a, Pastoral Development Network, Overseas Development Institute. Available online: www.odi.org.uk/pdn/papers/paper33a.html (accessed February 22, 2007).

Ostrom, E. (1990) *Governing the Commons: The Evolution of Institutions for Collective Action*, Cambridge: Cambridge University Press.

Ostrom, E. (2002) “Common-pool resources and institutions: toward a revised theory”, in B. Gardner and G. Rausser (eds) *Handbook of Agricultural Economics: Volume 2*, Amsterdam: Elsevier Science.

Oxfam (2006) *Climate Change and the Sahel Drought: Stories and Case Studies*, Oxfam, Oxford. Available online: www.oxfam.org.uk/what_we_do/issues/climate_change/story_turkana.htm (accessed February 24, 2007).

Pantuliano, S. (2002) *Sustaining Livelihoods across the Rural-Urban Divide: Changes and Challenges Facing the Beja Pastoralists of North Eastern Sudan*, Pastoral Land Tenure Series 14, International Institute for Environment and Development. Available online: www.iied.org/pubs/pdf/full/9163IIED.pdf (accessed February 22, 2007).

Scoones, I. (ed.) (1994) *Living with Uncertainty: New Directions in Pastoralist Development in Africa*, International Institute for Environment and Development, Intermediate Technology Publications, London.

Solomon, T.B., Snyman, H.A., and Smit, G.N. (forthcoming) “Cattle-rangeland management practices and perceptions of pastoralists towards rangeland degradation in the Borana zone of southern Ethiopia”, *Journal of Environmental Management*.

Sserunkuuma, D. and Olson, K. (2001) “Private property rights and overgrazing: an empirical assessment of pastoralists in Nyabushozi County, western Uganda”, *Economic Development and Cultural Change*, 49: 769–92.

Swallow, B. (1994) *The Role of Mobility within the Risk Management Strategies of Pastoralists and Agro-Pastoralists*, Gatekeeper Series 47, International Institute for Environment and Development, London.

Thompson, G. and Wilson, P. (1994) “Ejido reforms in Mexico: conceptual issues, potential outcomes”, *Land Economics*, 70: 448–65.

Thompson, M. and Homewood, K. (2002) “Entrepreneurs, elites, and exclusion in Maasailand: trends in wildlife conservation and pastoralist development”, *Human Ecology*, 30: 107–38.

Thornton, P.K., Kruska, R.L., Henninger, N., Kristjanson, P.M., Reid, R.S., Atieno, F., Odero, A.N., and Ndegwa, T. (2002) *Mapping Poverty and Livestock in the Developing World*, International Livestock Research Institute, Kenya. Available online: www.ilri.cgiar.org/InfoServ/Webpub/Fulldocs/Mappoverty/index.htm (accessed February 22, 2007).

Toulmin, C. (1994) “Tracking through drought: options for destocking and restocking”, in I. Scoones (ed.) *Living with Uncertainty: New Directions in Pastoralist Development in Africa*, International Institute for Environment and Development, Intermediate Technology Publications, London.

Toulmin, C. (2005) “Securing land and property rights in sub-Saharan Africa: the role of local institutions”, in T. Bigg and D. Satterthwaite (eds) *How to Make Poverty History: The Central Role of Local Organizations in Meeting the MDGs*, International Institute for Environment and Development, London. Available online: www.iied.org/Gov/mdgs/documents/mdg3/ch2_28pp.pdf (accessed February 22, 2007).

van den Brink, R., Bromley, D., and Chavas, J. (1995) "The economics of Cain and Abel: agro-pastoral property rights in the Sahel", *Journal of Development Studies*, 31: 373–99.

World Bank et al. (2003) *Poverty and Global Climate Change: Reducing the Vulnerability of the Poor through Adaptation*, World Bank, Washington, DC. Available online: lnweb18.worldbank.org/ESSD/envext.nsf/46ByDocName/KeyThemes VulnerabilityandAdaptationPovertyandClimateChange (accessed February 22, 2007).

Zadeh, L. (1965) "Fuzzy sets", *Information and Control*, 8: 338–53.

4 Application of partition function games to the management of straddling fish stocks

Pedro Pintassilgo and Marko Lindroos

The formation of coalitions among fishing States is a central issue in straddling stock fisheries. In this chapter, this is approached through a game in partition function form. The game is applied to a fishery represented by the classical Gordon–Schaefer bioeconomic model and to the North Atlantic bluefin tuna fishery. Both cases show very pessimistic prospects regarding the cooperative management of the fish resources. This study concludes that the breakdown of cooperative agreements on straddling stock fisheries is a very likely scenario if countries can adopt free-ride strategies.

4.1 Introduction

This chapter addresses straddling stock fisheries through coalition games. Straddling fish stocks are a special category of internationally shared fishery resources that straddle exclusive economic zones (territorial seas claimed by coastal States) and the adjacent high seas.[1] These species, usually targeted by both coastal States and distant water fishing States, became increasingly disputed after the establishment of exclusive economic zones by the United Nations Convention on the Law of the Sea (United Nations 1982). Lack of cooperation between coastal States and distant water fishing States has led to the overexploitation of many stocks worldwide.

Munro (1999) refers to the Alaska pollock fishery in the Bering Sea as a paradigmatic example of noncooperative behavior among fishing States and consequent stock overexploitation. A significant part of this straddling fish stock, one of the largest groundfish resources in the world, straddles the exclusive economic zone of the United States of America and a high seas enclave between the United States and Russian exclusive economic zones commonly known as the Doughnut Hole. The fishing pressure on the high seas portion of the stock exerted by distant water fishing States after 1984 resulted in the depletion of the stock in the early 1990s. The increasing tension between the United States and distant water fishing States led the US coastguards to seize several foreign fishing vessels allegedly using the Doughnut Hole as a base for fishing operations within United States waters. In 1992, after the stock crash, all States involved in the fishery reached an agreement to impose a fishing moratorium in the Doughnut Hole. Another classical example, given by Munro (1999), concerns the groundfish stocks on the Grand Banks of Newfoundland. The lack of cooperation between Canada and the European Union led to the overexploitation of the stocks and to conflicts that reached a climax in 1995 when Canadian authorities arrested a Spanish vessel in international waters.

The economic and biological overexploitation of the stocks and the increasing conflicts among countries, often called "international fish wars", induced the United Nations to convene the United Nations Conference on Straddling Fish Stocks and Highly Migratory Fish Stocks (1993–95). In 1995 the conference adopted the United Nations Fish Stocks Agreement[2] (United Nations 1995), which entered into force in 2001. The core of the agreement consists of placing regional fisheries management organizations (RFMOs) as the basic cells for the management of these marine resources. These organizations should integrate both the coastal States and the distant water fishing States effectively interested in the fishery. According to Munro (2006), the United Nations Fish Stocks Agreement left a few problems unsolved that may undermine cooperative agreements achieved under the aegis of RFMOs. One problem is the possibility of prospective new members wanting to join an RFMO and share the harvest following recovery of the stock, thus reaping the benefits of stock management without having borne any of the cost of the investment (free riding). Another problem is the one posed by nonmembers who do not follow an RFMO management regime and thus behave noncooperatively when exploiting the fishery resources in their exclusive economic zones or in the high seas. Munro (2006) refers to the Commission for the Conservation of Southern Bluefin Tuna as an example where both these problems have surfaced.

The economics of straddling stock fisheries is based on coalition games. In this chapter the main studies on coalition games applied to the management of straddling fish stocks are reviewed. The core aim of the chapter is to explore the potential of partition function games, a class of coalition games recently introduced into the fisheries literature. In these games the payoff for each player is defined by a partition function that depends not only on the coalition the player belongs to but also on the way the other players form coalitions. Partition function games have been suggested for cases in which coalition formation produces externalities that affect nonmembers (Yi 1997). This is a typical situation in the management of straddling fish stocks under RFMOs. The more RFMO member players the better for nonmembers, as they can adopt free-rider strategies. Thus, partition function games are an appropriate framework to address free-rider problems faced by RFMOs. In the present chapter a partition function game on a straddling stock fishery is defined. In addition, the stability of cooperative agreements and the equilibrium of the game are analyzed.

The chapter is structured as follows. Section 4.2 presents a brief revision of coalition games on straddling fish stocks. Section 4.3 discusses the relevance of partition function games and defines the main concepts. Section 4.4 presents the partition function game used in the chapter, and is followed by two applications: section 4.5, a straddling stock fishery represented by the classical Gordon–Schaefer model, and section 4.6, the North Atlantic bluefin tuna fishery. Section 4.7 concludes with a discussion of the main findings and policy recommendations.

4.2 Coalition games

A straddling stock fishery usually involves many countries or fleets. The analysis of games in which the number of players exceeds two requires analysis of coalitions. A coalition means a subset of the set of players. Two or more countries are considered to form a coalition if they ratify (or sign) a multilateral agreement on the particular fishery.

Three types of coalition scenarios may result. If all parties concerned sign the agreement the situation is denoted as full cooperation, and a grand coalition is said to be formed. If some countries are left outside the agreement the situation is denoted as partial cooperation and outsiders may act as free riders. Finally, in the case of noncooperation there are no agreements between the countries, and each is only interested in maximizing individual benefit from the fishery.

Based on the three possible outcomes described above a characteristic function of the game can be established. The characteristic function assigns a value to each possible coalition. The value in the case of straddling fish stocks is, generally, interpreted as the net present value of the fishery to a certain coalition.

The value for coalition members depends on the particular behavior of nonmembers. The different assumptions about their behavior give rise to the α, β, and γ characteristic functions. The α and β characteristic functions assume that the nonmembers jointly try to punish the coalition members. The problem with this assumption in the context of international fisheries games is that the punishments decrease the payoff of the nonmembers. Therefore, the γ-characteristic function, introduced by Chander and Tulkens (1995), is typically preferred. The assumption made by the authors is that nonmembers play as singletons and adopt individually best-reply strategies against the coalition. This results in a Nash equilibrium between the coalition and the nonmembers.

An example of a γ-characteristic function based on a game between three players exploiting Norwegian spring-spawning herring (Lindroos and Kaitala 2000) is given in Table 4.1. The players are Norway, Iceland, and the European Union (EU).

In Table 4.1, the first value (US$6,800 million) corresponds to the payoff of the grand coalition formed by all the players. This payoff is calculated by maximizing the joint net present value of profits over a period of 50 years. Under the full cooperative case, the stock level in the long run is the highest

Table 4.1 Example of a characteristic function

Coalition	*Payoffs (net present value)*[a]
Norway, Iceland, EU (grand coalition)	6,800
Norway, Iceland	3,000
Norway, EU	2,800
Iceland, EU	2,700
Norway	750
Iceland	350
EU	140

a. Approximate values in 10^6 US$ (constant prices of 1998).

compared to the partial and noncooperative cases. The next three values are the payoffs of the two-player coalitions under partial cooperation. These values result from the Nash equilibrium of the game among a two-player coalition, maximizing its joint benefits, and a single country (free rider). The last three values give the noncooperative equilibrium payoffs. Clearly, countries gain by joining together; for example, a coalition of Norway and Iceland would receive $3,000 million compared to $1,100 million, the sum of their payoffs under noncooperation. Whether a grand coalition can be formed is a more complex problem. In this case, the conditions when the grand coalition would be stable need to be defined. This is discussed in detail later in section 4.4. The following sections present an overview of coalition game literature (see also Lindroos et al. 2006 for a detailed and more technical review).

Kaitala and Lindroos (1998) introduced coalition games to the economic analysis of straddling stocks and fisheries economics in general by using a three-player model. Their main result is that there is a partial cooperation equilibrium stock level that is higher than the noncooperative stock level (Clark 1980), though lower than the cooperative stock level (Clark and Munro 1975). They also applied Shapley value and nucleolus as cooperative solutions to the sharing of full cooperative benefits. Contrary to previous results, uneven distribution of benefits was suggested because of coalitional bargaining power differences.

Lindroos (2004a) extended the Kaitala–Lindroos model (1998) by allowing for restricted coalition formation in a four-player game. This was analyzed in the context of straddling stock negotiations where distant water fishing States negotiated with coastal States. The main result was that, by joining together, distant water fishing States gained compared to unrestricted negotiations. Arnason et al. (2000) studied the case of Atlanto-Scandian herring and found that Norway was a crucial country for any coalition to be stable. Their findings relate to those of Lindroos (2004a), who considered the possibility of veto countries. The effect of uncertainty and selectivity was studied by Lindroos (2004b). He showed that the safe minimum economic stock level that would maximize the possibilities for cooperation was higher than the safe minimum biological level (precautionary level).

Kennedy (2003) applied the concept of coalition-proof equilibrium to the Northeast Atlantic mackerel fishery. Sang Kwon (2006), in one of the most recent applications of coalition games to straddling stocks, extended the Levhari–Mirman (1980) model to include coalitions.

4.3 Partition function games

Characteristic function games have been applied to straddling stock fisheries since the late 1990s. Nonetheless, as observed by Greenberg (1994), the framework of a characteristic function approach, although sufficiently general to encompass many contributions of coalition formation theory, is not fully satisfactory. Most importantly, it ignores the possibility of externalities among coalitions, that is, the effects that coalition mergers have on the payoffs of players who belong to the other coalitions.

> *Definition 1:* Externalities are present, in a game in coalition form, if and only if there is at least a merger of coalitions that changes the payoff of a player belonging to a coalition not involved in the merger. If the merger increases (decreases) the payoff of the player, the externality is considered as positive (negative).

In the context of straddling fish stock management through RFMOs, externalities are generally present. In fact, as these organizations tend to adopt conservative management strategies, nonmembers are typically better off when more players become members, as free-rider strategies can be adopted. Therefore, when a player joins an RFMO it generally creates a positive externality for nonmembers.

According to Yi (1997), the formation of economic coalitions with externalities has opened a new strand of literature on noncooperative game theory. Most studies (for example Bloch 1996; Yi 1996) are centered on finding the equilibrium number and size of coalitions and share a common two-stage game framework. In the first stage players form coalitions, whereas in the second stage coalitions engage in noncooperative behavior. The coalition payoffs are represented by a partition function (definition 3). This function assigns a value to each coalition as a function of the entire coalition structure (definition 2). Therefore, it captures the externalities across coalitions that are assumed to be absent in the characteristic function.

> *Definition 2:* A coalition structure $C = \{S_1, S_2, \ldots, S_z\}$ is a set of coalitions that altogether integrates all the players, with each player belonging to only one coalition.

> *Definition 3:* The partition function $\Pi\ (S_k;\ C)$ yields the payoff of coalition S_k, which is an element of the coalition structure $C = \{S_1, S_2, \ldots, S_z\}$. The payoff of player i, which is a member of coalition S_k, is given by the per-member partition function $\pi_i\ (S_k;\ C)$.

Recently, games in partition function form have been applied to internationally shared fish stocks. Pintassilgo (2003) models straddling stock fisheries as a partition function game and derives general results regarding the stability of coalition structures and the equilibrium of the game. The author explores, in particular, the case in which positive externalities are present. An application to the North Atlantic bluefin tuna fishery is undertaken using an age-structured, multigear bioeconomic model. Pham Do and Folmer (2003) study the feasibility of partial cooperation and its impacts on fishing effort through a game in partition function form. Using a static model, general results on the coalition structure in the competitive equilibrium are derived.

4.4 Stylized game

In this section, the game framework used by Pintassilgo (2003) is described. This will be applied to specific bioeconomic fishery models in the following sections. Assume that an RFMO is established with the purpose of managing and conserving a straddling fish stock. Consider a two-stage game and a finite number of players. In the first stage, each player decides whether to become a member of the RFMO or to remain a nonmember. As nonmembers typically adopt noncooperative behavior and do not join together, it is assumed that they will act individually as singletons. This is a standard assumption in the literature on international environmental agreements (Finus 2001). In particular, Chander and Tulkens (1997) adopt it in modeling an economy with multilateral externalities. The authors assume that when a coalition is formed, the players outside the coalition adopt only individual best-reply strategies. According to this assumption, coalition payoffs are described by a γ-characteristic function.

Regarding the rule of coalition formation, the simultaneous-move open membership game (Yi 1997) is assumed. Under this rule, the membership of each coalition is open to all players who are willing to follow its strategies. This game is designed to model an institutional environment in which players are allowed to form coalitions freely, as long as no player is excluded from joining any coalition. Thus, any player can choose to be either a member or a nonmember of the RFMO. This assumption is clearly within the legal framework set by the United Nations Fish Stocks Agreement. According to Article 8(3), any State with a real interest in the fisheries concerned should not be precluded from becoming a member of the regional RFMO (United Nations 1995). Furthermore, the agreement appears to leave open the possibility of nonmembers fishing in their exclusive economic zones or in the high seas (Munro 1999).

In the second stage it is assumed that the grand coalition members act cooperatively by choosing the fishing strategy that maximizes their aggregate payoff. For all the other coalition structures, each coalition chooses the strategy that maximizes its own payoff given the behavior of the remaining players. This noncooperative behavior leads to a noncooperative solution for each coalition structure, assumed to be unique. Thus, the coalition payoffs in the second stage can be defined as a partition function. Given the partition function, which yields the equilibrium payoffs of the second-stage game, the equilibrium coalition structures of the first-stage game are the Nash equilibrium outcomes of the open membership game of coalition formation.

An important issue that emerges by setting a coalition approach to a straddling stock fishery is the coalition structure stability. As the United Nations Fish Stocks Agreement calls for cooperative management, special interest is centered on the stability of the grand coalition. According to Greenberg (1994), since the merger of players into an RFMO tends to create positive externalities for nonmembers, the analysis of stability based on single-player deviations emerges naturally. Moreover, in the context of positive externalities, Yi (1997) refers to the concept of stand-alone stability as being particularly useful in characterizing equilibrium coalition structures. This concept is defined as follows.

> *Definition 4:* A coalition structure is stand-alone stable if and only if no player finds it profitable to leave its coalition to form a singleton (one-player) coalition, holding the rest of the coalition structure constant (including its former coalition). In the case of the grand coalition, this occurs when no player is interested in leaving the cooperative agreement to adopt free-rider behavior.

The following proposition, presented by Pintassilgo (2003), establishes a condition under which a given coalition structure is not stand-alone stable.

> *Proposition 1:* A sufficient condition for a coalition structure not to be stand-alone stable is that the sum of the payoffs of the singleton coalitions, resulting from unilateral deviations from any of its coalitions, exceeds the value of that coalition.

This proposition provides a tool to determine those cases in which no sharing rule can make a coalition structure stand-alone stable. Specifically, it can be used to assess the possibility of a stable cooperative agreement.

4.5 Application 1: a classical bioeconomic model

The game setting established in the previous section is now applied to a straddling fish stock fishery represented by the classical Gordon–Schaefer bioeconomic model (Gordon 1954). Because of its simplicity, this static model is commonly used in game-theoretic approaches to internationally shared fish resources (for example Ruseski 1998; Lindroos 2002).

Consider n ex ante symmetric players[3] (fleets/countries) in a straddling fish stock fishery. The fish stock dynamics follow the standard Schaefer model (Schaefer 1954), which can be represented through the following equations:

$$\frac{dX}{dt} = G(X) - \sum_{i=1}^{n} H_i \tag{4.1}$$

$$G(X) = rX\left(1 - \frac{X}{k}\right) \tag{4.2}$$

$$H_i = qE_iX \tag{4.3}$$

where X presents fish stock biomass; t the time; $G(X)$ the logistic growth function; r the intrinsic growth rate of fish; k the carrying capacity of the ecosystem; H_i the harvest of player i; q the catchability coefficient; and E_i the fishing effort of player i.

According to (4.1), the variation of the stock in time is given by the difference between stock growth and total harvest. Stock growth is defined by a logistic function (4.2). This is an inverted U-shaped function. Stock growth increases with the stock level up to a maximum value, often designated as maximum sustainable yield. As the stock continues to increase, the growth starts to decrease, and upon reaching zero, the stock stabilizes at the carrying capacity of the ecosystem. Thus, for low levels the fish multiply, but once they begin to compete for food, growth reduces and the stock tends to the level that can be sustained by the environment. The harvest function (4.3) indicates that the harvest of each player increases with stock level and fishing effort—an aggregate measure of the inputs devoted to harvesting, such as days at sea.

Equations (4.1) to (4.3) can be used to determine the equilibrium, or steady state, stock level that corresponds to a given fishing effort that is constant through time. This steady state relation is given by:

$$X = \frac{k}{r}\left(r - q\sum_{i=1}^{n} E_i\right). \tag{4.4}$$

As expected, (4.4) indicates a negative relation between the equilibrium stock level and the players' fishing effort.

Assuming constant price and cost per unit of effort, the aggregate economic rent from the fishery is:

$$ER = p\sum_{i=1}^{n} H_i - c\sum_{i=1}^{n} E_i \tag{4.5}$$

where ER denotes the aggregate economic rent; p the price; c and the cost per unit of effort.

Formalizing the economic rent, or economic profit, is the most common way to introduce the fishery economic dimension in bioeconomic modeling. Herein, the standard assumption that agents maximize their profits is adopted.

The Gordon–Schaefer model is still the main reference in theoretical approaches to fisheries bioeconomics. This aggregated model has shown significant potential to demonstrate fundamental economic principles in fisheries management. In particular, it has been the principal bioeconomic model used in addressing noncooperative management of internationally shared stocks (Munro 1999). Nonetheless, the scope of application of this model to empirical studies is very limited, because of its simplicity.

Consider the two-stage framework presented in the previous section. The per-member partition function is determined assuming that each coalition chooses the fishing strategy that maximizes its steady state economic profit, given the behavior of other coalitions. Furthermore, as players are ex ante symmetric, an equal sharing of coalition payoffs is assumed.

In the case of full cooperation, all players adhere to the RFMO. This grand coalition maximizes its steady state profit (4.5). The corresponding aggregate fishing effort and the payoff received by each member are given respectively by:

$$E = \frac{r}{2q}(1-b) \tag{4.6}$$

$$\pi = \frac{1}{n}(pqk-c)\frac{r}{4q}(1-b) \tag{4.7}$$

where $b = \frac{c}{pqk}$. This is usually designated an "inverse efficiency parameter", as it increases with the cost per unit of effort and decreases with price and catchability coefficient. Therefore the higher b, the lower players' efficiency.

As expected the payoff decreases with the number of players and the cost per unit of effort, and increases with the remaining parameters (price, catchability coefficient, intrinsic growth rate of fish, and carrying capacity of the ecosystem).

In the partial cooperation scenario, a coalition of two or more RFMO members plays against singleton nonmembers. In the Nash equilibrium, the fishing effort of each coalition is the following:

$$E = \frac{r}{(m+2)\,q}(1-b) \tag{4.8}$$

where m denotes the number of nonmembers. Hence, $n-m$ represents the number of RFMO members, and $m+1$ the number of coalitions.

The per-member payoff of the coalition formed by the players that adhere to the RFMO is given by:

$$\pi = \frac{(pqk-c)\,r\,(1-b)}{(n-m)\,(m+2)^2\,q} \tag{4.9}$$

The singleton nonmembers receive:

$$\pi = \frac{(pqk - c)\, r\, (1 - b)}{(m + 2)^2\, q} \tag{4.10}$$

From (4.9), it can be concluded that the payoff of the RFMO members decreases with the number of players, for a given number of nonmembers. Regarding the payoff of nonmembers, (4.10) indicates that it decreases with the number of coalitions.

Finally, under noncooperation all the players are nonmembers of the RFMO and behave as singletons. In the Nash equilibrium, the fishing effort and the payoff of each player are given, respectively, by:

$$E = \frac{r}{(n + 1)\, q}(1 - b) \tag{4.11}$$

$$\pi = \frac{(pqk - c)\, r\, (1 - b)}{(n + 1)^2\, q} \tag{4.12}$$

The payoff of each singleton decreases with the number of players. Moreover, it converges to zero as the number becomes infinitely large.

Having determined the values of the per-member partition function of this game, the next step is to explore the properties of this function.

> *Proposition 2:* The per-member partition function game presents positive externalities.[4]

This proposition indicates that the merger of coalitions increases the payoff of players who belong to the other coalitions. Having established the presence of positive externalities in the game, its consequences will now be analyzed in terms of coalition structure stability and equilibrium. The following proposition addresses the coalition structure stand-alone stability for the present straddling stock fishery game.

> *Proposition 3:* In a two-player game both the grand coalition and the coalition structure formed by singletons are stand-alone stable. In a game with three or more players the only stand-alone stable coalition structure is the one formed by singletons.

From this proposition it can be concluded that, aside from the unusual case of two players, no cooperative agreement is stand-alone stable. This is because of the presence of free-rider incentives associated with positive externalities. The free-rider behavior of nonmember fleets, which tends to occur in the presence of externalities, has been pointed out as one of the main factors behind the difficulties faced by RFMOs in the cooperative management of straddling fish stocks. A paradigmatic case is the Commission for the Conservation of Southern Bluefin Tuna. In order to protect itself against free riders, this RFMO has been trying to encourage nonmembers to adhere to the organization and has implemented a trade information scheme with the aim of deterring illegal, unreported, and unregulated fishing by effectively denying access to markets for the species.

Finally, the equilibrium of the game will be considered. This is a very relevant aspect since it is the expected scenario in the straddling stock fishery after all strategic adjustments have been made.

> *Proposition 4:* In a two-player game the grand coalition, formed by the two players, is the Nash equilibrium coalition structure. If the number of players is three or more, the Nash equilibrium is the coalition structure in which all players act as singletons, that is, complete noncooperation.

Proposition 4 indicates that the cooperative management of a straddling fish stock under an RFMO is a very unlikely outcome. This result emphasizes the importance of dealing effectively with free-rider behavior by nonmembers.

4.6 Application 2: North Atlantic bluefin tuna fishery

In this section the game in partition function form is applied to the eastern stock of the North Atlantic bluefin tuna. This straddling fish stock, whose management falls under the aegis of the International Commission for the Conservation of Atlantic Tunas (ICCAT), has been pointed out as one of the most typical cases of failure to implement a cooperative agreement (Pintassilgo 2003).

4.6.1 The fishery

The North Atlantic bluefin tuna is a large oceanic pelagic and also the largest of the tunas. Its normal fork length, in adult stage, is between 1.6 and 2.4 meters, but can reach more than 3 meters. Like many other tunas, the bluefin tend to be found in schools of similar-sized individuals. This species spawns in two main areas: the West Atlantic (Gulf of Mexico and Florida Straits) and the Mediterranean Sea (around the Balearic Islands and in the southern Tyrrhenian Sea). Based on the spawning areas, ICCAT defines two stocks for management purposes: the West Atlantic and the East Atlantic.[5] Hereafter, the analysis is centered on the eastern stock, which is distributed roughly from the Canary Islands to southern Iceland and in all the Mediterranean Sea.

In the Northeast Atlantic and Mediterranean Sea, the bluefin tuna is harvested by a large number of fleets from European Union member coastal States (Cyprus, France, Greece, Italy, Malta, Portugal, Spain), other coastal States (Algeria, Croatia, Libya, Morocco, Turkey), and distant water fishing States (China, Japan, Republic of Korea, Taiwan, United States). A variety of fishing gears are used, including purse seine, longline, trap, and baitboat. The different fishing gears target different quality and size specimens, which have different market values. Large-size and high-quality bluefin tunas are absorbed by the Japanese sashimi market, where prices are highest.

According to recent stock assessments, the eastern bluefin tuna stock is severely depleted and current reported catches (28,889 tonnes in 2004) cannot be sustained (ICCAT 2005). Furthermore, ICCAT suspects that there has been an increasing underreporting of catches over recent years. One of the main factors contributing to this situation is the high number of fleets involved in the fishery, both members and nonmembers of ICCAT. Another factor is the high price of the bluefin tuna in the Japanese market. This market has a strong and selective demand towards large-size and high-quality specimens, for which it is virtually the only consumer. The depletion of the southern bluefin tuna stock in the Pacific Ocean also contributes to the fishing pressure on the North Atlantic bluefin tuna.

4.6.2 Partition function game

Assume that all the countries participating in the bluefin tuna fishery in the East Atlantic and Mediterranean are represented in the RFMO, ICCAT, by one of the following players: European Union (EU), other coastal States (OCSs), and distant water fishing States (DWFSs). Consider the two-stage game framework presented in section 4.4. In the first stage, the coalition structure is determined by the players' choice to join ICCAT or play outside as singletons. In the second stage, the coalitions choose their fishing strategy.

The optimal fishing strategy for the coalition formed by ICCAT members is assumed to be the one that maximizes the net present value of profits over a 25-year period, given the behavior of the nonmembers. Nonmembers are assumed to follow market behavior by adjusting effort according to the sign of profits.[6] The coalition strategies and payoffs were computed by using an age-structured, multigear bioeconomic model (Pintassilgo and Costa Duarte 2002). For all the coalition structures, a unique equilibrium[7] payoff vector was obtained. The resulting payoff matrix (Pintassilgo 2003) is shown in Table 4.2. In representing the coalition structures, all the players that belong

Table 4.2 Coalition payoffs for bluefin tuna fishery

	Payoffs (net present value)[a]				
Coalition structure	*Total*	*ICCAT*	*EU*	*OCS*	*DWFS*
(EU, OCSs, DWFSs)	1,291.7	1,291.7	—	—	—
(EU, OCSs), (DWFSs)	991.8	154.2	—	—	837.6
(EU, DWFSs), (OCSs)	976.4	224.0	—	752.4	—
(OCS, DWFSs), (EU)[b]	7.6	21.4	–13.8	—	—
(EU), (OCSs), (DWFSs)	7.6	—	–13.8	–5.0	26.4

a. Values in 10^6 US$ (constant prices of 1995).

b. If ICCAT is composed of OCSs and DWFSs, there is no conservative strategy that can yield higher payoff for these players than in the case in which the three players act noncooperatively as singletons.

to the same coalition are placed within the same parentheses. As it is assumed that nonmembers act as singleton coalitions, all the coalitions with more than one member belong to ICCAT. Moreover, it is also assumed that ICCAT members will never be singletons, as no single player can gain by adopting a conservative strategy.

This game presents positive externality characteristics. In fact, Table 4.2 shows that when the EU joins the OCSs the payoff of the DWFSs increases from US$26.4 million to $837.6 million. A similar external effect occurs when the EU joins the DWFSs. The stability of the cooperative agreement will now be analyzed.

> *Proposition 5:* There is no sharing rule that makes the cooperative agreement stand-alone stable.

This can easily be shown through proposition 1. The values of the singleton coalitions, resulting from unilateral deviations from the grand coalition, are –$13.8 million, $752.4 million, and $837.6 million for the EU, OCSs, and DWFSs, respectively. As the sum of these values ($1,576.2 million) exceeds the payoff of the grand coalition ($1,291.7 million), it can be concluded that there is no sharing rule that can make the grand coalition stand-alone stable. Since stand-alone stability is a necessary condition for a coalition structure to be a Nash equilibrium, the grand coalition cannot be a Nash equilibrium outcome of this game.

What is the Nash equilibrium of the game? It must be a stand-alone stable coalition structure. Using definition 4, it can be shown that there are only three coalition structures that can potentially follow this property: {(EU, OCSs), (DWFSs)}; {(EU, DWFSs), (OCSs)}; and {(EU), (OCSs), (DWFSs)}. Whether any is a Nash equilibrium depends on the particular sharing rule adopted for the division of the coalition payoffs. By using a variable sharing rule, endogenously determined, that can differ among coalition structures, Pintassilgo (2003) shows that there is no Nash equilibrium coalition structure for this game.

From these results, a successful cooperative agreement on the management of this fishery seems an unlikely outcome. However, as shown by Pintassilgo (2003), if DWFSs are not allowed to adopt free-rider behavior then there are sharing rules that make the grand coalition stand-alone stable. Thus, the results indicate that legal restrictions, prohibiting fishing to fleets that do not follow the regulations set by RFMOs, are essential in order to sustain cooperative agreements.

4.7 Conclusion and policy implications

The management of straddling fish stocks has become a major international problem since the drafting of the United Nations Convention on the Law of the Sea (United Nations 1982). According to Munro (1999), the segment of the convention that addresses the management of these fishery resources "proved in retrospect to be seriously inadequate". In 1995, the United Nations Fish Stocks Agreement was adopted in order to supplement the convention. However, some problems still remained to be solved; for example, the threat to the cooperative management of RFMOs posed by prospective new members and nonmembers (Munro 2006). These problems are still clearly faced by such RFMOs as the Commission for the Conservation of Southern Bluefin Tuna and ICCAT.

Coalition games have been the main tool used by economists to address the management of straddling fish stocks. This chapter explores the potential of partition function games, a particular type of coalition game recently introduced into the fisheries literature. The main advantage of this approach, compared to traditional characteristic function games, is that it captures externalities among coalitions, which are generally present in straddling stock fishery games.

In the chapter, a game in partition function form is defined and applied in order to address the stability of cooperative agreements and the equilibrium coalition structures. The first application is a straddling stock fishery represented through the classical Gordon–Schaefer bioeconomic model. The results show that the game exhibits positive externalities. Moreover, in the case of three or more players, no cooperative agreement is stand-alone stable and the only Nash equilibrium coalition structure is the one formed by singletons. Thus, complete noncooperation is the most likely outcome. The second application is the bluefin tuna fishery in the Northeast Atlantic and Mediterranean Sea. In this application a fairly disaggregated bioeconomic model is used. It is also concluded that positive externalities are present and there is no sharing rule that can make the grand coalition stable.

Both games show very pessimistic prospects regarding the cooperative management of straddling fish stocks under RFMOs. This conclusion is in line with the results obtained by Yi (1997) for classical economic coalitions characterized by the presence of positive externalities, such as output cartels and coalitions formed to provide public goods. The author concludes that an open membership game rarely supports the grand coalition as a Nash equilibrium, and equilibrium coalition structures are often very fragmented.

According to the results obtained, the breakdown of cooperative agreements on straddling stock fisheries is a very likely scenario if countries can free ride on cooperative agreements. Thus, in order to protect cooperation, the legal regime must prevent those who engage in noncooperative behavior from having access to the resource. As Munro (2006) stresses, if the RFMO regime is to prosper, unregulated fishing by nonmembers must be eliminated. This is clearly on the agenda of the Food and Agriculture Organization of the United Nations (FAO) and is specifically addressed in its International Plan of Action to Prevent, Deter, and Eliminate Illegal, Unreported, and Unregulated Fishing (FAO 2001). This is a voluntary instrument that recommends, among other measures, that States should develop and implement national plans of action to achieve the objectives of the International Plan of

Action, and adopt trade restrictions on fish and fish products derived from illegal, unreported, and unregulated fishing, including import and export controls or prohibitions. According to the International Plan of Action, institutional and policy strengthening of RFMOs is a key element in addressing the problem of illegal, unreported, and unregulated fishing.

In 2003 a group of fisheries ministers and director-generals of international nongovernmental organizations created the High Seas Task Force with the aim of tackling "the root causes" of illegal, unreported, and unregulated fishing in the high seas. The final report (High Seas Task Force 2006) established a set of proposals intended to enhance enforcement, thereby increasing the risks to those engaged in illegal, unreported, and unregulated fishing operations, and making such fishing less profitable. Among the most prominent measures proposed by the High Seas Task Force are the improvement of the exchange of knowledge derived from monitoring, control, and surveillance activities through the development of a global information system on high seas fishing vessels; providing guidance to RFMOs in order to disseminate best practices in implementation of international fishery instruments; setting guidelines on flag State performance and measures in order to promote port State controls; and supporting developing countries in overcoming illegal, unreported, and unregulated fishing.

A key idea that emerges from this chapter is that the success of the management of straddling fish stocks under RFMOs, prescribed by the United Nations Fish Stocks Agreement, is at serious risk unless the gaps in this document are filled with complementary measures that empower RFMOs to combat noncooperative behavior.

Notes

1. This broad definition includes what, in the terminology of the Food and Agriculture Organization of the United Nations, is called highly migratory fish stocks (mainly the six major tuna species). According to Munro (2006), there is no meaningful difference between straddling fish stocks and highly migratory fish stocks as far as economic analysis is concerned.
2. This is the most common abbreviated designation of the agreement and is used throughout the chapter. The full title is: Agreement for the Implementation of the Provisions of the United Nations Convention on the Law of the Sea of 10 December 1982 Relating to the Conservation and Management of Straddling Fish Stocks and Highly Migratory Fish Stocks.
3. The assumption of ex ante symmetric players is dominant in the literature on partition function games (for example Yi 1997; Finus 2001). Allowing for asymmetric players raises significantly the level of complexity of the analysis because the results depend on the particular rule chosen to share the benefits from cooperation.
4. The proofs of this and the following propositions are provided in the Appendix to this chapter.
5. Nonetheless, recent data on bluefin tuna releases and recoveries show significant mixing between the two stocks (ICCAT 2006).
6. This is a very common way to model noncooperative behavior in fisheries (for

example Amundsen et al. 1995). One of its main advantages is that it can incorporate the natural restrictions on the effort adjustment.

7. The uniqueness of the equilibrium payoffs was one of the main factors behind the option for a particular modeling of nonmembers' strategies rather than the use of the traditional Nash equilibrium. In fact, the complexity of the bioeconomic model raised the problem of nonuniqueness of the Nash equilibrium.

References

Amundsen, E., Bjørndal, T., and Conrad, J. (1995) "Open-access harvesting of the Northeast Atlantic minke whale", *Environmental and Resource Economics*, 6(2): 167–85.

Arnason, R., Magnusson, G., and Agnarsson, S. (2000) "The Norwegian spring-spawning herring fishery: a stylized game model", *Marine Resource Economics*, 15: 293–319.

Bloch, F. (1996) "Sequential formation of coalitions in games with externalities and fixed payoff division", *Games and Economic Behavior*, 14: 90–123.

Chander, P. and Tulkens, H. (1995) "A core-theoretic solution for the design of cooperative agreements on transfrontier pollution", *International Tax and Public Finance*, 2(2): 279–93.

Chander, P. and Tulkens, H. (1997) "The core of an economy with multilateral environmental externalities", *International Journal of Game Theory*, 26: 379–401.

Clark, C. (1980) "Restricted access to common property fisheries resources: a game-theoretic analysis", in P. Liu (ed.) *Dynamic Optimization and Mathematical Economics*, New York: Plenum Press.

Clark, C. and Munro, G. (1975) "Economics of fishing and modern capital theory: a simplified approach", *Journal of Environmental Economics and Management*, 2: 92–106.

FAO (Food and Agriculture Organization of the United Nations) (2001) *International Plan of Action to Prevent, Deter and Eliminate Illegal, Unreported and Unregulated Fishing*, Rome: FAO.

Finus, M. (2001) *Game Theory and International Environmental Cooperation*, Cheltenham, UK: Edward Elgar.

Gordon, H.S. (1954) "The economic theory of a common property resource: the fishery", *Journal of Political Economy*, 62: 124–42.

Greenberg, J. (1994) "Coalition structures", in R.J. Aumann and S. Hart (eds) *Handbook of Game Theory with Applications to Economics, Volume 2*, Amsterdam: North-Holland.

High Seas Task Force (2006) *Closing the Net: Stopping Illegal Fishing on the High Seas*, Governments of Australia, Canada, Chile, Namibia, New Zealand, and the United Kingdom, World Wide Fund for Nature (WWF), World Conservation Union (IUCN), and the Earth Institute at Columbia University. Available online: www.high-seas.org (accessed February 20, 2007).

ICCAT (International Commission for the Conservation of Atlantic Tunas) (2005) *Report for the Biennial Period 2004–2005, Part II, Volume 2*, 81–8, Madrid.

ICCAT (International Commission for the Conservation of Atlantic Tunas) (2006) *Statistical Bulletin*, 35: 136–7, Madrid.

Kaitala, V. and Lindroos, M. (1998) "Sharing the benefits of cooperation in high sea fisheries: a characteristic function game approach", *Natural Resource Modeling*, 11: 275–99.

Kennedy, J. (2003) "Scope for efficient multinational exploitation of North-East Atlantic mackerel", *Marine Resource Economics*, 18: 55–80.

Levhari, D. and Mirman, L.J. (1980) "The great fish war: an example using a dynamic Cournot-Nash solution", *Bell Journal of Economics*, 11: 322–34.

Lindroos, M. (2002). "Coalitions in fisheries", paper presented at the XIVth Annual Conference of the European Association of Fisheries Economists, Faro, Portugal.

Lindroos, M. (2004a) "Restricted coalitions in the management of regional fisheries organizations", *Natural Resource Modeling*, 17: 45–70.

Lindroos, M. (2004b) "Sharing the benefits of cooperation in the Norwegian spring-spawning herring fishery", *International Game Theory Review*, 6: 35–54.

Lindroos, M. and Kaitala, V. (2000) "Nash equilibria in a coalition game of the Norwegian spring-spawning herring fishery", *Marine Resource Economics*, 15: 321–40.

Lindroos, M., Kronbak, L., and Kaitala, V. (2006) "Coalitions in fisheries games", in T. Bjørndal, D. Gordon, R. Arnason, and U.R. Sumaila (eds) *Advances in Fisheries Economics: Festschrift in Honour of Professor Gordon R. Munro*, London: Blackwell.

Munro, G. (1999) "An economic review of the United Nations Agreement for the Implementation of the United Nations Convention on the Law of the Sea of 10 December 1982 Relating to the Conservation and Management of Straddling and Highly Migratory Fish Stocks", in *Proceedings from the Conference on the Management of Straddling Fish Stocks and Highly Migratory Fish Stocks, and the UN Agreement*. Papers on Fisheries Economics 38, Centre for Fisheries Economics, Norwegian School of Economics and Business Administration.

Munro, G. (2006) "International allocation issues and the high seas: an economist's perspective", paper presented at the Sharing the Fish Conference 2006, Fremantle, Australia, 26 February–2 March 2006.

Pham Do, K. and Folmer, H. (2003) *International Fisheries Agreements: The Feasibility and Impacts of Partial Cooperation*, CentER DP 2003–52, Tilburg University, Netherlands.

Pintassilgo, P. (2003) "A coalition approach to the management of high seas fisheries in the presence of externalities", *Natural Resource Modeling*, 16: 175–97.

Pintassilgo, P. and Costa Duarte, C. (2002) "Optimal management of the Northern Atlantic bluefin tuna", *Marine Resource Economics*, 17: 47–67.

Ruseski, G. (1998) "International fish wars: the strategic roles for fleet licensing and effort subsidies", *Journal of Environmental Economics and Management*, 36: 70–88.

Sang Kwon, O. (2006) "Partial international coordination in the great fish war", *Environmental and Resource Economics*, 33: 463–83.

Schaefer, M. (1954) "Some aspects of the dynamics of populations important to the management of commercial marine fisheries", *Bulletin Inter-American Tropical Tuna Commission*, 1: 25–56.

United Nations (1982) *United Nations Convention on the Law of the Sea*, United Nations Doc. A/Conf.62/122.

United Nations (1995) *United Nations Conference on Straddling Fish Stocks and Highly Migratory Fish Stocks: Agreement for the Implementation of the Provisions of the United Nations Convention on the Law of the Sea of 10 December 1982*

Relating to the Conservation and Management of Straddling Fish Stocks and Highly Migratory Fish Stocks, United Nations Doc. A/Conf./164/37.

Yi, S.-S. (1996) "Endogenous formation of customs unions under imperfect competition: open regionalism is good", *Journal of International Economics*, 41: 151–75.

Yi, S.-S. (1997) "Stable coalitions with externalities", *Games and Economic Behavior*, 20: 201–37.

Appendix

Proof of proposition 2

Note that as all players are ex ante identical a coalition can be identified by its size. Thus, following Yi (1997), $C = \{S_1, S_2, \ldots, S_z\}$ is hereafter written as $C = \{n_1, n_2, \ldots, n_z\}$ where n_i is the size of coalition S_i.

The presence of positive externalities in this game, according to definition 1, can be proved by showing that the merger of coalitions increases the payoff of all the players who belong to the other coalitions. Formally, this can be written as:

$$\pi\,(n_i; C) < \pi\,(n_i; C') \tag{4.A1}$$

where $\{n_i\} \subset C$, $\{n_i\} \subset C'$, and $C' \backslash \{n_i\}$ can be derived from $C \backslash \{n_i\}$ by merging coalitions in $C \backslash \{n_i\}$.

As it is assumed that all nonmembers of the RFMO behave as singletons, (4.A1) can be proven by showing that nonmembers are better off when additional players join the organization. Consider the general coalition structure $C = \{1, \ldots, 1, n - m\}$, $m \in \{1, \ldots, n - 1\}$. Note that it includes all noncooperative coalition structures. Let $C' = \{1, \ldots, 1, n - m + l\}$, $l \in \{1, \ldots, m - 1\}$. Thus, the condition (4.A1) can be written as:

$$\pi\,(1; \{1, \ldots, 1, n - m\}) < \pi\,(1; \{1, \ldots 1, n - m + l\}) \tag{4.A2}$$

Using the per-member partition function (equation (4.10)):

$$\frac{(pqk - c)\, r\, (1 - b)}{(m + 2)^2\, q} < \frac{(pqk - c)\, r\, (1 - b)}{(m - l + 2)^2\, q} \tag{4.A3}$$

This yields:

$$\left(\frac{m - l + 2}{m + 2}\right)^2 < 1 \tag{4.A4}$$

which is always verified.

Proof of proposition 3

A two-player game will first be considered. The per-member payoff of the grand coalition and singletons, respectively, are given by:

$$\pi\ (2;\{2\}) = \frac{(pqk - c)\ r\ (1 - b)}{8q} \qquad (4.A5)$$

$$\pi\ (1;\{1,1\}) = \frac{(pqk - c)\ r\ (1 - b)}{9q} \qquad (4.A6)$$

The grand coalition is stand-alone stable as the per-member payoff exceeds that of noncooperation. In addition, the coalition structure formed by singletons is stand-alone stable by definition.

Take now the case of a game with three or more players. Consider first the grand coalition, and assume that this coalition structure is stand-alone stable. Using definition 4:

$$\pi\ (n;\{n\}) \geq \pi\ (1;\{1,n - 1\}) \qquad (4.A7)$$

Applying the per-member partition function (equations (4.7) and (4.10)):

$$\frac{1}{n}(pqk - c)\frac{r}{4q}(1 - b) \geq \frac{(pqk - c)\ r\ (1 - b)}{9q} \qquad (4.A8)$$

Simplifying the inequality:

$$n \leq \frac{9}{4} \qquad (4.A9)$$

Thus, the grand coalition is not stand-alone stable in a game with three or more players.

Consider now the coalition structures with at least two coalitions. This can be represented by $C = \{1,\ldots,1,n - m\}$, $m \in \{1,\ldots,n - 1\}$. Assume that C is stand-alone stable. Thus:

$$\pi\ (n - m;\{1,\ldots,1, n - m\}) \geq \pi\ (1;\{1,\ldots,1, n - m - 1\}) \qquad (4.A10)$$

Using the per-member partition function (equations (4.9) and (4.10)):

$$\frac{(pqk - c)\ r\ (1 - b)}{(n - m)\ (m + 2)^2\ q} \geq \frac{(pqk - c)\ r\ (1 - b)}{(m + 3)^2\ q} \qquad (4.A11)$$

Rewriting the inequality:

$$n - m \leq \left(\frac{m+3}{m+2}\right)^2 \tag{4.A12}$$

As $1 < \left(\frac{m+3}{m+2}\right)^2 < 2, \forall\, m \in \{1,\ldots,n-1\}$, then $n - m \leq 1$.

Hence, the only stand-alone stable coalition structure is the one formed by singletons.

Proof of proposition 4

Consider $n = 2$. From proposition 3 the grand coalition $C = \{2\}$ is stand-alone stable. Thus, $C = \{2\}$ is a Nash equilibrium coalition structure, because for the grand coalition there is equivalence between stand-alone stability and Nash equilibrium. The coalition structure $C\{1,1\}$ is not a Nash equilibrium as each player becomes better off by joining the other and establishing a cooperative agreement. This is shown in the proof of proposition 3.

Consider $n \geq 3$. From proposition 3, $C = \{1,\ldots,1\}$ is the only stand-alone stable coalition structure and therefore the only candidate to be an equilibrium. $C = \{1,\ldots,1\}$ is a Nash equilibrium if and only if:

$$\pi(1;\{1,\ldots,1\}) \geq \pi(2;\{1,\ldots,1,2\}) \tag{4.A13}$$

Using the per-member partition function (equations (4.12) and (4.9)):

$$\frac{(pqk-c)\,r\,(1-b)}{(n+1)^2\,q} \geq \frac{1}{2}\,\frac{(pqk-c)\,r\,(1-b)}{n^2\,q} \tag{4.A14}$$

This yields:

$$\left(\frac{n}{n+1}\right)^2 \geq \frac{1}{2} \tag{4.A15}$$

This inequality is verified for $n \geq 3$. Thus, in the case of three or more players the coalition structure formed by singletons is the only Nash equilibrium of the game.

5 To negotiate or to game theorize

Evaluating water allocation mechanisms in the Kat basin, South Africa

Ariel Dinar, Stefano Farolfi, Fioravante Patrone, and Kate Rowntree

Governments and developing agencies promote participatory approaches in solving common pool resource problems such as in the water sector. Two main participatory approaches have been applied separately, namely negotiation and mediation. This chapter applies the role-playing game that is a component of the companion modeling approach—a negotiation procedure—and cooperative game theory (Shapley value and the nucleolus solution concepts), which can be mirrored as a mediated mechanism, to a water allocation problem in the Kat basin[1] in South Africa. While the absolute results of the two approaches differ, the negotiation and cooperative game theory approaches provide similar shares of the benefit allocated to the players from various cooperative arrangements. By evaluating the two approaches, the chapter provides useful tips for future extension of both the role-playing game and the cooperative game theory applications.

5.1 Introduction

Common pool resources are those resources to which a community rather than a private individual has right of access, and which are managed jointly by that community. Examples of such resources (and relevant studies of the conflicts that may arise and the management arrangements among the community members who may have free access) include grazing land (Goodhue and McCarthy 2008), watershed (basin) resources (White and Ford Runge 1994), fisheries (Munro 1979), and irrigation water (Ostrom 1992; Aadland and Kolpin 1998, 2004).

Common pool resource disputes are increasing because of increase in competition and deterioration of resource quality. Cooperative and participatory arrangements have long been in the center of public interest, especially regarding the mechanisms used by communities to share and manage the resource. Governments and development agencies promote participatory approaches that allow the stakeholders to agree on the arrangements by which the common property resource is managed, allocated, and used. Negotiation approaches have been used by communities as a cooperative mechanism for solving common pool resource problems (Farrington and Boyd 1997; Steins and Edwards 1999; Ostrom et al. 1994). In parallel, cooperative game theory, which involves an agreed-upon mediated mechanism that, among other things, explores the conditions supporting specific mediated solutions, has been applied with success to many common pool resource problems (see review of relevant literature by Parrachino et al. 2006).

Although the two approaches of negotiation and cooperative game theory come from different directions and are based on different assumptions, they complement each other in that they are both based on the principles of fairness that were suggested by Carraro et al. (2005b: 39): "The perception of fairness plays a crucial role in determining how a surplus is divided, and the potential allocation rules must be perceived as 'equitable' and 'envy-free' by all parties." These principles of fairness together with the efficiency principle lead, under certain circumstances, to stable cooperative outcomes.

Negotiation approaches focus on the resolution of conflicts originating from stakeholders' different perceptions and use of natural resources. Recent literature on environmental management and catchment or basin management in particular places strong emphasis on achieving negotiated settlements to such conflicts (Becu et al. 2003). In the water sector, there have been many applications of both negotiation and cooperative game theory approaches at various levels, from sectoral to international (Carraro et al. 2005a; Parrachino et al. 2006). So far, the literature has applied negotiation and cooperative game theory approaches separately to the evaluation of allocation issues and has not attempted to compare the solutions emerging from both approaches. Other authors argue that participation in knowledge sharing for water is a fundamental requirement for efficient and equitable use because of the cumulative effects of individual actions on the patterns of water use at the irrigation system and river basin scales (Lankford and Watson 2006). Although it is expected that, under certain conditions of interaction among the parties, negotiation and cooperative game theory approaches will lead to similar allocation solutions, there is still no clear empirical evidence that this is the case.

In this chapter a negotiation procedure and a cooperative game theory solution approach are applied to a water allocation problem in the Kat basin in South Africa. Simplifying assumptions are used to allow an evaluation, based on similar parameters, of the allocation solutions. The role-playing game that is applied is a component of the companion modeling approach (Barreteau et al. 2003), which has already demonstrated its usefulness in promoting discussion among stakeholders with contrasted and eventually conflicting viewpoints (Dray et al. 2006). The same allocation problem is then also formulated as a cooperative game theory problem, explaining and evaluating a couple of allocation solution concepts (Shapley value and the nucleolus); these allocations in the present case are contained in the core, which means that they enjoy a relevant stability property. An explanation is given for sources of differences between the role-playing game and the cooperative game. Although the approach is centered on a specific case, the general issue of stable allocation arrangements in a basin that is characterized by multiple externalities is of interest to many around the world. Therefore, this chapter aims to demonstrate the application and usefulness of the cooperative game theory and role-playing game methods.

The chapter is structured as follows. Section 5.2 provides a short description of the geographical, historical, political, and institutional aspects of the Kat basin. In section 5.3 the water allocation issues in that basin is discussed. Section 5.4 formulates the water allocation problem as a negotiation game and section 5.5 presents the cooperative game solution to the allocation problem. Section 5.6 evaluates and explains the differences, and section 5.7 provides possible extensions. The chapter is then concluded.

5.2 The Kat basin in South Africa[2]

The valley of the Kat River (Figure 5.1), a tributary of the Great Fish River, is situated in Eastern Cape Province, South Africa. Although the catchment has a relatively high rainfall, much of the climate of the 1,700 square kilometer Kat basin is subhumid to semiarid. The fertile valley land can be utilized only through irrigation, using water from the Kat River. Prior to 1969, irrigators relied on the natural flow in the river, but since 1969 water from the Kat Dam (24 million cubic meters storage capacity) has been available. While irrigation takes up the majority of the water in the basin, domestic water users (about 49,500 inhabitants in 2001) represent an increasingly important component of the demand for water in the basin.

Four groups of irrigators can be identified in the Kat basin: small-scale black farmers, often forming cooperatives; large-scale "emerging" black farmers;[3] white commercial farmers with scheduled water rights; and white commercial farmers without scheduled water rights. The main water-related stakeholders in the Kat basin are therefore (a) the four groups of irrigators; (b) domestic water users; and (c) the municipality of Nkonkombe. The South African Department of Water Affairs and Forestry, currently operating the Kat Dam, is considered the fourth important stakeholder in the system. The complex and contentious political history of the valley has given rise to power dynamics that have historically favored the white commercial farmers producing citrus, who controlled water use through the Kat River Irrigation Board.[4]

5.3 Water allocation issues in the Kat basin

Water sources in the Kat basin are currently almost exclusively from surface water. Some groundwater developments are foreseen in the near future and this could increase water availability in the basin by nearly 10 percent (DWAF 2001).[5] As mentioned above, decisions about water allocation strategies will become the responsibility of the recently established Water

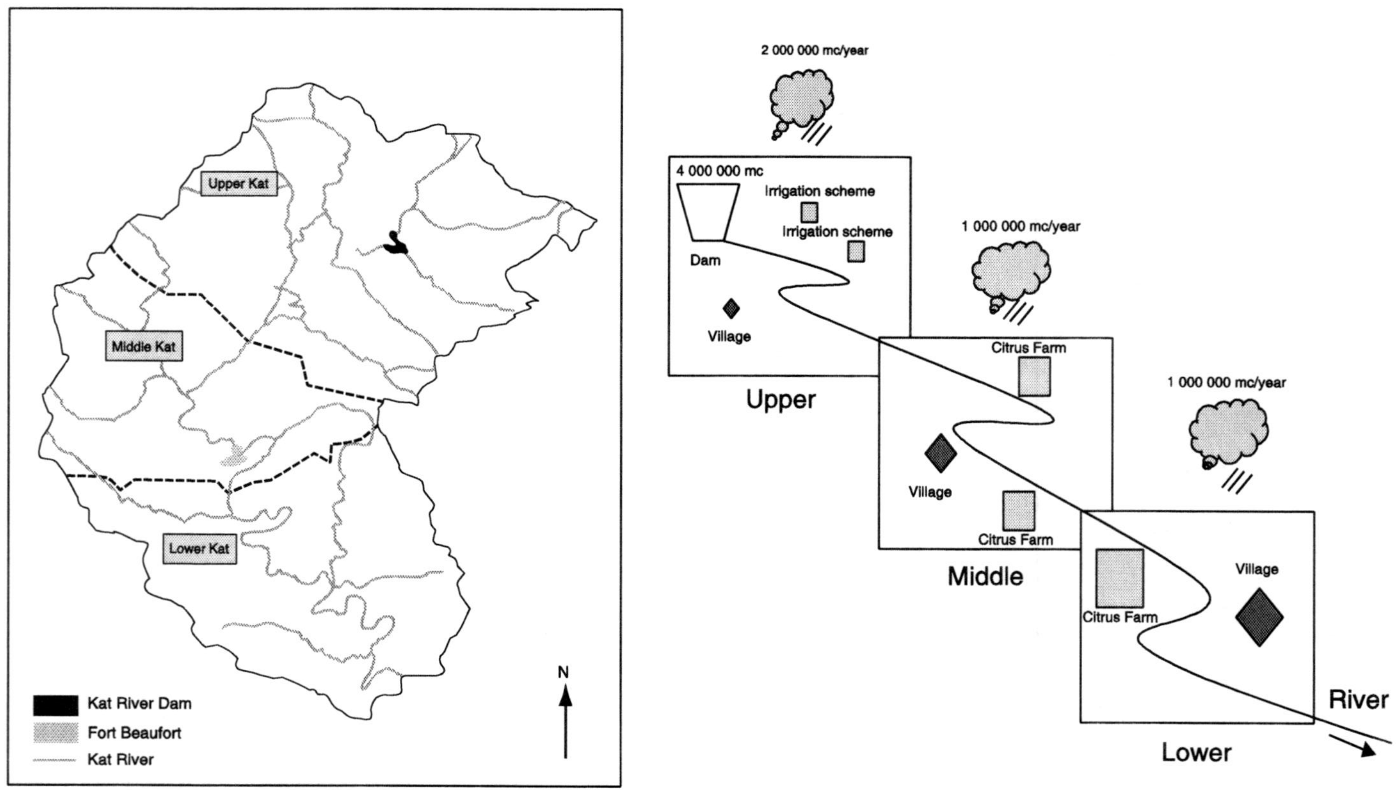

Figure 5.1 Kat basin and stylized subbasins in the role-playing game (Source: Left panel is based on map prepared by Alistair McMaster (McMaster 2002)).

Note: For the conversion factors used see Table 5.1.

Users Association (WUA), which represents the various groups of stakeholders in the basin.

Irrigated agriculture is at present the largest water user in the basin, accounting for 60 percent of the total water requirements. The other use requirements in the basin are the environmental flow requirement, known as the ecological Reserve[6] (12 percent), domestic uses (13 percent), afforestation (10 percent), and losses arising from alien vegetation (5 percent). The recent history of South Africa has led to the coexistence of different groups of irrigators in the basin. These farmers are located in specific portions of the basin[7] (Figure 5.1), namely smallholders and emerging farmers in the upper subbasin (U), emerging and large-scale farmers in the middle subbasin (M), and large-scale farmers in the lower subbasin (L) (Farolfi and Rowntree 2005).

Domestic water consumption per capita is low by modern standards, mainly because of striking rural poverty and limited services (Farolfi and Abrams 2005). A crucial issue for the National Water Resource Strategy of South Africa is the protection of environmental and ecological needs, which is translated into the definition of and respect for the ecological Reserve in each subbasin. Therefore, it is taken for granted in this study that the WUA that represents the water users respects the ecological Reserve.

The Kat Dam is certainly the main tool for water supply management in the basin. The dam is currently operated by the Department of Water Affairs and Forestry based on a mechanism of water licenses and scheduled and nonscheduled users. It is expected that soon the WUA will be responsible for the management of the dam. Therefore, in this chapter it is assumed that it is the WUA and not the Department of Water Affairs and Forestry that decides on dam water allocation. Future demands were identified (Farolfi and Abrams 2005) to include increase in citrus area; increase in small-scale irrigation schemes; increase in domestic uses, particularly in rural areas; and more tourism, game farms, and possibly golf courses. These will be addressed in future analyses.

5.4 Formulating water allocation decisions as a negotiation process

To facilitate discussions about water allocation strategies within the Kat River WUA, stakeholders' representatives agreed to take part in a process of companion modeling (Barreteau et al. 2003) consisting of an iterative and participatory development of a simulation model, KatAware (Farolfi and Bonté 2005), which illustrates alternative scenarios of water allocation in the basin (Farolfi and Rowntree 2005).

An important component of companion modeling is the use of role-playing games to facilitate stakeholders' comprehension of the developing model and to allow modelers to better understand the stakeholders' strategies and behavior. For this purpose, a role-playing game was constructed based on the model KatAware, which is being developed with the WUA following the companion modeling approach.

5.4.1 Structural details of the role-playing game

In the role-playing game, the Kat basin (Figure 5.1) is divided into three subbasins corresponding to the three mentioned areas: upper, middle, and lower. For the purpose of this game, two smallholding irrigation schemes (20 hectares each)[8] are located in the upper subbasin (*U*), two citrus farmers (30 hectares each) are located in the middle subbasin (*M*), and a citrus farmer (40 hectares) is located in the lower subbasin (*L*). Farmers have irrigated land on which they produce cabbages if they are smallholders or citrus if they are large-scale farmers. Domestic water users live in three villages: one in *U* (3,000 inhabitants), one in *M* (5,000 inhabitants), and one in *L* (15,000 inhabitants). An average amount of rainfall equivalent to 2 million cubic meters per year falls on *U*, whilst annual rainfall equivalent to 1 million cubic meters per year falls on *M* and on *L*. A dam with a storage capacity of 4 million cubic meters is located in *U*. A WUA exists in the basin and is responsible for water management and allocation according to the principles of social equity, environmental sustainability, and economic efficiency, as indicated in the water legislation of South Africa. All players are members of the WUA.

To simplify the game while still reflecting its reality, data representing the actual basin are transformed into values that are relatively proportional to the real ones, and only eight players plus the WUA are left to interact (Table 5.1).[9] The primary goal of the game is to allow for the available water resource of the basin as a whole to be managed in a sustainable way, taking into consideration the above-mentioned principles. At the same time, the goal for each player is to maximize his or her individual economic gain and if they are a village manager, to also maximize villagers' satisfaction, within the context of the group goal (see additional discussion in section 5.5). The game is set to allow sessions of up to seven simulated years. This choice is a compromise between playability and a time span sufficient to provide elements of discussion on midterm consequences of water allocation. The role-playing game session presented here spans six years.

Table 5.1 Actual and transformed values for use in role-playing game of main variables in Kat basin

Variable[a]	*Actual values*	*Values in role-playing game*
Dam capacity (m^3)	24,000,000	4,000,000
Natural runoff (m^3)	13,500,000	3,300,000
Domestic consumption (m^3)	1,500,000	580,000
Irrigation consumption (m^3)	11,000,000	1,064,000
Cabbage area (ha)	180	40
Citrus area (ha)	1,300	100
Inhabitants in basin	49,000	23,000
Annual outflow	1,600,000	550,000

a. m^3 = cubic meters; ha = hectares.

At the beginning of the game farmers receive the number of hectares corresponding to their farm (or irrigation scheme) and for each hectare a symbol corresponding to their production (cabbage or citrus). As the game progresses, farmers also receive an amount of money corresponding to their previous year's profit and a number of workers indicating how many permanent or seasonal employees they hired the previous season (see Appendix at end of this chapter). Every year farmers may decide to increase or reduce their irrigated area. They can also decide to change their crop (cabbage to citrus or vice versa). If they decide to plant new citrus, they can choose a more advanced irrigation technology (drip), which would cost more but will save water. Cabbage producers can decide to have one cycle, two cycles, or three cycles of cabbage production per year on their fields. Budgets and water consumption data for citrus and cabbage are provided to farmers at the beginning of the game. Village managers receive, and pay for, bulk water from the WUA, which manages the entire water in the basin and provides water services (including water distribution) to the households of each village. They start with a given ratio of water sources for the households of their village. These water sources are river water, collective tap, and in-dwelling tap. Each water source has a different cost (investment + operating cost) that has to be added to the cost of the bulk water the managers "buy" from the WUA. Village managers can charge their inhabitants a per capita tariff for the water services they provide, and this corresponds to their annual income. The households derive a certain level of satisfaction (utility) from their income that can be spent on consumption goods. Because households have different levels of effort associated with the various water sources they are provided with, they also obtain different levels of utility from each of the three sources of water.

The village manager's objective is twofold: to maximize his or her profit resulting from the difference between the tariff collection and the water provision cost + bulk water cost; and at the same time to maximize the sum of households' utility (see profit function in Appendix). Elements of budgets and utility values for households are provided to local village managers at the beginning of the game.

A number of factors that vary annually, such as rainfall, market prices, and population dynamics, can influence players' strategies. Variations in rainfall and prices over time in the presented role-playing game session are illustrated in the following section.

5.4.2 Negotiation results

This section illustrates some outcomes of a role-playing game session held in the Kat basin in November 2005. The setup of the game and the players participating in the session are presented in Figure 5.1 (panel to the right of the map). Table 5.2 shows the initial values of the exogenous factors controlled by the game operators, and the final values at the end of year 6 of the negotiation game. In Tables 5.3, 5.4, and 5.5, year 1 is the initial state and was set by the game facilitators; years 2 to 6 were actually played.

The game facilitators introduced a general trend of increasing water scarcity. This stress was produced by a combination of lower rainfall and increasing population in the basin. Some marginal changes (mainly reductions) affected crop prices. A relatively low level of uncertainty was introduced in the session, corresponding to a small difference between expected (forecasted) and actual exogenous factors stakeholders faced.

Clear differences in behavior and strategies among players were observed for the different sectors and in the three subbasins (Tables 5.3 and 5.4). In subbasin *U* the two irrigation schemes opted first for an intensification of their cabbage production (from two to three cycles per year). At the end of the game session the second irrigation scheme decided to reduce the cultivated area by 50 percent. In subbasin *M*, the two citrus farmers adopted two very different strategies: one farmer opted first for diversification (cabbage in addition to citrus) and then abandoned citrus, whilst the other farmer kept the citrus area constant but also planted an equivalent area of cabbage. In subbasin *L*, the large citrus farm adopted quite a conservative strategy consisting of reducing the planted area of citrus by 25 percent and not moving to cabbage. All new citrus plantations in the three farms were equipped with advanced irrigation technologies consisting of drip systems, more costly in terms of investment, but water saving.

Table 5.4 shows the dynamics in the village managers' decisions regarding water services and tariffs for their households. As a general trend, better water provision was introduced in all villages, and this was accompanied by an increase in water tariffs required from the households. In some cases the increase in domestic water tariffs was perceived as too high by local residents (village *L*), affecting negatively their utility. On the other hand, this water tariff increase in village *L* triggered a significant increase in the village manager's profit.

It was clear that the WUA gave priority to the domestic uses of water, not

Table 5.2 Exogenous factors in role-playing game session: initial, final, and difference values

Variable	*Initial*	*Final*	*Difference %*
Rainfall upper (m^3)	2,000,000	1,400,000	–30
Rainfall middle (m^3)	1,000,000	600,000	–40
Rainfall lower (m^3)	1,000,000	600,000	–40
Population upper (inhabs.)	3,000	3,500	17
Population middle (inhabs.)	5,000	5,500	10
Population lower (inhabs.)	15,000	16,000	7
Market price citrus (R/tonne)[a]	2,000	2,000	0
Market price cabbage (R/bag)[a]	6	5	–17

a. R = rand, the South African currency. US$1 = 6 rands at the time of the experiment.

Table 5.3 Strategies and outcomes for the five farms during role-playing game session: initial and final values

Farm[a]	*Initial*	*Final*	*Difference (%)*
Irrigation scheme 1 (U)			
Citrus old technology (ha)	0	0	0.0
Citrus new technology (ha)	0	0	0.0
Cabbage (ha)	20	20	0.0
Cycles cabbage	2	3	50.0
Total ha	20	20	0.0
Employment (n)	51	77	50.1
Profit (R)	64,208	250,000	289.4
Irrigation scheme 2 (U)			
Citrus old technology (ha)	0	0	0.0
Citrus new technology (ha)	0	0	0.0
Cabbage (ha)	20	10	–50.0
Cycles cabbage	2	2	0.0
Total ha	20	10	–50.0
Employment (n)	51	25	–51.0
Profit (R)	64,208	250,000	289.4
Citrus farm 1 (M)			
Citrus old technology (ha)	30	0	–100.0
Citrus new technology (ha)	0	5	
Cabbage (ha)	0	30	
Cycles cabbage	0	1	
Total ha	30	35	16.7
Employment (n)	46	46	0.0
Profit (R)	829,300	3,290,000	296.7
Citrus farm 2 (M)			
Citrus old technology (ha)	30	0	–100.0
Citrus new technology (ha)	0	30	
Cabbage (ha)	0	30	
Cycles cabbage	0	1	
Total ha	30	60	100.0
Employment (n)	46	84	82.6
Profit (R)	829,300	740,000	–10.8
Citrus farm 3 (L)			
Citrus old technology (ha)	40	0	–100.0
Citrus new technology (ha)	0	30	
Cabbage (ha)	0	0	0.0
Cycles cabbage	0	0	0.0
Total ha	40	30	–25.0
Employment (n)	62	44	–29.0
Profit (R)	1,105,700	2,710,000	145.1

a. Key:
U = upper subbasin, *M* = middle subbasin, *L* = lower subbasin,
ha = hectares, R = rand.

Table 5.4 Strategies and outcomes for the three villages during role-playing game session: initial and final values

Village[a]	*Initial*	*Final*	*Difference (%)*
Village 1 (U)			
Population (inhabs.)	3,000	3,500	16.7
Share of river source	0.8	0.0	–100.0
Share of collective tap	0.2	0.2	0.0
Share of in-dwelling tap	0.0	0.8	
Water tariff (R/m^3)	1.0	2.0	100.0
Satisfaction index	40.6	41.7	2.8
Manager's profit (R)	20,500	420,000	1,948.8
Village 2 (M)			
Population (inhabs.)	5,000	5,500	10.0
Share of river source	0.8	0.0	–100.0
Share of collective tap	0.2	0.2	0.0
Share of in-dwelling tap	0.0	0.8	
Water tariff (R/m^3)	1.0	1.7	70.0
Satisfaction index	40.6	42.89	5.7
Manager's profit (R)	34,180	300,000	777.7
Village 3 (L)			
Population (inhabs.)	15,000	16,000	6.7
Share of river source	0.1	0.0	–100.0
Share of collective tap	0.4	0.0	–100.0
Share of in-dwelling tap	0.5	1.0	100.0
Water tariff (R/m^3)	1.5	2.0	33.3
Satisfaction index	42.7	41.9	–1.8
Manager's profit (R)	128,130	2,110,000	1,546.8

a. Key:
U = upper subbasin, *M* = middle subbasin, *L* = lower subbasin,
m^3 = cubic meters, R = rand.

hampering any initiative by the local managers to increase water provision. Respect of an ecological Reserve set at 500,000 cubic meters per year in drought years and 750,000 cubic meters per year in normal years was another WUA priority, arising from recognition of legal requirements. Agricultural uses were more controlled and the release of new water licenses to farmers was less automatic, especially when the dam reserve became scarce (in the last three years of the game session).

The water allocation policy of the WUA allowed positive results (profit) in terms of economic outputs for four farms out of five. Cabbage was more profitable than citrus because of a relatively steady rise in market price (excluding the final two years) and, more importantly, because no investment is required for new plantations. Farm 4, in the *M* subbasin, registered the worst performance, paying the cost of heavy investment in new hectares planted with citrus combined with lower market prices in years 3 and 4. In addition, the session was too short to allow the farmer to recover the investment through production from the new citrus plants (in the role-playing game, citrus takes two years after plantation to become productive).

Job creation was generally positive for all farms. The water shortage provoked by the WUA's decision to stop releasing water in the last year had very negative impacts on job creation, particularly in the *M* and *L* subbasins, where citrus is cultivated.

Table 5.5 shows also the dynamics of water consumption in the three subbasins. At year 1 *L* is the most water consuming (large village and large citrus farm) followed by *M* and *U*. Subbasin *U* consistently increased its water consumption during years 2 to 4 because of the intensification of cabbage production. The increases in water consumption in the remaining subbasins stemmed from higher domestic demand. At year 5, water consumption in *U* decreased because of a change in strategy in one of the two irrigation schemes from three to two cycles of cabbage production per year. In year 6 the WUA decided to stop releasing water from the dam in order to allow it to refill.

The increasing water demand in the three subbasins is partially compensated by water releases from the dam decided by the WUA (Table 5.5). During the first four years of the game the WUA opted for the use of the dam water to satisfy users' water demand and to provide a water flow in the river able to maintain ecosystem functioning (the ecological Reserve).

At the end of year 5, as the dam level reached 1.3 million cubic meters, the WUA decided to stop suddenly and completely water releases. This decision contributed to an improvement in the dam water quantity, but had an immediate and dramatic consequence on the socioeconomic and environmental indicators in the basin. It is worth mentioning that in the role-playing game farmers could not pump groundwater nor extract water from the river against the allocation rules. Therefore, the WUA decision to stop water releases at year 6 had an amplified effect compared to reality.

Table 5.5 allows a comparison of general socioeconomic and environmental trends by subbasin. Job creation is linked to the area cropped and to the intensity of production (cycles of cabbages on the same area); therefore, it closely follows the dynamics of water consumption. As a general trend, annual job creation is positive in all subbasins (except for the last two years). Profit is more sensitive to water availability and during the first years of the game it is (negatively) influenced by high investments in the citrus farms.[10] This is also why the profits of *M* and *L* decline dramatically during the game compared with that of *U*, which increases. The net present value figures show clearly the dramatic impact of the WUA decision at year 5 on profit generation for the three subbasins. Again, *M* and *L*, where citrus farms are located, suffer because of the water shortage. Annual profit values are obtained by first calculating the net present value of the stream of profits for each subbasin and then spreading it over the game duration, using the capital recovery multiplier of 0.20336. A 6 percent discount rate, reflecting the real interest rate in South Africa, and six years are used.

Table 5.5 Profit, job creation, and water consumption in the three subbasins during role-playing game session

Variable[a]	*Year 1*	*Year 2*	*Year 3*	*Year 4*	*Year 5*	*Year 6*	*Annual/average profit value*[b]
				Rands			
Profit *U*	148,924	191,235	258,121	328,450	142,437	–127,311	173,371
Profit *M*	1,692,874	1,089,725	332,920	1,047,169	1,196,990	–949,883	840,854
Profit *L*	1,233,926	688,012	337,207	1,330,596	1,585,825	–333,844	875,872
Total profit	3,075,724	1,968,972	928,248	2,706,215	2,925,252	–1,411,038	1,890,097
				Rands			
Agric. profit *U*	128,416	168,774	184,898	228,657	40,474	–234,938	101,819
Agric. profit *M*	1,658,694	1,081,258	256,006	991,829	1,132,383	–1,038,061	785,101
Agric. profit *L*	1,105,796	525,873	143,245	833,148	1,031,818	–906,511	522,400
				Number of employees (rounded)			
Employment *U*	102	140	179	191	108	102	137
Employment *M*	92	119	133	142	156	130	129
Employment *L*	62	76	77	84	85	44	71
				Cubic meters			
Annual water cons. *U*	376,893	557,852	754,369	886,709	568,552	280,028	570,734
Annual water cons. *M*	464,526	612,535	657,732	800,720	843,208	564,019	657,123
Annual water cons. *L*	784,335	994,810	998,942	1,148,847	1,193,328	658,260	963,087
Dam level[c]	4,000,000	3,674,800	2,774,144	1,814,240	1,368,704	2,361,648	2,665,589
Ecological Reserve[c]	1,700,000	1,500,000	750,000	850,000	950,000	350,000	1,016,667

a. *U* = upper subbasin, *M* = middle subbasin, *L* = lower subbasin.
b. Annual profit values are calculated using a capital recovery multiplier coefficient,[16] calculated at 6 percent discount rate.
c. Dam level and ecological Reserve figures are for the whole basin.

Finally, it is worthwhile noticing that the decision to stop completely water releases from the dam had a negative impact also on the ecological Reserve, with flow well below the limit of 500,000 cubic meters per year at year 6. On average, the ecological Reserve was kept at about 1 million cubic meters per year, corresponding to 40 percent less than its level at the beginning of the negotiation game session, but 33 percent more than the limit set for wet years and 100 percent more than the limit set for dry years.

5.5 Formulating water allocation decisions as a cooperative game

The cooperative game theory model will introduce several assumptions. It is assumed that the players are rational price takers[11] and profit maximizers. They will engage in cooperative arrangements only if it can improve their economic situation compared with the status quo.

5.5.1 Physical characteristics and economic behavior of players in the cooperative game

The basin includes three players (each with several water activities)—upper, middle, and lower subbasin—designated *U*, *M*, and *L*, as is described in the right panel of Figure 5.1 (using $i = 1$ for *U*, $i = 2$ for *M*, and $i = 3$ for *L*). There is water storage (a dam) in the *U* subbasin, and an outlet of the river beyond the *L* subbasin. There is a determinable rainfall quantity that falls on the area of each subbasin and ends up in the river. There are no losses of water and all the rainfall can be used as a source for the water activities (this assumption can also be modified by having a fraction of the rainfall available for use, assuming losses and evaporation). So, each player refers to the amount of rainfall on its area as water in the river available for use. There are also groundwater sources not yet fully developed, so they are not included. Players can also use water from the reservoir that is released (flushed) by the WUA upon a request from the player (if supplies last). The WUA can refuse to supply water from the reservoir if the amount in the reservoir is below a given level.

The WUA in this game is the authority that oversees and regulates the players' behavior. From the point of view of cooperative game theory, the WUA could decide on an allocation of water that respects agreed-on principles (for example social equity, environmental sustainability, and economic efficiency). It is assumed that the players obey the WUA rules of behavior. The objective of each player is to maximize annual profits subject to water availability, prices and costs, and the WUA rules of behavior.

A rule of behavior respected by the players is that no player extracts water that runs in the river that does not belong to that player. Such water includes the ecological Reserve that flows through the segments of the river running through *U*, *M*, and *L*, and water that was released from the dam at the request of a given player and has to run in the river through the "territory" of other players. While such behavioral requirements may look naïve in the light of cheating attempts on the part of the users, one can include monitoring and enforcement costs associated with operating such a system in a real-world situation.

The cooperative game played is an annual one. Therefore, constraints for each player include (a) fixed citrus land; (b) limited land for cabbage; (c) fixed number of inhabitants consuming water; and (d) given amount of rainfall that can be utilized via the flow of the river. The cooperative game uses the values applied in the role-playing game. Constraints or rules for the WUA include (a) a given amount of water in the storage at the beginning of the year; (b) a given amount of water required in the storage at the end of the year; (c) a given amount to be released from the Kat to the Fish River; (d) a given amount to be left in the river for local benefits; (e) a minimum annual amount of water per inhabitant to fulfill human needs; and (f) a given amount of available laborers to work in citrus or cabbage operations. It is assumed that these laborers can move freely between the three subbasins. Additional assumptions represent the linkage between the water use and water flow in the river. These linkages are expressed in the equations of the cooperative game theory model in the Appendix.

The study investigates the likelihood of cooperation among the three players, *U*, *M*, and *L*. Cooperative game theory introduces the concept of the characteristic function for each set of coalitional arrangements among the players in the subbasin. A characteristic function is the best outcome of a coalition. Further assumptions are needed. As indicated above, in this analysis a one-year game period is considered. The essential aspect of this choice is that investment decisions will not be considered.[12] This is a clear limitation, but in the game players are not in a position to make optimal investment decisions (especially because of the short time span allowed). Hence, the one-year time span used in the cooperative game theory approach should offer an appropriate benchmark as far as evaluation of the role-playing game and the cooperative game is concerned.

The status quo individual coalitions will first be considered. The status quo is represented by each player (subbasin) working on their own to maximize their utility (profits) from the available water they have, subject to the individual constraints of each player and the rule of behavior constraints imposed by the WUA. There is no special treatment of the water activities and the land constraints. However, in the maximization problem for each player in the case of the status quo, there is a need to add several assumptions to deal with the subbasin constraints: labor, minimum ecological flow in the Kat River, and a given amount be released for the Fish River.[13] A total amount of available labor in each subbasin and a total amount of environmental flow to be released to the Fish River are assumed. The total labor available to each player will be based on the relative total available land of that player (indicating the potential employment ability of that player). The allocation of the environmental flow amount will be made on the basis of the total amount of water used by that player. The minimum flow in the Kat River will be simply deducted from the amount of water available for each player.

The next step involves calculation of the characteristic functions of the partial coalitions. It is assumed that all permutations are possible, even that between player 1 (*U*) and player 3 (*L*). However, for these players it will simply be assumed that what they can achieve in cooperating with each other is just the sum of what they can get separately. The difference between the calculation of the value of the other partial coalitions and the individual coalitions will be that the decision on water allocation and the total amounts of the minimum flow and the environmental flow can be made jointly rather than individually. Additionally, the allocation of the rule values for labor, minimum flow, and environmental flow will follow the pattern suggested in the case of the individual coalition calculations.

The final step involves calculation of the characteristic function of the grand coalition. In this case the labor constraint is at the subbasin level, the minimum flow constraint is also at the subbasin level, and the environmental flow constraint can be met at the outlet of the Kat when it leaves the area of player 3 (*L*).

The variables will now be introduced (remember that $i = 1,2,3$ is equivalent to $i = U, M, L$). F_{ia} is the (natural) available flow (with probability greater than or equal to a). In this model, $a = 1$ in subbasin i. S_i is the stream entering subbasin i ($S_1 = 0$). E_i is the ecological Reserve constraint for part i (flow leaving part i). Since the ecological Reserve level that is of interest is the quantity that leaves the subbasin, each subbasin i is expected to release to the next one, and L is expected to release to the environment that same quantity, denoted by E. This is another simplifying assumption, as the level of E in each subbasin i could quite easily be a decision variable. C_i: water for domestic use in part i. $W(D)$ is the additional water available from the dam.

A player, or a coalition, will use its available water to maximize its revenue. Actually, it is assumed that each player solves an optimization problem to maximize the use of the available water via allocation among all possible water uses. The plan that maximizes the returns is called the characteristic function of the coalition or player. A very general exposition of the optimization problem (generalized to each coalition) is as follows:[14]

$$\underset{\substack{\textit{water from dam, water fee,}\\ \textit{water from river}}}{\mathit{Max}} \left\{ \begin{array}{l} \textit{(land under citrus); (land, water intensity,} \\ \textit{cycles of cabbage); (water consumption in} \\ \textit{villages by source); (environmental release)} \end{array} \right\}$$

Subject to:

(1) *environmental Reserve outflow*

(2) *rain availability*

(3) *water in river*
(4) *land*
(5) *labor force*
(6) *other institutional and technical constraints*

Such an optimization problem is solved to yield solutions to the following coalitions (individuals, partials, and grand coalition): $\{U\}$, $\{M\}$, $\{L\}$, $\{U, M\}$, $\{M, L\}$, and $\{U, M, L\}$. In the solution for each coalition, the characteristic functions are denoted by: $v(\{U\})$, $v(\{M\})$, $v(\{L\})$, $v(\{U, M\})$, $v(\{M, L\})$, and $v(\{U, M, L\})$. Note that according to the simplifying assumption $v(\{U, L\})$ will be replaced by $v(\{U\}) + v(\{L\})$.

Using various game theory solution concepts, allocations of payoffs are made among the three players. For demonstration the Shapley value solution concept is used (Shapley 1953):

$$\phi_i = \frac{1}{n}\sum_{s=1}^{n}\frac{1}{c(s)}\sum_{\substack{i \in s \\ |S| = s}} [v(S) - v(S \setminus \{i\})],\ i = U, M, L$$

where $c(s)$ is the number of coalitions of size s containing player i:

$$c(s) = \binom{n-1}{s-1} \equiv \frac{(n-1)!}{(n-s)!\,(s-1)!}$$

Another cooperative game theory solution concept, the nucleolus (Schmeidler 1969), is also used (for a definition of the nucleolus see standard references, such as Owen 1995). The nucleolus always lies in the core, provided that it is nonempty. In these results, the values of the nucleolus differ very little from those of the Shapley value.

5.5.2 *Cooperative game solution results*

Based on the model calculations (see Appendix at end of this chapter), the following are the characteristic values of the Kat cooperative game (in rands):

$$
\begin{aligned}
v(\{U\}) &= 336{,}060 \\
v(\{M\}) &= 1{,}758{,}946 \\
v(\{L\}) &= 1{,}185{,}693 \\
v(\{U, M\}) &= 2{,}341{,}140 \\
v(\{U, L\}) &= 1{,}521{,}753 \\
v(\{M, L\}) &= 2{,}944{,}639 \\
v(\{U, M, L\}) &= 3{,}552{,}913
\end{aligned}
$$

The resulting Shapley allocation is:

$$\phi_U = 467{,}820.33$$
$$\phi_M = 1{,}890{,}706.33$$
$$\phi_L = 1{,}194{,}386.33$$

with

$\phi_U + \phi_M + \phi_L = v(\{U, M, L\})$, since the Shapley value provides an efficient allocation,
$\phi_i \geq v(\{i\})$, $i = U, M, L$, which suggest individual rationality, and
$\phi_U + \phi_M \geq v(\{U, M\})$, $\phi_U + \phi_L \geq v(\{U, L\})$, and $\phi_M + \phi_L \geq v(\{M, L\})$, which suggests group rationality (otherwise stated, the Shapley value lies in the core).

The payoff is distributed among the three players such that *U*, *M*, and *L* get 13 percent, 53 percent, and 34 percent of the total cooperative profits, respectively. *U* is clearly the main beneficiary from the cooperative game allocation, increasing its share in the cooperative payoff by 39 percent compared with the noncooperation payoff, while *M* and *L* gained 7 percent and 1 percent, respectively. Cooperative game theory assumes utility transfer in the form of payments (or compensations).

For the nucleolus, the allocation is 465,647 for *U*, 1,888,533 for *M*, and 1,198,733 for *L* (all allocations in rands). Since the core is nonempty (the Shapley value, as seen, lies in it), the nucleolus is also in the core. Since the allocations provided by the nucleolus are so close to those provided by the Shapley value (the differences are much smaller than variations because of the approximations used or to the assumptions done), the same comments in the next section will apply for both.

5.6 Evaluation of the negotiation and cooperative game theory allocations

This section attempts to evaluate the results from the role-playing game and the cooperative game allocations. The basic features of both approaches that lead to the solutions obtained will be reviewed. Some extensions will then be suggested that may be likely to bring the results of the two closer. Table 5.6 displays the annualized allocations of profits by the role-playing game, the Shapley value, and the nucleolus.

Any comparative evaluation of the negotiation approach (role-playing game) and cooperative game outcomes has to be subject to several caveats. First, the role-playing game is based on an empirical/real negotiation framework and the cooperative game is based on an axiomatic model. Second, the role-playing game has a dynamic nature that cannot be captured with the present version of the cooperative game model. And third, the main differences (between the role-playing game and the cooperative game) in the calculations of the profits of the players lead to possible discrepancies between the total and individual profits. While the calculation of profits to the *U*, *M*, and *L* players in the case of the cooperative game is a result of an optimization

Table 5.6 Allocation of annual (and annualized) profits according to the role-playing game, Shapley value, and the nucleolus in the Kat game

	Role-playing game	*Shapley value*	*Nucleolus*
Player		*Rands*	
U	173,371	467,820	465,647
M	840,854	1,890,706	1,888,533
L	875,872	1,194,386	1,198,733
Total	1,890,097	≅3,552,913	3,552,913

process that takes into account a very strict set of variables, the role-playing game process incorporates "real" players that take into consideration many more factors than the algorithm used in the cooperative game. Just these three caveats may explain possible differences in total basin outcomes.

As for the behavior of individual players, observations from the role-playing game session suggest that both irrigators and village managers aim at improving their respective indicators of performance (profit for irrigators; profit + residents' satisfaction for village managers) without necessarily maximizing them. This might be because of a lack of information on the possible alternative strategies they could adopt during the game session and exemplifies a behavior called "satisficing", where satisficing is an alternative to optimization in cases where there are multiple and competitive objectives and one gives up the idea of obtaining a "best" solution (Simon 1992). Players therefore adopt year-after-year strategies of incremental improvement of their indicators. These strategies take into account external factors and must be discussed within (and cleared by) the WUA before they can be put in practice. In addition, willingness to reach an improved state does not correspond necessarily to an improved state for all the players through the role-playing game session: lack of play skills or external factors whose dynamics are worse than forecast can be the causes of performances less positive than expected. There is another reason behind the difference in the results: in cooperative game theory there are less "players" than in the role-playing game approach. In the cooperative game, each of the three players corresponds to a group of players in the role-playing game approach.[15] For example, farm 1 and farm 2 and the village in *U* are merged and transformed into player *U* in the cooperative game. Because of the optimization approach used in cooperative game theory, this leads to an optimal coordination and allocation among these players. This is a condition that cannot be taken for granted in the role-playing game approach. A further caveat is that *UV* (and similarly other village managers) is assumed not to have any strategic role in the cooperative game theory approach.

As can be seen from Table 5.6, the basin profit (outcome) based on the role-playing game outcomes is 1.891 million rands on an annualized basis for each of the six years. The annual profit based on the cooperative game characteristic function calculations is 3.553 million rands. The difference arises from economic decisions that are based on different algorithms and assumptions. While the total payoff at the basin level may be different in the role-playing game and cooperative game procedures, because of use of different assumptions, a more useful insight is obtained from the distribution of the payoff among the players. In this case, the shares of the three subbasins *U*, *M*, and *L* in the basin total profit were 9 percent, 45 percent, and 46 percent respectively in the role-playing game based on the annualized allocation, and 13 percent, 53 percent, and 34 percent respectively in the cooperative game solution. Note the equal share of profit for *M* and *L* in the role-playing game solution, compared with a lower share to *L* (compared to *M*) in the cooperative game solution.

5.7 Conclusion and policy implications

This chapter has developed a framework that helps to evaluate cooperative game theory and role-playing game solutions to a problem of water allocations and payoff distribution among competing uses. Such a framework is useful for several reasons. First, it allows the analysts to assess the nature of the assumptions made during the calculations or negotiation session. Second, it creates feedback loops between the cooperative game and the role-playing game that may assist further development of the tools. And third, it may suggest complementary roles for each approach under different conditions that the parties in the allocation problem face.

The role-playing game results indicate a clear willingness of the WUA members to interact and cooperate for better management of water resources in the basin. Domestic water provision and the protection of the ecological Reserve are priorities for the local stakeholders, and as a consequence the improvement of individual profits is subject to these objectives. These findings have important policy implications, as the WUA will be responsible for water allocation in the near future. The WUA attitude during the role-playing game session will be discussed and possibly translated into a water management plan that the WUA will submit for approval to the Department of Water Affairs and Forestry.

With the present level of analysis, it is believed that the role-playing game approach has the potential to support management of common pool resources. Interest groups that are associated with the resource have a reliable and flexible tool, allowing them to respond to various states of nature and changes in prices and other institutional reforms that could be imposed by the regulator. Cooperative game theory provides a fast and dirty approximation to the details that the role-playing game is capable of obtaining. It is believed that with the proposed improvements in both approaches, synchronization with and reflection of real-life policies would be better captured.

Acknowledging the overly simplified optimization procedure in the case of cooperative game theory, it is suggested that the baseline scenario in the cooperative game will be modified in order to address new constraints and scenarios that have been considered in the role-playing game session. This will include the dynamic nature of the allocation problem, and various structures to consider the environmental flow needs. Outstanding issues that have an intrinsic dynamic characteristic are investment decisions and dam management. The latter issue, in particular, calls for due consideration on the part of the WUA, taking into account its (already mentioned) future responsibilities. Future sessions of the role-playing game will take cooperative game theory results into consideration, such as allowing for negotiation among subbasins, and integrating WUA and subbasin procedures in order to eliminate unilateral decisions, such as the decision to stop releasing water from the dam.

As was indicated in the introduction, there is already a sizeable number of contributions using individual negotiation and bargaining theory and using cooperative game theory to analyze common pool resource management and allocation. The approach used in this chapter offers the possibility of creating a bridge between these two theories.

Notes

1. The term "basin" is used here to mean the total area drained by a river and its tributaries. Other terms for this area used in the literature include "catchment", a hydrological term referring to the area from which a river catches or collects its water; and "watershed", a term that properly refers to the high ground forming a dividing line between two river basins, but is often used as synonymous with "basin".
2. Modified from Farolfi and Rowntree 2005.
3. "Emerging" citrus farmers are black farmers who after 1994 had the right to occupy and develop citrus farms located during the apartheid period in the Ciskei Bantustan and previously owned and managed by the public Bantustan administration.
4. The Irrigation Board was transformed recently into a more participatory Water Users Association including representatives of all main groups of stakeholders in the Kat basin. This organization is now responsible for defining the water business plan indicating water allocation strategies and resource management for the basin. While the difference between the Irrigation Board and the Water Users Association may not be essential for cooperative game theory, it is very important for the negotiation process, as will be discussed later in the chapter.
5. Developing groundwater resources in the basin would help irrigators overcome droughts. However, in this chapter groundwater development is not considered because of the need for external financial support.
6. Defined in the National Water Act as "the quantity and quality of water required to protect the aquatic ecosystems of the water resource in order to secure ecologically sustainable development and use of the resource".
7. These portions of the basin correspond to the three voting areas identified to nominate the Kat River Valley Water Users Association representatives (Figure 5.1).
8. 1 hectare = 2.5 acres.
9. The conversion factors for the different variables range between 2.1 (Inhabitants in basin) and 13 (Area citrus). The arbitrary choice of the conversion factors arose from the tradeoff between having a role-playing game that represents the reality and a role-playing game that is "playable". It may be noted that players

validated the role-playing game representation of the Kat during the first session. The only criticism was for the disproportionately high importance of domestic water consumption in the role-playing game compared to the real one. This issue was addressed in a second version of the game by changing the conversion factor of "Inhabitants in basin" from 2.1 to 4.9.

10. In this game, profit = total income less total costs. If a farmer invests in citrus plantations, therefore, that farmer's annual income during the first years of new orchards is constant (no production) whilst the costs increase. It was noticed by citrus farmers during the game debriefing that this is not really how they see things because an investment is calculated as a positive asset in their budget, whereas here it is a negative (cost) item. They suggested calling "cash flow" what is called "profit" in the game outcomes.
11. Price takers face given prices rather than affecting the prices in the market.
12. It should be stressed that the formulation of this cooperative game theory problem allows consideration of longer time horizons by creating for each player and coalition optimization problems that span over T years but what the player considers is the expected annualized value.
13. It is assumed that the ecological Reserve serves both for environmental health in the Kat River and for releases into the Fish River's estuary.
14. A multiyear problem will be developed in future work.
15. We are indebted to a referee for having suggested giving due emphasis to this fact.
16. The capital recovery multiplier equation used is $[r \cdot (1 + r)^t]/[(1 + r)^t - 1]$, where r is discount rate (0.06), and t is number of years of the game (6).

References

Aadland, D. and Kolpin, V. (1998) "Shared irrigation costs: an empirical and axiomatic analysis", *Mathematical Social Sciences*, 35: 203–18.

Aadland, D. and Kolpin, V. (2004) "Erratum to 'environmental determinants of cost sharing' ", *Journal of Economic Behavior and Organization*, 55: 105–21.

Barreteau, Olivier, et al. (2003) "Our companion modelling approach", *Journal of Artificial Societies and Social Simulation*, 6(1). Available online: http://jasss.soc.surrey.ac.uk/6/2/1.html (accessed January 5, 2007).

Becu, N., Perez, P., Walker, A., Barreteau, O., and Le Page, C. (2003) "Agent based simulation of a small catchment water management in northern Thailand: description of the CATCHSCAPE model", *Ecological Modelling*, 170: 319–31.

Carraro, C., Marchiori, C., and Sgobbi, A. (2005a) "Application of negotiation theory to water issues", Policy Research Working Paper 3641, World Bank, Washington, DC.

Carraro, C., Marchiori, C., and Sgobbi, A. (2005b) "Advances in negotiation theory: bargaining, coalitions and fairness", Policy Research Working Paper 3642, World Bank, Washington, D.C.

Dray, A., Perez, P., Jones, N., Le Page, C., D'Aquino, P., and Auatabu, T. (2006) "The AtollGame experience: from knowledge engineering to a computer-assisted role playing game", *Journal of Artificial Societies and Social Simulation*, 9(1). Available online: http://jasss.soc.surrey.ac.uk/9/1/6.html (accessed January 5, 2007).

DWAF (Department of Water Affairs and Forestry) (2001) "Fish to Tsitsikamma water management area: water resources situation assessment", WMA 15 Report P WMA 150000/00/0203.

Farolfi, S. and Abrams, M. (2005) "Water uses and their socio-economic impact in the Kat River catchment: a report based on primary data", unpublished Water Research Commission Report, Project K5/1496, Institute for Water Research, Rhodes University, Grahamstown.

Farolfi, S. and Bonté, B. (2005) "KatAware: a negotiation-support tool for participatory water resource management in the Kat River Valley (prototype model)", unpublished Water Research Commission Report, Project K5/1496, Institute for Water Research, Rhodes University, Grahamstown.

Farolfi, S. and Bonté, B. (2006) "KatAware: from prototype to version 1—development of the model following the iterative process of companion-modelling", unpublished Water Research Commission Report, Project K5/1496, Institute for Water Research, Rhodes University, Grahamstown.

Farolfi, S. and Rowntree, K. (2005) "Accompanying local stakeholders in negotiation processes related to water allocation through simulation models and role-playing games: an experience from South Africa", paper presented at conference on End-User Ownership and Involvement in Integrated Water Resource Management, Cairo, Egypt, 13–17 November 2005. Available online: www.empowers.info/page/704 (accessed January 5, 2007).

Farrington, J. and Boyd, C. (1997) "Scaling up the participatory management of common pool resources", *Development Policy Review*, 15(4): 371–91.

Goodhue, R.E. and McCarthy, N. (2008) "Traditional grazing rights in sub-Saharan Africa and the role of policy", this volume, Chapter 3.

Lankford, B. and Watson, D. (2006) "Exploring metaphor in natural resource gaming: insights from the river basin game", in O. Barreteau, M. Etienne, C. Le Page, and P. Perez (eds) *Symposium Issue of Simulation and Gaming*, SAGE, USA.

McMaster, A. (2002) "GIS in participatory catchment management: a case study in the Kat River Valley, Eastern Cape, South Africa", unpublished MSc thesis, Rhodes University, Grahamstown.

Munro, G. (1979) "The optimal management of transboundary renewable resources", *Canadian Journal of Economics*, 12(8): 355–76.

Ostrom, E. (1992) *Crafting Institutions for Self-Governing Irrigation Systems*, San Francisco: Institute for Contemporary Studies.

Ostrom, E., Gardner, R., and Walker, J. (1994) *Rules, Games, and Common-Pool Resources*, Ann Arbor, MI: University of Michigan Press.

Owen, G. (1995) *Game Theory*, San Diego: Academic Press.

Parrachino, I., Dinar, A., and Patrone, F. (2006) "Cooperative game theory and its application to natural, environmental, and water resource issues: 3. Application to natural and environmental resources", Policy Research Working Paper 4074, World Bank, Washington, DC.

Schmeidler, D. (1969) "The nucleolus of a characteristic function game", *SIAM Journal of Applied Mathematics*, 17: 1163–70.

Shapley, L.S. (1953) "A value of n-person games", *Annals of Mathematical Studies*, 2(28): 307–18.

Simon, H.H. (1992) "Methodological foundations of economics", in J.L. Auspitz et al. (eds) *Praxiologies and the Philosophy of Economics*, New Brunswick, NJ: Transaction Publishers.

Steins, N.A. and Edwards, V.M. (1999) "Platforms for collective action in multiple-use common pool resource management", *Agriculture and Human Values*, 16(3): 241–55.

White, T.A. and Ford Runge, C. (1994) "Common property and collective action: lessons from cooperative watershed management in Haiti", *Economic Development and Cultural Change*, 43: 1–20.

Appendix: The formulation of the profit function in the KatAware model and the characteristic function in the cooperative game

The assumptions used in the calculation of the annual profit in the role-playing game and in the formulation of the characteristic function in the cooperative model are provided here.

Since the cooperative game model and the role-playing game differ in the way they calculate payoffs and players, different symbols will be used for the same variables in the calculations of the characteristic function and in the profit function.

Calculation of the profit functions in the KatAware model

In the role-playing game, irrigators' annual profit per hectare is calculated by subtracting a unitary production cost from the income resulting from the produced crop multiplied by the crop price for that year:

$$\Pi_{ij} = Q_{ij}P_i - C_{ij},$$

where Π is annual profit per hectare (in rands); Q is annual crop yield per hectare (in tonnes for citrus and in bags for cabbage) using technology j. However, in the empirical calculations no distinction is made between the yields obtained by each technology; P is price per unit of crop, net of marketing and harvesting cost; C is annual cost per hectare (in rands); i = crop (citrus or cabbage); j = irrigation technology (traditional or innovative).

A crop yield response to available water is introduced in the model following a linear relationship:

$$Q = mx + k,$$

where x is irrigation water (in cubic meters per hectare); m is an irrigation coefficient with a value of 0.00434 for citrus and 0.2057 for cabbage; and k is an intercept having a value of 14.62 for citrus and 1,800 for cabbage.

The profit function for a village manager is:

$$\pi = \max_{y \cdot x_i} \left\{ y \cdot x_i \cdot N[p \cdot q_{x_i} - w_{x_i} - \tau \cdot q_{x_i}] \right\}$$

where $y \cdot x_i$ is the proportion of water source for villagers ($0 < y \cdot x_i < 1$); $i = 1$ for river, 2 for collective tap, 3 for in-dwelling tap (to each source an amount per capita of consumption is fixed); p is tariff that villagers pay to the village manager for water; q_{x_i} is water consumption per capita per month for each water source; w_{x_i} is cost of water source (rands per capita per year); τ is water price paid to the Department of Water Affairs and Forestry; N is number of inhabitants in the village.

The formulation of the characteristic function in the cooperative game

The physical model

Denote by D^0 the water stock behind the dam at the beginning of the planning period. R_i (i = 1,2,3) is the amount of rainfall on the area of player i (1 = U, 2 = M, and 3 = L are used interchangeably). The variable D^0 is also interpreted to be the amount of water available for use by player i, and also the amount of water in the river that player i can use. D_i is the request for flashes of water from the dam operator by player i. R_1^D is the amount of water stored behind the dam during the planning period. As noticed, only rainfall in the part of the area of player 1 can be dammed. D will be the amount of water behind the dam that can be used during the planning period.

Several relationship hold so far:

$$R_1^D \leq R_1 \tag{5.A1}$$

$$D = D^0 + R_1^D \geq \sum_i D_i \tag{5.A2}$$

Denote by W_i the total demand for water, during the planning period, by player i and by E the demand (allocation) for environmental needs where the basin ends and the Kat River becomes a tributary of the Fish River. This occurs after player L's area.

Here several more relationships are added:

$$W_i \leq S_i, \tag{5.A3}$$

where, S_i is the supply of water available to player i as follows:

$$S_1 = D_1 + R_1 - R_1^D \tag{5.A4}$$

$$S_2 = D_2 + R_2 + (S_1 - W_1) \tag{5.A5}$$

$$S_3 = D_3 + R_3 + (S_2 - W_2) \tag{5.A6}$$

It is now necessary to include more specific relationships representing the total water demand by each player. Given the water use patterns by each player, there are agricultural uses (citrus and cabbage), and domestic uses (in the villages and urban centers). This amount is imposed on the subbasin and is subject to a policy decision.

For the purposes of this study, let V_i^v, V_i^u, C_i^C, and C_i^B be the amounts of water used by the village, the urban center, the citrus, and the cabbage, respectively, for player *i*. The total demand for water by player *i* is:

$$W_i \; V_i^v + V_i^u + C_i^C + C_i^B. \tag{5.A7}$$

Additional constraints

The WUA imposes minimum flow (MF_i) to keep the river flowing in each subbasin. This minimum flow amount is deducted from each player's amount of available water S_i.

$$S_i^{MF} = S_i - MF_i \tag{5.A8}$$

In addition, the WUA imposes the environmental flow constraint. That constraint is handled in the following way: a total of *EF* has to leave the Kat River to the Fish River; this amount has to fulfill the following relationship:

$$EF \leq \sum_i EF_i \tag{5.A9}$$

The optimization process in each subbasin

The villages. Assume that each village has a given population N_i^r, that the annual water consumption per person is v_i^r, the cost of providing each unit of water is P^{r_i}, and that the utility per inhabitant in the village from being provided a unit of water is u^{r_i}. It is assumed that the utility is linear in money. For simplicity assume that the ratio is 1:1. The total "benefits" (U_i^v) from providing water to the village is therefore:

$$U_i^r = V_i^r \cdot (u^{r_i} - P^{r_i}),$$

where $V_i^r = N_i^r \cdot v_i^r$,
with V_i^v being the amount of water consumed in village *i*.

A constraint is introduced on the minimal amount per year that a village inhabitant should be receiving:

$$v_i^r \geq \underline{v} \tag{5.A10}$$

The citrus industry. Assume that citrus is grown with three factors of production, namely land, water, and labor. Since this is a perennial crop and it can be assumed there will be no investment in new plantation, the area of citrus in each subbasin will be fixed at L_i^C. Since in this model the area is *fixed*, the decision growers make is how much water per hectare to apply; they may opt not to irrigate at all, giving a minimum yield (Farolfi and Bonté 2006).

The payoff for citrus production in subbasin i is:

$$F_i^C = L_i^C \cdot [Y_i^C\,(C_i^C) \cdot P^C - C_i^C \cdot P^{C_i^C} - B_i^C \cdot P^B],$$

where $Y_i^C\,(C_i^C)$ is the citrus production function. It has a positive intercept at $C_i^C = 0$; P^C is the price per unit of citrus produced, which is a function of the amount of water applied per hectare, C_i^C; and $P^{C_i^C}$ is the cost of water charged to the citrus operation in subbasin i. Note that different water charges are allowed per crop and subbasin; B_i^C is the labor per hectare of citrus; and P^B is the cost per unit of labor, assuming the same for the entire Kat basin.

The cabbage industry. Production is very similar to that of citrus except that, under the set of assumptions used here, the land for growing cabbage is not fixed (however, for realism a constraint of 60 hectares is imposed on the extension of land that can be used for cabbage; this constraint turns out to be binding only for player U and for the subcoalition $\{U, M\}$). In the case of cabbage, the growers do not vary the amount of water per hectare, but decide only the area to be cultivated with cabbage with a given amount of water per hectare. The payoff per hectare of cabbage is:

$$F_i^B = L_i^B \cdot \left[C_i^B \cdot P^B - C_i^B \cdot P^{C_i^B} - B_i^B \cdot P^B\right],$$

where P^B is the cost per unit of cabbage produced (cost of labor has the same symbol); C_i^B is the amount of water applied per hectare of cabbage; $P^{C_i^B}$ is the cost of water charged to the cabbage operation in subbasin i; B_i^B is the labor per hectare of cabbage. Cabbage growers decide on the area they plant with cabbage L_i^B.

The objective function of subbasin i. Subbasin i maximizes payoff from the three activities, subject to physical and institutional constraints:

$$Y_i = \underset{V_i^V + C_i^C + C_i^B}{Max} \left\{U_i^u + U_i^v + F_i^C + F_i^B\right\},\ i = 1, 2, 3 \qquad (5.A11)$$

s.t. the relevant constraints in (5.A1)–(5.A10).

The characteristic functions

The value of the characteristic function for each individual coalition is actually a solution to a linear programming problem that is unique to the coalition at stake. In the case of the individual coalitions, (5.A11) is solved subject to relevant constraints in (5.A1)–(5.A10), including imposed rules of allocation of water from the dam, allocation of the minimum flow, and allocation of the environmental flow among players *1, 2, 3*.

Then there is the possibility of subcoalitions. Clearly a coalition of {1, 2} and a coalition of {2, 3} can be envisioned. As said, a coalition of {1, 3} is less obvious. Such coalitions can be included on the premise that the WUA enforces rules and water transfers that are respected by its members.

6 Cooperation and equity in the river-sharing problem

Stefan Ambec and Lars Ehlers

This chapter considers environments in which several agents (countries, farmers, cities) share water from a river. Each agent enjoys a concave benefit function from consuming water up to a satiation level. Noncooperative extraction is typically inefficient and any group of agents can gain if they agree on how to allocate water with monetary compensations. This chapter describes which allocations of water and money are acceptable to riparian agents according to core stability and several criteria of fairness. It reviews some theoretical results and then discusses the implementation of the proposed allocation with negotiation rules and in water markets. Lastly, it provides some policy insights.

6.1 Introduction

Water is essential to life. It is consumed for a variety of purposes, from domestic to agricultural and industrial uses. Because of population growth, the development of irrigated agriculture, and industrialization, demand has tremendously increased, and water has become a locally scarce resource in many regions on earth. The so-called tragedy of the commons has considerable relevance to water resources: free-access (or decentralized) extraction leads to inefficiencies, and increasing benefits to all users requires centralized planning and cooperation of the economic agents (farmers, firms, cities, countries) who share a water resource. In practice, concretization of such coordination may take many forms, from international agreements signed by sovereign countries to allocation rules or water markets established by communities of farmers. The process is often facilitated by local authorities, in many cases with the involvement of users.

This chapter deals with the issue of coordinating water management along rivers. It investigates the incentives of riparian agents to agree to share water efficiently. It examines what kind of agreement is acceptable. The definition of "acceptability" is twofold. First, the river-sharing agreement (or allocation) should be stable in the sense that no users or group of users are better off designing another river-sharing agreement. Second, it should be perceived as fair according to certain justice principles.

This issue is tackled using cooperative game theory and the axiomatic theory of justice. The chapter describes the cooperative game induced by a river-sharing problem, and analyzes the stable river-sharing agreements in this cooperative game. Next, it considers standard axiomatic principles of fair division and adapts them to the river-sharing problem.[1] It posits fair sharing rules for total welfare.

The chapter reviews several important theoretical papers on the river-sharing problem in an informal and simple way, without formal proofs. The goal is to provide the intuition of these results and their policy implications.

Note that the focus here is on the direct benefits of water consumption. The model ignores other benefits for which water is not directly consumed, such as flood control, navigation, recreation activities, biodiversity, and hydropower production.[2]

Rivers are common water resources that possess several interesting features. First, each agent can only consume water entering the river upstream of its location. Therefore, agents have unequal access to the resource, depending on their location on the river. Upstream agents have a first mover advantage in water extraction. Yet, as river flow increases downstream, this advantage is offset by the lower amount of water available upstream. Second, the welfare achieved by cooperation depends on the locations of the cooperating agents. In order to increase welfare, a group of agents exchanges water. This can only be done from upstream agents to downstream agents and, preferably, by neighbors: water exchange among distant agents is subject to extraction or free riding by those located in between.

International river-sharing agreements are numerous worldwide. For instance, the Nile Treaty, signed in 1929, specifies a sharing rule for the Nile River water flow between Egypt and Sudan. The Columbia River Treaty specifies a sharing rule for the costs and benefits from flood control and hydropower production between Canada and the United States (Barrett 1994). In the case of the Syr Darya River, the upstream country, Kyrgyzstan, agreed to increase summer discharges to supply the downstream country, Uzbekistan, in exchange for fossil fuel transfers (Abbink et al. 2005). Similarly, the Lao People's Democratic Republic and Thailand signed an agreement on developing hydropower production on one of the tributaries of the Mekong River inside the former country. It specifies a payment in hard currency from Thailand to the Lao People's Democratic Republic in exchange for electricity supply (Barrett 1994). In the United States, states sign interstate river compacts that prescribe a fixed or a percentage allocation of water (Bennett et al. 2000).

Governments or farmers themselves set up rules to encourage efficient exploitation of water for irrigation, including water pricing, subsidies, and water markets (Ostrom 1990; Dinar et al. 1997). Such rules lead to an allocation of water and can result in a redistribution of the benefits arising from water extraction. Because different users (for example farmers, urban dwellers) use water for different purposes and thus derive different values from an additional unit of water, there is an impetus for moving water from lower-value to higher-value uses. During this process, the seller of a certain volume of water obtains monetary compensation from those who buy it. In general, farmers have to pay for water consumption. The money collected is then spent on maintenance or transferred to some users through subsidies. For instance, Thomas et al. (2004) provide evidence that French farmers receive on average four times more subsidy than the amount they pay to the water agencies.

These monetary transfers, whether they are direct (peer-to-peer in a water market) or not (through centralized water pricing, taxes, or subsidies), and the allocation of water comprise the total benefit from consuming water, defining a particular distribution of the total welfare. When water is exclusively consumed to irrigate crops, the farmer's total welfare is simply the value of total production, though it could include the monetary equivalent of the utility consumers derive from consuming water.

This chapter proceeds as follows. Section 6.2 introduces a general model to address the issue of cooperation and equity in river sharing. Section 6.3 examines the optimal allocation of water in this model. Section 6.4 shows how noncooperative free-access extraction leads to an inefficient allocation of water. Therefore, any movement towards Pareto optimality requires cooperation and monetary compensation mechanisms that are acceptable to all agents. The transfers are only acceptable if the allocation of water and money belongs to the core of the cooperative game (section 6.5) or is perceived as fair (section 6.6). Section 6.7 presents a negotiation game that lead to an efficient allocation of water and fair and stable transfer schemes in the subgame perfect equilibrium. Section 6.8 posits property rights that lead to efficient water allocation and fair transfers in competitive water markets. The concluding section discusses the policy implications of the analysis of the river-sharing problem.

6.2 General framework

The river-sharing problem is represented by the following stylized model. Agents are ranked according to their location along the river and numbered from upstream to downstream: $i < j$ means that i is upstream of j and j is downstream of i. There are n agents. The set of agents is denoted by $N = \{1, \ldots, n\}$. Agent i's benefit or production function from consuming a level x_i of water is $b_i(x_i)$. Benefits are measured in monetary terms. The function b_i is strictly concave and differentiable for every x_i and $i \in N$. It is increasing up to a satiation level y_i. Above this level, the agent infers a loss from over-consumption. Indeed, above satiation, the cost of extraction and sanitation exceeds the benefit from consumption; or, even worse, the agent suffers from flooding. Marginal benefits are strictly decreasing and positive up to the satiation level. Above that, they are strictly decreasing and negative. It is also assumed that the marginal benefit at 0 (no water) is high enough to avoid corner-type solutions (no water consumption by some agents) in the efficient water allocation program. This means that water is indispensable for a user, as it is very costly (or deadly) to be deprived of water. Examples of such benefit functions are quadratic forms such as

$$b_i(x_i) = a_i x_i - b_i \frac{x_i^2}{2}$$

where $y_i = \frac{a_i}{b_i}$, provided that the marginal benefit at zero a_i is high enough that everybody should be supplied with water. Our assumptions on the benefit functions are broadly consistent with the production function for irrigated crops, as represented, for example, by Griffin (2006: 19), except for the marginal benefit at 0.

The amount of water entering the river at the location of the most upstream agent 1 is $e_1 > 0$. In addition, several tributaries might enter the river after the location of agent 1. Denote by $e_i \geq 0$ the amount of water flowing from tributaries located between agents $i - 1$ and i for $i = 2, \ldots, n$. The highest total amount of water available at location i (without extraction upstream) is thus $E_i = e_1 + e_2 + \ldots + e_{i-1} + e_i > 0$. A river-sharing problem is formally defined by a set of agents N, a vector of (strictly concave and single-peaked) benefit functions $(b_1, \ldots, b_n)$, and a vector of water inflows along the river $(e_1, \ldots, e_n)$.

6.3 Noncooperative extraction

Under noncooperative extraction, the river-sharing problem defines the following sequential game. Player 1 chooses how much to consume x_1 under the constraint that this level does not exceed the amount available e_1. Then player 2 selects x_2 from the remaining water $e_1 - x_1 + e_2$, that is, the water left by the upstream agent 1 added to the water flowing between 1 and 2. And so forth until agent n. Of course, guided by selfishness, each player i maximizes its benefit function b_i when choosing x_i subject to the constraint $x_i \leq e_1 - x_1 + \ldots + e_{i-1} - x_{i-1} + e_i$.

In the subgame perfect equilibrium of this game, each player i extracts the maximum between its satiation level y_i and the amount of water available at its location $e_1 - x_1 + \ldots + e_{i-1} - x_{i-1} + e_i$. The remaining water is left in the river to be consumed by the downstream agents.

The noncooperative equilibrium is in general inefficient in the sense that the payoff of players can be increased with another allocation of water through monetary compensations. In other words, there is often room for Pareto improvement by allowing transfers between agents.

To understand that, consider simply a river shared by two agents ($n = 2$) with no tributaries ($e_2 = 0$). The upstream agent 1 consumes the minimum of y_1 and e_1, enjoying a benefit $b_1(\min\{e_1, y_1\})$. If $e_1 \geq y_1$ then the downstream agent 2 gets nothing. If $e_1 > y_1$ and $e_1 - y_1 < y_2$, agent 2 consumes $e_1 - y_1$ and enjoys $b_2(e_1 - y_1)$. This case is represented in Figure 6.1.

Now reduce the upstream agent's extraction by ε, as in Figure 6.1, to

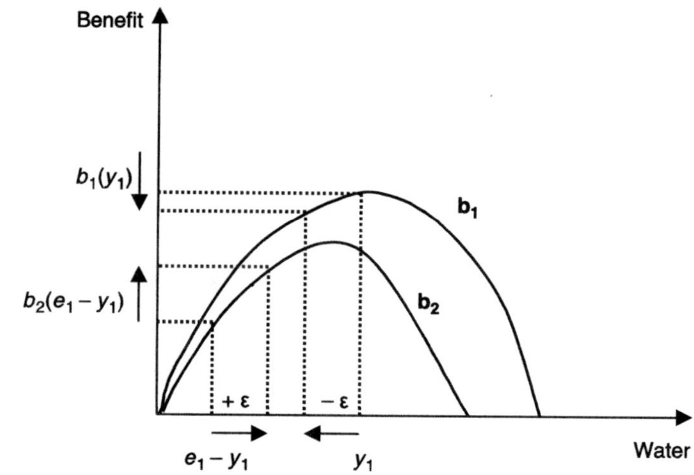

Figure 6.1 Example of inefficient noncooperative extraction with Pareto improving reallocation of water.

increase the downstream agent's consumption by the same amount, thus benefiting the latter. Because of the strict concavity of the benefit functions (or because of diminishing marginal benefits), the increase in benefit to the downstream agent is larger than the loss of benefit to the upstream agent. Formally, for $\varepsilon > 0$ small enough $b_2(e_1 - y_1 + \varepsilon) - b_2(e_1 - y_1) > b_1(y_1) - b_1(y_1 - \varepsilon)$. In Figure 6.1, the vertical downward arrow is shorter than the vertical upward arrow. Of course, the same argument applies if the upstream agent consumes e_1 so there is no water left downstream (and also for other assumptions on the number of agents and water inflows). This change of water allocation is Pareto improving if the upstream agent is somehow compensated by the downstream agent for its loss of benefit. It therefore requires some sort of transfer from agent 2 to agent 1. But which transfer?

With two agents, the problem is easy. Let us denote (x_1^*, x_2^*) the efficient allocation of water (where x_i^* stands for the water consumed by agent *i*), which, by definition, maximizes the sum of the two agent's benefits, $b_1(x_1) + b_2(x_2)$ subject to the resource constraint $x_1 + x_2 \leq e_1$. The two parties will agree to change their water consumption to (x_1^*, x_2^*), provided that 2 pays a transfer t to 1, if they are both better off doing so. The upstream agent 1 accepts if $b_1(x_1^*) + t \geq b_1(y_1)$ while downstream agent 2 agrees if $b_2(x_2^*) - t \geq b_2(e_1 - y_1)$. The transfer must therefore satisfy the two acceptability constraints $b_2(x_2^*) - b_2(e_1 - y_1) \geq t \geq b_1(y_1) - b_1(x_1^*)$. With more than two agents, the transfers need to be acceptable not only to individual agents but also to any group (or coalition) of agents. Otherwise, a group of agents is better off refusing the agreement and signing its own agreement on the part of the river it controls. Technically speaking, the transfer scheme defines a distribution of welfare. To be acceptable, this distribution of welfare should be in the core of the cooperative game in the sense that no coalition of agents is better off forming its own river-sharing agreement.

Furthermore, the transfer should be perceived as fair by riparian agents. Indeed fairness is often invoked as the main principle for sharing life-essential natural resources such as water. Several fair ways to share the benefit of river water extraction will be considered.

This chapter focuses on transfer schemes that implement efficient allocations of water while being acceptable to n riparian agents (in ways to be defined) for any $n > 1$ (for example more than two agents). It then discusses how to implement them. First, the optimal allocation of water in the river-sharing problem is analyzed.

6.4 Efficient extraction

To simplify the analysis without loss of generality, assume $e_i \leq y_i$ for every i. Indeed, if $e_i > y_i$ for one i, then since agent i will never choose to consume more than y_i, the next downstream agent $i + 1$ can rely on $e'_{i+1} = e_{i+1} + e_i - y_i$. So the river-sharing problem can always be redefined with water supplies $e'_i = y_i$ and $e'_{i+1} = e_{i+1} + e_i - y_i$ and so forth.

The efficient allocation of water is the allocation that maximizes the total benefit from water extraction. Formally, it is the vector $\boldsymbol{x} = (x_1, \ldots, x_n)$ that maximizes the sum of all b_i, that is, $b_1(x_1) + \ldots + b_n(x_n)$. Such an allocation should be feasible in the sense that what is extracted up to any location i should not exceed what is available up to that point, that is, $x_1 + \ldots + x_i \leq E_i = e_1 + \ldots + e_i$ for $i = 1, \ldots, n$.

The above feasibility conditions define n resource constraints. The solution to this problem is formally described in Ambec and Sprumont 2002 and Kilgour and Dinar 2001. An informal description is provided here.

Basically, the efficient allocation divides the set of agents N into K subsets of consecutive agents $N_1 = \{1, \ldots, i\}$, $N_2 = \{i + 1, \ldots, j\}, \ldots, N_K = \{h + 1, \ldots, n\}$. In a given subset N_k, the allocation of water equalizes marginal benefits among all members. Across subsets, the marginal benefits decrease. Agents consuming their satiation level have marginal benefit 0 and they must belong to N_K. The intuition is as follows. Since the marginal benefits are decreasing, the efficient way to share the total amount of water flowing down the river $E_n = e_1 + \ldots + e_n$ is to equalize marginal benefits across agents when possible. This marginal benefit is positive as soon as water becomes scarce (that is, if not everybody can consume its satiated level y_i) and is equal to the shadow cost of the water.

In the special case of all water inflows coming from the source e_1 (i.e., $e_2 = e_3 = \ldots = e_n = 0$ or $e_1 = E_n$) and with the same benefit functions, the total amount of water e_1 is optimally shared equally among all users, each of them getting $\frac{e_1}{n}$. But, in general, agents are heterogeneous. Then those with higher marginal benefits (for example the more productive farmers) obtain more than the others. Still, with only one source of water e_1, the water is shared such that the marginal benefits of all agents in the river are equal to the shadow cost of the unique resource constraint $x_1 + \ldots + x_n = e_1$.

Now, if the water picks up volume along its course, there is more water downstream than upstream. The total amount of water available at the downstream end E_n is efficiently shared among riparian agents if marginal benefits are equalized. The condition might not be feasible because of the lack of water at some location along the river. At say location j, there might not be enough water to achieve this goal. For instance, if agents are endowed with identical benefits, it might not be possible to assign $\frac{E_n}{n}$ to agent j and, therefore, to those upstream of j. This means that water is more scarce at j than downstream. Therefore the shadow value is higher. Then it is efficient to make the agents upstream of j, including j, share the total inflow up to j $E_j = e_1 + \ldots + e_j$, those downstream of j relying on the water flowing from tributaries $j + 1, \ldots, n$. This defines the subset N_K in which agents' marginal benefits are equal. Again, efficiency prescribes that E_j is shared so as to equalize marginal benefit among agents $1, \ldots, j$. But this might not be feasible at say $i < j$, in which case marginal benefits are equalized among agents i and j and all others in between. This defines the subset N_{K-1}. And so on until the source of the river is reached.

To sum up, the efficient allocation defines the subset of consecutive agents or portions of the river N_k in which the total water inflow is shared so as to equalize marginal benefits among agents and to the shadow value of water. This shadow value decreases strictly moving downstream from one portion of river N_k to the next one N_{k+1}.

For instance, suppose that $n = 3$, $b_1(x) = 20x - x^2 = b_3(x)$, and $b_2(x) = 8x - x^2$. The satiated consumption levels are $y_1 = y_3 = 10$ and $y_2 = 4$. Suppose first that $(e_1, e_2, e_3) = (9, 3, 6)$. Then efficiency is achieved when the total amount of water $e_1 + e_2 + e_3 = 18$ is divided such as to equalize marginal benefits among agents. Formally, (x_1^*, x_2^*, x_3^*) satisfies $20 - 2x_1^* = 8 - 2x_2^* = 20 - 2x_3^*$ and the binding resource constraint at the end of the river $x_1^* + x_2^* + x_3^* = 18$. This leads to $x_1^* = x_3^* = 8$ and $x_2^* = 2$. The shadow cost of the resource is then 4 at any location on the river. Now suppose that $(e_1, e_2, e_3) = (6, 4, 8)$, that is, most water flow comes from a downstream tributary and not the source, but still the total amount of water to be shared is $e_1 + e_2 + e_3 = 18$. Then agent 1 cannot consume 8 units of water so as to equalize all marginal benefits. The resource constraint at that agent's location, formally $x_1 \leq e_1$, is binding and therefore $x_1^* = e_1 = 6$. The next two agents share the rest of the water flow $e_2 + e_3 = 4 + 8 = 12 = x_2^* + x_3^*$ so as to equalize marginal benefits, that is, $8 - 2x_2^* = 20 - 2x_3^*$, which leads to $x_2^* = 3$ and $x_3^* = 9$. Therefore the set of agents N is divided into two subsets $N_1 = \{1\}$ and $N_2 = \{2,3\}$ whose agents share the water they control. The marginal benefit of agent 1 in N_1 is 8. This is the shadow value of water at its location, and is higher than the marginal benefits of agents 2 and 3 in N_2, or the shadow value of water downstream, which is 2.

6.5 Cooperation

The efficient allocation of water $\boldsymbol{x}^* = (x_1^*, \ldots, x_n^*)$ described above generally requires that upstream users refrain from extracting water from the river. If nobody (for example a regulator) can oblige them to act thus, they will accept only if they receive (monetary) compensation, which should at least cover the loss from consuming less water than available. But, since such compensation would be financed by downstream agents, it should not exceed the gain in benefit of the downstream agents. In the two-agent example in Figure 6.1, the upstream agent would accept a reduction in consumption of ε if compensated by at least the amount represented by the vertical downward arrow. On the other hand, the maximum compensation of the upstream agent acceptable to the downstream agent is represented by the vertical upward arrow. With more than two agents, the arguments should apply not only to individuals, but also to groups. Any group of agents should be compensated at least for the total of the amounts that the agents could obtain if operating individually. These acceptability conditions of transfers, which implement the efficient allocation $\boldsymbol{x}^*$, are based on the notion of the core in cooperative game theory. They require defining formally what a group of agents can achieve by itself in the river-sharing problem—the value or characteristic function, in the jargon of cooperative game theory.

Consider a group of agents or coalition S. The total benefit or welfare that a coalition can achieve alone is called the "value" of that coalition. In the river-sharing problem, it is the total benefit of the best way to share the water the coalition can rely on. In other words, it is the group's benefit from the efficient allocation of water available. Yet what is available to a coalition still has to be defined. A coalition can, of course, rely on the flow from tributaries they control, or e_i for every $i \in S$. But they might also receive inflow of water from outside the coalition. The amount of external water flow depends on how the agents outside the coalition behave, in particular whether they cooperate (by forming a coalition) or not (by playing noncooperatively). To understand that, consider a river shared by three agents. Consider the middle agent 2, located between 1 and 3. How much water can 2 expect to rely on at its location? It will depend on whether the other two agents 1 and 3 cooperate or not. If they do not, 1 consumes the maximum of the water inflows e_1 in the river at its location, with a satiation consumption y_1. This maximum is equal to e_1 (recall that by assumption $e_i \leq y_i$ for every i). Thus the amount of water available at 2's location is just the inflow from the tributary it controls e_2. The maximal benefit that 2 achieves is therefore $b_2(e_2)$ when agents 1 and 3 act noncooperatively, that is, when they belong to different coalitions. Now, if 1 and 3 group together, then it might be in their interest to pass some water from 1 to 3 through 2 (even if 2 consumes part of this water flow). This is particularly likely if there are no tributaries between 2 and 3 ($e_3 = 0$), when the most downstream agent 3 can rely on is water left in the river by upstream agents 1 and 2. Agent 3 thus consumes 0 if agent 1 does not leave any water running down the river (which is very costly or deadly). But if agent 1 leaves a sufficient amount of water, at least more than $y_2 - e_2$, then even if 2 consumes up to its satiation level y_2 (which it would do), 3 enjoys a positive consumption level of water. The marginal benefit of these first units of water being high, 3 might overcome the loss of benefit by 1 even if there is some water lost to agent 2. Therefore, passing some water between 1 and 3 sometimes increases the total benefit of {1,3} even if it is at the cost of supplying 2, an outsider of {1,3}. Then 2 consumes its satiation level y_2 and obtains its highest benefit $b_2(y_2)$, while the coalition {1,3} loses $y_2 - e_2$ from e_1 in this transfer.

The above argument can easily be illustrated with the example introduced in section 6.4. Suppose $n = 3$, $b_1(x) = 20x - x^2 = b_3(x)$, $b_2(x) = 8x - x^2$ and $(e_1,e_2,e_3) = (9,3,6)$. Consider agent 2. How much water can it expect to rely on at its location? What benefit can it achieve on its own? If agents 1 and 3 do not cooperate, then agent 1 consumes $e_1 = 9$ units of water flow coming from the source and, therefore, agent 2 relies only on $e_2 = 3$ units, enjoying a benefit of $b_2(e_2) = b_2(3) = 15$. If agents 1 and 3 do cooperate, in other words if 1 and 3 are in the same coalition, then agent 1 might supply agent 3 with some of its 9 units of water. In this case, agent 2 would extract the water flowing down at its location up to its satiated level $y_2 = 4$. Since $e_2 = 3$, it would consume up to 1 unit of the water transferred by agent 1 to agent 3. To reach agent 3, the water flow left by agent 1 should therefore exceed 1 unit. By transferring water downstream, the coalition {1,3} loses $y_2 - e_2 = 1$ unit of water and, therefore, can rely on $e_1 + e_3 - 1 = 9 + 6 - 1 = 14$ units of water. Since they have the same benefit function, the efficient way to share these 14 units is to divide the amount equally $x_1 = x_3 = 7$ (which is feasible because $7 \leq e_1$). In doing so, the coalition {1,3} enjoys a benefit of $b_1(7) + b_3(7) = 196$, which is higher than its benefit without transferring water, which is $b_1(e_1) + b_3(e_3) = b_1(9) + b_3(6) = 99 + 84 = 183 < 196$. Therefore, despite the loss of 1 unit of water, if agents 1 and 3 cooperate, agent 1 leaves 3 units of water flowing down the river but only 2 units reach agent 3. Agent 2 consumes its satiated level $y_2 = 4$, enjoying a benefit of $b_2(4) = 16$, which is strictly more than the 15 units it gets if 1 and 3 do not cooperate.

As shown above, the maximal benefit or value of a coalition S depends on the coalition structure of the other agents $N\backslash S$.[3] At one extreme, all members outside S can act cooperatively by forming a single coalition $N\backslash S$. Similarly, as above, they might pass some water through some agents that belong to S. At the other extreme, all members outside S form singletons. They act noncooperatively and thus pass no water through subsets of S. Between those two extremes, one can think of other more or less coarse coalition structures or partitions of $N\backslash S$. In general, a partition of N defines a sequential game in which agents cooperate within coalitions but not between coalitions. Broadly speaking, such a game is similar to that described in the noncooperative extraction section (6.3) except that the players are consecutive subcoalitions.[4,5] Moreover, those who belong to the same coalition cooperate while the others do not. This game is formally described and analyzed in Ambec and Ehlers 2006.

As suggested by the above example and proved in Ambec and Ehlers 2006, the value of a coalition is higher if outside members cooperate than if they do not. The basic idea is that if people outside S cooperate, they might pass some water through members of S, while they are unlikely to do so if they act noncooperatively. When computing how much welfare they can achieve by their own, the members of a coalition must have expectations about how the others will behave. They might thus expect to get some water from outside the part of the river they control. The pessimistic view is that outsiders do not cooperate at all. They form singletons in the partition of the sequential game. They thus never leave any water from their inflows to downstream agents, including members of S (recall that we have normalized to $e_i \leq y_i$ but they might leave what exceeds their peak consumption y_i if the amount of water coming from the upstream agents is sufficiently large). Call the value function the "cooperative value" when members outside cooperate, and the "noncooperative value" when they do not cooperate.[6]

Going back to the problem of searching for acceptable contributions and compensations, denote by $\boldsymbol{t} = (t_1, \ldots, t_n)$ the allocation of "money" or transfer scheme, where t_i denotes the compensation assigned to agent i (which is negative in the case of a contribution). The allocation should be budget balanced: it must sum up to 0 or less. The transfer scheme $\boldsymbol{t}$ and the water allocation $\boldsymbol{x}^*$ yield a payoff or utility $b_i(x_i^*) + t_i$ to any agent i for every $i \in N$. A transfer scheme defines a distribution of the maximal total welfare. The transfer scheme $\boldsymbol{t}$ or the allocation $(\boldsymbol{x}^*,\boldsymbol{t})$ is acceptable in the sense of the core if every group of agents S obtains at least what it can achieve on its own, that is, its value. Formally, the sum of the payoffs $b_i(x_i^*) + t_i$ of the agents belonging to S is at least $v(S)$ for any $S \subset N$. The core defines the lower bounds on agents' payoffs or, equivalently, on transfers t_i (given the optimal allocation of water $\boldsymbol{x}^*$). They depend on the value function v under consideration. The cooperative (or noncooperative) core lower bounds are the ones defined using the cooperative (or noncooperative) values of coalitions.

Since the cooperative value characteristic function is greater than or equal to the noncooperative function, the noncooperative lower bounds are easier to satisfy. Ambec and Ehlers (2006) show that, in any river-sharing problem, the transfer scheme that assigns to every agent its marginal contribution to its predecessors satisfies the noncooperative core lower bounds, yielding a so-called downstream incremental distribution. Formally, denoting $Pi = \{1, \ldots, i\}$ the set of predecessors of i (including i) and $P^\circ i = \{1, \ldots, i-1\}$ the set of strict predecessors of i, the downstream incremental distribution assigns to any agent i the payoff $b_i(x_i^*) + t_i = v(Pi) - v(P_i^\circ)$. It thus prescribes a transfer scheme $\boldsymbol{t}^d$ with $t_i^d = -b_i(x_i^*) + v(Pi) - v(P_i^\circ)$ for every $i \in N$. Furthermore, other transfer schemes might also satisfy the noncooperative core lower bounds. In the particular case where benefit functions are always increasing (so y_i goes to infinity), Ambec and Sprumont (2000) show that then the cooperative game is convex. This implies that the noncooperative core lower bounds might be satisfied by many transfer schemes, including the one that assigns to any agent its marginal contribution to the coalition composed of its followers $v(Fi) - v(F^{\circ}i)$, where $Fi = \{i, \ldots, n\}$ and $F^{\circ}i = \{i+1, \ldots, n\}$, namely the upstream incremental distribution; and also including the well-known Shapley value, which is the barycenter of the core in convex games.

The cooperative core lower bounds are less easy to satisfy. If the river is shared among two or three agents, then there is always a distribution satisfying them. With four agents or more, the cooperative core lower bounds might not be satisfied in some river-sharing problems. Ambec and Ehlers (2006) provide an example in which this is indeed the case with four agents. The logic is the following. If one of the two middle agents 2 and 3 is alone, it obtains its satiation benefit $b_i(y_i)$ because the remaining agents pass some water through its location. As a consequence, agent 2's and 3's payoff should be higher than $v(\{2\}) = b_2(y_2)$ and $v(\{3\}) = b_3(y_3)$, respectively. Moreover, each of the agents at the extremes of the river, 1 and 4, should get at least their stand-alone benefits, $v(\{1\}) = b_1(e_1)$ and $v(\{4\}) = b_4(e_4)$, respectively. However, the total benefit of the efficient water allocation $b_1(x_1^*) + b_2(x_2^*) + b_3(x_3^*) + b_4(x_4^*) = v(\{1,2,3,4\})$ is strictly lower than $b_1(e_1) + b_2(y_2) + b_3(y_3) + b_4(e_4)$. Therefore, here it is impossible to distribute the benefit from $\boldsymbol{x}^*$ while giving every agent more than its stand-alone cooperative core lower bound $v(\{i\})$.[7]

To sum up, the set of transfers that are acceptable in the sense of the core (in that the members of any group obtain at least what they would get on their own) depends on the cooperative behavior of members outside the group. If they cooperate then this set might be empty, meaning that no (budget-balanced) transfer is acceptable. As a consequence, the agents might fail to implement the efficient allocation of water $\boldsymbol{x}^*$. If they do not cooperate, then existence is guaranteed, and the set of transfers might be quite large. The next section reviews some fairness principles that may be used to select transfer schemes in this set.

6.6 Equity

While efficiency is defined by the application of the Pareto principle, there are many ways to define fairness or equity, depending on how people determine judgments that can be translated formally into axioms. The section starts by defining three axioms inspired by different judgments on what is fair. It then posits transfer schemes that satisfy the defined axioms while implementing the efficient allocation of water $\boldsymbol{x}^*$. Yet some fairness axioms or criteria might not be compatible, thereby implying that no transfers satisfy all of them. The first two axioms, equal sharing individual rationality and envy-free, are standard in fair division problems. The third, the aspiration upper bounds, is a solidarity axiom that particularly suits the river-sharing problem.

Note that the noncooperative core (or the noncooperative core lower bounds) can be seen as a fairness principle by itself. Without well-defined property rights for water, an agent or group of agents may claim property rights on the water it controls. For instance, in international river disputes, the principle of absolute territorial sovereignty grants to a country the right to water originating in its territory. It is then fair that the agent or group of agents obtains at least the benefit from consuming the water that it claims to own (Ambec and Sprumont 2002).

The first fairness principle is equal sharing individual rationality. It is based on the oldest axiom of the literature, on fair division, often taken as the definition of fairness (Steinhaus 1948; Moulin 1991). It stipulates that any agent should get at least the benefit of an equal division of the resources. Like the core lower bounds, it thus defines a lower bound on agents' payoffs.

The equal sharing individual rationality axiom can be easily adapted to the river-sharing problem when all water originates from one source, that is, if $0 = e_2 = e_3 = \ldots = e_n$ and, therefore, $e_1 = E_n$, for instance when agents are farmers located along a canal devoted to irrigation and linked to a single water pool. In this case, an equal sharing of water means that everybody is entitled to claim at least $\frac{e_1}{n}$. In term of benefits or payoffs, it means that every agent $i \in N$ should obtain at least $b_i\left(\min\left\{\frac{e_1}{n}, y_1\right\}\right)$, assuming free disposal. The transfer scheme $\boldsymbol{t}$ is thus equal sharing individual rational if:

$$b_i(x_i^*) + t_i \geq b_i\left(\min\left\{\frac{e_1}{n}, y_1\right\}\right) \text{ for every agent } i \in N. \tag{6.1}$$

With more than one tributary, one way to adapt the axiom is to assume that any agent i can claim an equal share of all tributaries located upstream (including e_i). Formally, agent i has the right to a consumption level $C_i = \frac{e_1}{n} + \frac{e_2}{n-1} + \ldots + \frac{e_1}{n+1-i}$. The transfer scheme $\boldsymbol{t}$ is equal sharing individual rational if $b_i(x_i^*) + t_i \geq b_i(\min\{C_i, y_i\})$ for every agent $i \in N$.

A second fairness principle also central to the axiomatic theory of justice is no envy (or envy-freeness), also called superfairness (Baumol 1986). According to the standard definition, an agent does not envy another agent if its payoff is higher with its assigned consumption bundle (here water and money) than it would be with the other agent's bundle. An allocation satisfies no envy if no agent envies the bundle assigned to another agent (Varian 1974).

The no envy principle can easily be defined in the river-sharing problem when, again, all water inflow comes from one point, that is, $0 = e_2 = e_3 = \ldots = e_n$. In this case, all agents can claim to consume the water level assigned elsewhere in the river to another agent. Under free disposal, an allocation $(\boldsymbol{x}^*, \boldsymbol{t})$ satisfies no envy (or is superfair) if

$$b_i(x_i^*) + t_i \geq b_i\left(\min\{x_j^*, y_i\}\right) + t_j \text{ for every } i,j \in N. \tag{6.2}$$

Similarly, a transfer scheme $\boldsymbol{t}$ implements $\boldsymbol{x}^*$ without envy if condition (6.2) is satisfied.

With tributaries ($e_h > 0$ for $h > 1$), the problem is that an agent located upstream might not be able to consume the level of water assigned to a downstream agent it envies because of lack of water at its location. One way to deal with this problem is to restrain envy-freeness to feasible water allocations; or, more precisely, to consider the feasible level of water the closest to the one the agents might envy. Formally, to consider $E_i = e_1 + \ldots + e_i$ as the alternative allocation for i if the bundle (x_j, t_j) it might envy is such that $x_j < E_i$. So, in the general case, an allocation $(\boldsymbol{x}^*, \boldsymbol{t})$ satisfies feasible no envy (or a transfer scheme $\boldsymbol{t}$ implements $\boldsymbol{x}^*$ without feasible envy) if

$$b_i(x_i^*) + t_i \geq b_i\left(\min\left\{x_j^*, y_i, E_i\right\}\right) + t_j \text{ for every } i,j \in N. \tag{6.3}$$

Obviously if condition (6.2) holds, then so does condition (6.3). Therefore, if an allocation (or a transfer scheme) satisfies no envy according to the original definition, it also satisfies feasible no envy.[8]

Concerning the two above fairness axioms, Ambec (2006) provides a general result in the case of only one source e_1, assuming that the benefit functions are single crossing. Denote λ the shadow value of water with the efficient allocation, which is also the marginal benefit of agents at $\boldsymbol{x}^*$:

$$\frac{\partial b_i(x_i^*)}{\partial x} = \lambda \text{ for every } i \in N. \tag{6.4}$$

The transfers $t_i^e = \lambda\left(\frac{e_1}{n} - x_i^*\right)$ for $i = 1, \ldots, n$ implement the efficient allocation of water $\boldsymbol{x}^*$ while satisfying both equal sharing individual rationality and no envy. Furthermore, when the number of agents is large and agents are sufficiently heterogeneous, these transfers are unique.

One way to achieve the transfer scheme $\boldsymbol{t}^e$ is to price water or tax extraction at λ and to distribute the money collected equally. Then every agent i will extract water up to an equalized marginal benefit λ, thus selecting x_i^*. The total amount of money collected is thus $\lambda(x_1^* + \ldots + x_n^*) = \lambda e_1$, where the last equality arises from the binding resource constraint. Each agent obtains a share $\frac{\lambda e_1}{n}$ of the money collected λe_1. Therefore, each agent i obtains in the end $b_i(x_i^*) - \lambda x_i^* + \lambda\frac{e_1}{n} = b_i(x_i^*) + t_i^e$. It will be clear later that another way to allocate money as in $\boldsymbol{t}^e$ is to define property rights for water in a competitive water market.

Although the result in Ambec 2006 relies on the one water source case, it can be adapted to the general river-sharing problem as follows. Consider the subsets $N_1, \ldots, N_K$ of N defined by the efficient allocation of water (see section 6.4). Denote λ_k the shadow value of water in the subset k for $k = 1, \ldots, K$. It is equal to the marginal benefit of agents in N_k and decreases strictly moving downstream from N_k to N_{k+1}. Denote also $E(N_k) = e_i + \ldots + e_j$ for any $N_k = \{i, \ldots, j\}$ for $k = 1, \ldots, K$ the total flow of water controlled by members of N_k. Notice that efficiency requires that the agents in N_k share $E(N_k)$ for $k = 1, \ldots, K$. Thus in a river-sharing problem with one source $E(N_k)$ shared by the agents in N_k, following Ambec (2006), $t^*_i = \lambda\left(\frac{E(N_k)}{|N_k|} - x^*_i\right)$ for every $i \in N_k$ (where $|N_k|$ denotes the number of agents in N_k) satisfies equal sharing individual rationality and no envy in the subset N_k for $k = 1, \ldots,$ K. Since the no envy conditions are more stringent than the feasible no envy ones, $\boldsymbol{t}^f = (t^f_1, \ldots, t^f_n)$ defined above satisfies equal sharing (of $E(N_k)$) individual rationality and feasible no envy among agents in N_k for $k = 1, \ldots, K$.

The third fairness principle is a solidarity axiom. It relies on the idea that, since water is scarce, everybody should make an effort. It starts by considering the welfare that an agent could achieve if it were alone on the river. In the absence of others, an agent i could consume up to the full water stream originating from upstream of its location $E_i = e_1 + \ldots + e_i$. Call the benefit from consuming up to E_i the agent's aspiration welfare. Formally, i's aspiration welfare is $b_i(\min\{E_i, y_i\})$. Of course, it is not possible to assign to every agent its aspiration welfare because the sum of the individuals' aspiration welfares exceeds the total welfare, that is, $\sum_{i \in N} b_i(\min\{E_i, y_i\}) > \sum_{i \in N} b_i(X^{*1})$. Therefore, by solidarity, no agent should end up with a welfare or payoff higher than its aspiration welfare, that is, $b_i(x^*_i) + t_i \leq b_i(\min\{E_i, y_i\})$ for every $i \in N$. In Moulin's (1991) terms, since the river-sharing problem exhibits negative group externalities, it is natural to ask that everyone takes up a share of these externalities. In addition, as argued in Ambec and Sprumont 2002, the aspiration uppers bounds can be seen as an interpretation of the unlimited territorial integrity principle often invoked in international river conflicts (see Godona 1985; Sadoff et al. 2003).

The above argument applies not only for individuals but also for coalitions. The aspiration welfare of a coalition $S \subset N$ is the highest welfare it could achieve in the absence of $N \backslash S$. It is denoted by $w(S)$ and formally defined in Ambec and Sprumont 2002. The aspiration welfare of a coalition S is the total benefit achieved by the members of coalition S if they share efficiently the full stream of water in the river. A transfer scheme that implements $\boldsymbol{x}^*$ satisfies the aspiration welfare upper bounds if any coalition welfare is lower that its aspiration welfare, formally,

$$\sum_{i \in S} b_i(x^*_i) + t_i \leq w(S) \text{ for every } S \subset N. \tag{6.5}$$

Ambec and Ehlers (2006) show that the unique transfer scheme that satisfies both the noncooperative core lower bounds and the aspiration welfare upper bounds is $\boldsymbol{t}^d$: the one that implements the downstream incremental distribution. It yields to every agent i its marginal contribution to its predecessors, that is, $b_i(x_i^*) + t_i^d = v(Pi) - v(P^\circ i)$ for every $i \in N$. The next section addresses the issue of implementing the downstream welfare distribution with negotiation rules.

6.7 Implementation with negotiation rules

In practice, agents often negotiate to decide how much water each of them is entitled to consume. They may also bargain over compensation, as in the case of the Columbia River (Barrett 1994) or the Syr Darya River (Abbink et al. 2005).[9]

To coordinate international river management, countries often join institutions or sign treaties with specific negotiation rules. For instance, the "principe d'approbation des Etats" (principle of approval by the States) included in the treaty founding the Organisation pour la Mise en Valeur du Fleuve Sénégal (OMVS), an international institution that manages the Senegal River, forbids any member from changing the water flow without the consent of all others. Another example of negotiation rules is the process leading to interstate river (or water) compacts to solve interstate river conflicts in the United States. These agreements are subject to congressional consent. In the case of disagreement, an allocation can be forced by the US Supreme Court (Bennett and Howe 1998; Bennett et al. 2000). International treaties or negotiation rules sometimes come close to explicit game forms. This section describes a game, proposed in Ambec and Sprumont 2000, that implements the downstream incremental distribution as a subgame perfect equilibrium of this game.[10]

The game gives priority lexicographically to the most downstream user n, $n - 1, \ldots, 2, 1$. At the first stage, agent n proposes an allocation of water and money $(\boldsymbol{x},\boldsymbol{t})$ to the other agents in the river. Of course $(\boldsymbol{x},\boldsymbol{t})$ should be feasible: $\boldsymbol{x}$ must satisfy the resource constraints at every location in the river and $\boldsymbol{t}$ must be budget balanced. If all accept, the allocation is enforced. If at least one refuses, agent n leaves the negotiation table, obtaining the bundle $(x_n, t_n) = (e_n, 0)$. Then the next upstream agent $n - 1$ proposes a (feasible) allocation of water $(x_1, \ldots, x_{n-1})$ and money $(t_1, \ldots, t_{n-1})$ for the river-sharing problem upstream. It is enforced if unanimously accepted. Otherwise, agent $n - 1$ gets $(e_{n-1}, 0)$ and leaves the negotiation table. And the game proceeds this way until the last stage (if reached), in which agent 2 proposes a feasible water allocation (x_1,x_2) and budget-balanced transfer scheme (t_1,t_2) to 1, who accepts or refuses. It is enforced if 1 agrees. Otherwise, 2 gets $(e_2,0)$ and 1 gets $(e_1,0)$. Straightforward backward induction shows that every subgame perfect equilibrium of this game implements the efficient allocation $\boldsymbol{x}^*$ and the transfer scheme $\boldsymbol{t}^d$ that yields the incremental welfare distribution.

6.8 Decentralization in water markets

For centuries markets for water have existed worldwide in irrigation communities (see Ostrom 1990 for case studies). Application of a market system is often recommended by economists to achieve efficiency because the inefficiency of free-access extraction arises from the lack of well-defined property rights for water. It thus seems natural to define property rights, leading to an efficient allocation of water on the premise that traders are price takers. But which rights? How should water be divided? Obviously, the assignment of property rights affects the payoffs of agents in the market through an allocation of money, leading to a transfer scheme in the river-sharing problem. As for transfers, an allocation of property rights is acceptable by agents if it is perceived as fair.

In the case of a one-source river ($e_2 = \ldots = e_n = 0$), it is easy to show that equal division (of the water e_1) leads to the equal sharing individual rational and envy-free transfer scheme $\boldsymbol{t}^e$. At the (competitive) market equilibrium, the agent's marginal benefits are all equal to the equilibrium price, which is then the shadow value of water λ. This equilibrium therefore implements $\boldsymbol{x}^*$. At this price, any agent i buys or sells the difference between its endowment $\frac{e_1}{n}$ and its efficient water consumption x_i^*. Agent i thus obtains $\lambda\left(\frac{e_1}{n} - x_i^*\right) = t_i^e$.

More generally, in any river-sharing problem, equally dividing the water controlled by the agents in the subsets N_k of N (described in section 6.4) for $k = 1, \ldots, K$ leads to a transfer scheme that satisfies no envy and equal sharing individual rationality among the members of N_k for $k = 1, \ldots, K$.

Notice that, in the one-source river-sharing problem, equally splitting water might violate aspiration welfare upper bounds. Indeed, Ambec (2006) shows that the three fairness axioms outlined in section 6.6 might not be compatible, and posits a transfer scheme that implements $\boldsymbol{x}^*$ while satisfying no envy, the aspiration welfare upper bounds, and the weaker requirement of individual rationality (agents' payoffs are nonnegative) in a one-source river-sharing problem. The scheme can be implemented by pricing water or taxing extraction at λ but without redistributing the money collected.

6.9 Conclusion and policy implications

This synthetic review of the river-sharing problem is now concluded by proposing some insights for public policies.

First, the analysis of the cooperative game helps to assess the potential gains from cooperation in the management of international or interstate rivers. There is no doubt that some form of transfer from downstream countries to upstream ones is needed to achieve efficiency. Such a transfer may take several forms, including of course direct monetary compensation, through a water market or fiscal transfers among states in a federal State, or compensatory payments through international treaties. But it could also take the form of a sharing rule of joint costs and benefits of utilities such as dams, canals, or hydropower plants, as for the Columbia River (Barrett 1994) and the Senegal River. Water can also be traded in exchange for other commodities, for example fuel supply on the Syr Darya River (Abbink et al. 2005) or electricity supply on the Mekong River between Thailand and the Lao People's Democratic Republic.

The likelihood of reaching an international river-sharing agreement depends on the country's expectations about the status quo in the case of disagreement. If a country expects that the others will cooperate by reaching an agreement among them, it might be tempted to free ride on the agreement. Otherwise, an agreement is feasible. Nevertheless, for rivers shared by two or three countries, which is the case for many of them, an agreement is possible for any expectations. With more than three countries, it might not be manageable. These results might therefore provide some support for partial agreements with only the few main countries (for example Thailand and the Lao People's Democratic Republic on the Mekong).

To illustrate the main argument of the chapter, let us consider as an example the section of the Nile River shared by Egypt, Sudan, and Ethiopia. The largest consumer, Egypt, is located downstream, whereas most of the flow (80 percent) originates from the most upstream country, Ethiopia. Egypt might be tempted to deal with Ethiopia to secure the water inflow exiting Ethiopian territory in exchange for some compensation. But then Sudan can free ride on the deal by extracting this increased supply of water entering its own territory without paying its cost (in the form of a compensation to Ethiopia). An inclusive agreement among the three countries must take into account this temptation to free ride by Sudan. With three countries or less, it is always possible to find such an inclusive agreement. But if one more country, for example Uganda, is included, no agreement might be acceptable by all. In this case, a partial agreement, for example between Egypt, Sudan, and Ethiopia, might be recommended.

Second, there are at least three reasons to expect or recommend the implementation of the downstream incremental distribution (or the transfer scheme $\boldsymbol{t}^d$) in international river agreements: (a) it assigns to every coalition of sovereign countries at least its noncooperative value (that is, the welfare that this coalition can achieve on its own); (b) it is a compromise between two conflicting fairness principles invoked during international river disputes, the absolute territorial sovereignty and the unlimited territorial integrity; and (c) it is the outcome of a game defined by simple negotiation rules. Basically, those rules assign more negotiation power to downstream countries than to upstream countries. If they adhere to the absolute territorial sovereignty and unlimited territorial integrity principles, countries might include these negotiation rules in international river-sharing agreements.

In practice, river-sharing treaties include international negotiation rules. For instance, the Indus Water Treaty establishes that a permanent Indus

Commission is required to meet regularly to discuss potential disputes and to plan cooperative arrangements for the development of the basin. In the case of disagreement, the matter may be taken up by intergovernmental negotiations or, failing these, arbitration (Barrett 1994). In the United States, states must follow specific negotiation procedures to sign an interstate river water compact. The disagreement outcome is a solution imposed by the federal government (Bennett and Howe 1998).

Third, the axiomatic analysis of equitable transfer schemes might shed light on how to divide water among farmers producing irrigated crops. The analysis deals with heterogeneous farmers (differing land size, crops) sharing the same pool of water. The only way to sustain no envy and equal sharing individual rationality is to split this pool equally, provided that the water market is competitive and production functions satisfy some regularity property. If not, or if farmers are reluctant to market, one way to implement an efficient allocation of water complying with no envy and equal sharing individual rationality is to price or tax water at its shadow cost (which nevertheless has to be estimated) and to redistribute equally the money collected.

Notes

1. Complementary to this approach is that of Tsur and Dinar (1995), who examine the equity properties of real-world pricing methods for irrigation water.
2. For a model of hydropower production with several agents, see Ambec and Doucet 2002.
3. The notation $N\backslash S$ refers to the set of agents in N outside S.
4. A coalition is connected or consecutive if for all $i,j \in S$ and all $k \in N$, $i < k < j$ implies $k \in S$.
5. The analysis of noncooperative games between coalitions goes back to Aumann and Dreze (1974) and leads to the literature on stable coalition structures (for example Bloch 1996; Ray and Vohra 1997).
6. This feature is common to cooperative games with externalities, such as the international pollution reduction game (Tulkens 1997).
7. For extensions of the Shapley value for a cooperative game with externalities, see Maskin 2003 and Macho-Stadler et al., forthcoming.
8. Sadoff et al. (2003) also mention no envy or superfairness as a fairness principle that can be applied to the river-sharing problem.
9. See Carraro et al. 2005 for a review of negotiation on water issues.
10. See Moore 1992 for an introduction to the theory of implementation through game forms in complete information environments.

References

Abbink, K., Moller, L.C., and O'Hara, S. (2005) "The Syr Darya River conflict: an experiment case study", Working Paper, University of Nottingham.

Ambec, S. (2006) "Sharing a common resource with concave benefits", Working Paper, GAEL, Grenoble.

Ambec, S. and Doucet, J. (2002) "Decentralizing hydropower production", *Canadian Journal of Economics*, 36(3): 587–607.

Ambec, S. and Ehlers, L. (2006) "Sharing a river among satiable countries", Working Paper, GAEL and Université de Montréal.

Ambec, S. and Sprumont, Y. (2000) "Sharing a river", Working Paper 2000–08, Université de Montréal.

Ambec, S. and Sprumont, Y. (2002) "Sharing a river", *Journal of Economic Theory*, 107(2): 453–62.

Aumann, R.J. and Dreze, J.H. (1974) "Cooperative games with coalition structures", *International Journal of Game Theory*, 3(4): 217–37.

Barrett, S. (1994) "Conflict and cooperation in managing international water resources", Policy Research Working Paper 1303, World Bank, Washington, DC.

Baumol, W.J. (1986) *Superfairness*, Cambridge, MA: MIT Press.

Bennett, L.L. and Howe, C.W. (1998) "The interstate river compact: incentives for non-compliance", *Water Resources Research*, 34(3): 485–95.

Bennett, L.L., Howe, C.W., and Shope, J. (2000) "The interstate river compact as water allocation: efficiency aspects", *American Journal of Agricultural Economics*, 82(4): 1006–15.

Bloch, F. (1996) "Sequential formation of coalitions in games with externalities and fixed payoff division", *Games and Economic Behavior*, 14: 90–123.

Carraro, C., Marchiori, C., and Sgobbi, A. (2005) "Application of negotiation theory to water issues", Working Paper 656.06, Fondazione Eni Enrico Mattei.

Dinar, A., Rosegrant, M.W., and Meinzen-Dick, R. (1997) "Water allocation mechanisms: principles and examples", Policy Research Working Paper 1779, World Bank, Washington, DC.

Godona, B. (1985) *Africa's Shared Water Resources*, London: Frances Pinter.

Griffin, R.C. (2006) *Water Resource Economics*, Cambridge, MA: MIT Press.

Kilgour, D.M. and Dinar, A. (2001) "Flexible water sharing within an international river basin", *Environmental and Resource Economics*, 18: 43–60.

Macho-Stadler, I., Pérez-Castrillo, D., and Wettstein, D. (forthcoming) "Sharing the surplus: an extension of the Shapley value for environments with externalities", *Journal of Economic Theory*.

Maskin, E. (2003) "Bargaining, coalitions, and externalities", manuscript, Princeton University.

Moore, J. (1992) "Implementation, contracts, and renegotiation in environments with complete information", in J.-J. Laffont (ed.) *Advances in Economic Theory*, Cambridge: Cambridge University Press.

Moulin, H. (1991) "Welfare bounds in the fair division problem", *Journal of Economic Theory*, 54: 321–37.

Ostrom, E. (1990) *Governing the Commons: The Evolution of Institutions for Collective Action*, Cambridge: Cambridge University Press.

Ray, D. and Vohra, R. (1997) "Equilibrium binding agreements", *Journal of Economic Theory*, 73: 30–78.

Sadoff, C.W., Whittington, D., and Grey, D. (2003) *Africa's International Rivers: An Economic Perspective*, World Bank, Washington, DC.

Steinhaus, H. (1948) "The problem of fair division", *Econometrica*, 59(2): 509–19.

Thomas, A., Feres, J.G., and Nauges, C. (2004) "Country case: France", in R. Seroa da Motta, A. Thomas, L. Saade Hazin, J.G. Feres, C. Nauges, and A. Saade Hazin (eds) *Economic Instruments for Water Management*, Cheltenham: Edward Elgar Publishing.

Tsur, Y. and Dinar, A. (1995) "Efficiency and equity considerations in pricing and allocating irrigation water", Policy Research Working Paper 1460, World Bank, Washington, DC.

Tulkens, H. (1997) "Cooperation vs. free riding in international environmental affairs: two approaches", CORE Working Paper 9752, Université Catholique de Louvain, Belgium.

Varian, H.R. (1974) "Equity, envy and efficiency", *Journal of Economic Theory*, 9: 63–91.

7 Negotiating over the allocation of water resources

The strategic importance of bargaining structure

Rachael E. Goodhue, Gordon C. Rausser, Leo K. Simon, and Sophie Thoyer

Stakeholder negotiation is an increasingly important policymaking tool. However, relatively little is understood about the relationship between the structure of the negotiating process and the effectiveness with which participating stakeholders can pursue their individual interests. In this chapter, the Rausser–Simon bargaining model is applied to a specific negotiation regarding the use of a water resource in order to illustrate the effects that decisions regarding negotiation structure have on the ultimate outcome of the bargaining process. The results highlight a number of aspects of the relationship between negotiation structure and bargaining power.

7.1 Introduction

Stakeholder participation is an increasingly important component of effective policymaking in a number of areas, including water policy. Negotiations that include stakeholder groups, as well as government representatives, have resolved thorny water allocation and quality issues. However, the inclusion of stakeholders does not guarantee a successful outcome. Designing policies that meet environmental and economic objectives and satisfy stakeholders' political requirements is a challenging undertaking. Policymakers have an important opportunity to enhance the likelihood that a negotiation is successful when designing the structure of the negotiation process. Unfortunately, relatively little is understood about the relationship between the structure of the negotiating process and the effectiveness with which participating stakeholders can pursue their individual interests.

This chapter analyzes how the "rules of the game" affect the outcome of a negotiation, and derives lessons for the design of stakeholder negotiation processes. To do so, the Rausser–Simon multilateral bargaining model (Rausser and Simon 1999) is applied to a specific negotiation process over water allocation in the basin of the Adour River in southwestern France. The objective of this negotiation was to relax the water supply constraints facing users by constructing one or more dams. While the government would fund dam construction costs, users were supposed to determine the number of dams built, and who would be responsible for funding the cost of ongoing operations. The Adour negotiation process is an interesting case to examine, because the negotiations ultimately deadlocked. By identifying some negotiation structure decisions made by the French government, the present study may contribute to an understanding of why the negotiations were unsuccessful.

The analysis provides insights for policymakers seeking to design stakeholder negotiation processes elsewhere by examining the consequences of design choices that influence the ultimate outcome of the negotiation, and its stability. It demonstrates that the structure of the negotiation, which is under the control of the policymaker, interacts with the preferences of the participants to determine an outcome. The analysis focuses on two specific negotiation design choices that are under the control of the policymaker initiating the negotiation process: first, the assignment of access to the negotiation process for specific interest groups; and second, the extent to which a group of individual stakeholders, whose interests are relatively similar compared to the overall set of stakeholder interests, can be represented at the bargaining table by a single negotiator. For the second design choice, an evaluation is made of how the outcome is affected by the definition of the "issue space" over which the interest groups negotiate. Fundamentally, both these design choices relate to the task of delineating the set of stakeholders for a specific negotiation. When these stakeholders are broadly defined ("the environment", "agriculture"), specific organized groups, often with more selective interests, must be assigned to represent these broader interests. The results highlight the effects of these design choices on bargaining outcomes for relatively similar stakeholder groups. The focus is on relatively similar stakeholder groups because these groups may be represented collectively by a single negotiating stakeholder when the structure of the negotiation process is defined, or else entitled to represent their own interests individually.

The following specific results were obtained. First, a stakeholder may obtain a better bargaining outcome when that stakeholder's access is reduced, provided that the redistribution increases the access of another stakeholder with similar interests but with a more favorable strategic location. Second, the interests of the coalition as a whole will usually, but not always, be advanced if the access allocated to its members is reassigned to a spokesperson representing their common interests. However, some coalition members may be adversely affected. The definition of the issue space affects how much the coalition's welfare improves. When the issue space allows greater room for coalition members to make bargaining proposals that further their own individual interests at the expense of the interests of other coalition members, the benefit from the spokesperson is greater.

This analysis contributes to the economic literature applying bargaining theory to negotiations regarding water allocation and water quality. This literature can be divided into three categories: negotiation support systems, applications of cooperative bargaining theory, and applications of noncooperative bargaining and game theory. Because of space considerations, this review is extremely brief. For a more thorough discussion, see Simon et al. 2007.

Carraro et al. (2005) survey the literature regarding negotiation support systems, a relatively new field of research, as well as the literature regarding water management applications of negotiation theory. Negotiation support systems are designed to facilitate specific negotiation processes by optimizing over multiple objectives and issues, given the system's physical constraints. In some cases, such as in Barreteau et al. 2003, the outcomes of scenarios generated by the negotiation support systems are presented to stakeholders in the negotiations. Based on their reactions, a new set of simulations is generated and analyzed. In this approach, the support system is used iteratively, and its outcomes more closely reflect stakeholder preferences over successive rounds. Another approach is explicitly interactive, as in Thiesse et al. 1998. Under this approach, the system begins with information regarding stakeholders' preferences, rather than collecting it over time. The model is used to generate alternatives based on preferences and the system's physical constraints. In this case, the model is intended to introduce mutually beneficial outcomes to negotiators early in the negotiation process.

Dinar et al. (2008) describe a recent example of the application of cooperative game theory to water negotiations. They compare the cooperative approach to a negotiated approach utilizing a negotiation support system for the case of water allocation in the Kat basin in South Africa. They find that the theoretical results of the cooperative game theory model are similar to those obtained utilizing a negotiation support system, which suggests that there may be complementary roles for the two approaches. Parrachino et al. (2006) and Dinar et al. (1992) provide extensive reviews of analyses that use cooperative game theory to analyze water negotiations.

While cooperative game theory assumes that players maximize their joint welfare, noncooperative game theory assumes that each player maximizes its own welfare, subject to the other players receiving sufficient utility to participate in the negotiation process. Noncooperative game theory has been used to model negotiations over water resources in other contexts. Adams et al. (1996) utilize the Rausser–Simon multilateral bargaining model to examine the three-way negotiations over water allocation and quality conducted in California in the early 1990s. Frisvold and Emerick (2008) model water transfers in the United States–Mexico border region as bargaining games. They find that the use of contingency contracts will increase water trades, and that increasing the efficiency of water use may reduce, rather than increase, water trades. Noncooperative models of water allocation negotiations also identify a role for negotiation support systems: an interactive model of the negotiation problem may aid in identifying mutually beneficial solutions to complex problems that may not be immediately obvious to self-interested players.

Such systems may also aid in clearly defining player preferences, which may increase the chance of reaching a negotiated solution that is satisfactory to all participants.

The Adour negotiation process is modeled here as a noncooperative bargaining problem for two reasons. First, the diversity of stakeholder preferences and the large number of stakeholder groups suggest that it is more plausible for each group to seek to maximize its own welfare, rather than the joint welfare of all groups. Second, the specification of the bargaining problem imposed by the French government does not provide any obvious incentives for joint welfare maximization. While stakeholders other than water users were included in the negotiations over the number and size of dams, only water users, in this case farmers, were required to pay operation costs, even though other stakeholders stood to benefit from the construction of larger, and hence more expensive, dams.

The remainder of the chapter is structured as follows. Section 7.2 introduces the Adour negotiation process. Section 7.3 introduces the modeling framework. Section 7.4 describes how the modeling framework is implemented in the context of the case study, and section 7.5 presents the results. The study concludes in section 7.6 with an application of the results to other negotiations over water resources and an extensive discussion of the implications of these findings for policymakers designing other stakeholder negotiations.

7.2 Adour River negotiations

The analysis in this chapter is based on a specific negotiation process regarding water allocation, water storage capacity, and water prices for users in the Adour catchment area in southwestern France. This example provides an excellent case study for analysis using the Rausser–Simon multilateral bargaining model, because a substantial amount of information about the context of the negotiations is available. This information includes details about the hydrology of the river, the use of the water in agriculture, specific minimum flow requirements at specific points along the river, the political economy of water allocation in the region, and the explicit rules governing the negotiation process specified by the French central government, including the set of issues contained in the negotiation, the definition of different stakeholder groups included in the negotiation process, and what would happen if the negotiation process proved unsuccessful.[1] The information on river hydrology allows modeling of the effects of surface water use in a given subbasin on the availability of water for use elsewhere and on residual flows, as well as the effects of the different potential dams; and the minimum flow requirements allow modeling of the physical constraints governing the negotiation outcome. The information on water use in agriculture allows modeling of agricultural production and profits as a function of the price and quantity of water available, which in turn enables modeling of farmers as profit-maximizing agents in the negotiation. Information on negotiation process rules allows implementation of a model of the negotiation process that incorporates much of the actual negotiation structure.

7.2.1 The Adour basin

The Adour catchment area includes three subbasins above its intersection with the Midouze. The upper subbasin is significantly larger than the middle and lower subbasins, with 18,997 hectares of agricultural land, compared to 14,363 hectares in the middle subbasin and 13,814 in the lower subbasin. Over the past thirty years, irrigation has become the primary water use in the region, with approximately 47,500 irrigated hectares. Irrigated area as a share of total agricultural area is relatively similar across subbasins: 61 percent in the lower subbasin, 56 percent in the middle subbasin, and 58 percent in the upper subbasin (Thoyer et al. 2001).

The pattern of agricultural activity differs across subbasins, so that the marginal value of water per hectare in agriculture varies across subbasins. Farmers who primarily grow corn account for 72 percent of all irrigated area in the upper subbasin, and for 65 percent of all irrigated area in the middle and lower subbasins. In the upper subbasin, beef and dairy cattle account for a larger share of total area than in the middle and lower subbasins. Perhaps the most critical difference is that in the middle subbasin farmers who produce corn and seed or vegetable crops account for 28 percent of irrigated area, while farmers in this category account for only 7 percent of irrigated area in the upper subbasin and 5 percent of irrigated area in the lower subbasin. Because seed and vegetable crops are relatively high valued, the marginal value of irrigation water per hectare is higher in the middle subbasin than in the other two (Thoyer et al. 2001).

Within each subbasin, farmers withdraw water for irrigation. Under the current system, farmers pay a fixed price for a fixed amount of quota per formally registered irrigated hectare. No additional land can be registered as irrigated land. Given the hydrology of the catchment area, water flows are insufficient to provide farmers with their quotas in approximately two years out of ten. When flows are deemed insufficient to meet minimum flow objectives, all irrigation is halted.

7.2.2 National water policy and the Adour negotiations

The Adour negotiations were part of a nationwide water policy reform. France's 1992 Water Law renationalized the water supply, and specified a new water management structure that devolved decisionmaking authority below the central government. While the implementation of policy was devolved, the structure of the devolution was tightly controlled by the center. All six of France's major hydrological basins were required to develop and implement a water development plan (schéma directeur d'aménagement et de gestion des eaux, or SDAGE), and water regulations were required to be negotiated at the smaller, catchment scale by stakeholders, under the supervision of local authorities, and with participation by local, regional, and national government representatives as schéma d'aménagement et de gestion des eaux (SAGE) projects. Negotiations for a SAGE project were to be undertaken by a local water commission (commission locale de l'eau), with membership specified by the 1992 Water Law: 50 percent of the membership to be elected representatives from the relevant political entities (regions, departments, towns), 25 percent from the central government, and 25 percent from users, consumers, and nongovernmental organizations. Except for farmers, who have formally elected representatives who can be appointed to the commission, the 1992 Water Law does not specify how seats should be apportioned among the different interest groups.

The Adour-Garonne SDAGE recommended the formation of an Adour SAGE to consider the construction of up to three dams in order to balance the irrigation needs of agriculture in the Adour catchment area above its intersection with the Midouze with the need to maintain water flows in the river. As part of the SDAGE, the three subbasins are separated by points at which specific minimum flow requirements are defined. The middle and upper subbasins are separated by a flow monitoring point at Estirac. The middle and lower subbasins are separated by a flow monitoring point at Aire sur Adour. The lower subbasin is separated from further downstream by a flow monitoring point at Audon.

Because of their different locations, the three dams would have different implications for residual water flows (the flow rate after deducting total agricultural water diversion rates), total water storage, and the pattern of water use for irrigation. Dams 1 and 3 are in the upper subbasin. Water stored behind either of these two dams can be used for irrigation purposes in any subbasin, or can be used to enhance residual water flows at every flow monitoring point. Dam 2 is in the lower subbasin. It can only be used for irrigation in that subbasin, and water stored behind it will only enhance residual water flows at Audon. In addition, technical considerations restricted the maximum possible storage capacity of dam 1 to approximately one fourth of the maximum possible storage capacity of the other two dams.

As mandated by the central government, stakeholder negotiations were required to determine whether or not any of the dams would be built, and how their operation would be funded by water users through the assignment of water quotas and water prices. The government specified that the negotiations would include elected representatives of agricultural producers; a group representing the interests of all nonagricultural water users; environmental groups; and a number of government representatives, including national and local environmental and agricultural agencies, elected local governments, and the Adour Basin Water Authority, which is a semipublic institution with responsibilities including infrastructure maintenance, irrigation water delivery, and monitoring of river flows.

In the Adour case, negotiations were not technically undertaken within the context of the formation of an official SAGE recommended by the SDAGE. Rather, the initial negotiations were undertaken as part of a low water flow management plan (plan de gestion des étiages) among the relevant parties. In the initial negotiations, stakeholders agreed to initiate and fund studies regarding future water consumption needs and supplies, and agreed on a total water consumption volume, and farmers agreed to fund a substantial portion of system management and maintenance costs. However, the second stage of negotiations deadlocked over the specifics of water allocation, including allocating quotas across farmers and managing limited supplies in times of drought (Thoyer et al. 2004). The failure to agree upon quotas and prices sufficient to fund the desired water storage capacity after reaching an initial agreement resulted in a failed SAGE: because the negotiations deadlocked, the SAGE was stillborn. Consistent with the central government's specification, the current water administration system remained in effect. Farmers pay for quotas. If necessary, all irrigation is halted in order to meet (or attempt to meet) residual flow requirements.

7.3 Conceptual framework: Rausser–Simon multilateral bargaining model

The noncooperative multilateral bargaining model developed in Rausser and Simon 1999 is designed to model the complex structure of multiplayer, multi-issue negotiations. It is a computational model that can be calibrated to the parameters of a specific bargaining problem, including the structure of the bargaining process itself. In contrast to most multilateral bargaining models in the literature, its solution is particularly sensitive to bargaining efforts made at the last minute before negotiations break down. Often, deadlines may induce negotiators to reach an eleventh-hour resolution, as in the case of an impending strike deadline leading a labor union and an employer to reach an agreement. Because the Rausser–Simon model is responsive to such effects, it is particularly well suited to evaluating changes in the structure of the negotiation process.

In the Rausser–Simon model, a finite number of players negotiate over a set of issues. Each possible outcome in this "issue space" is referred to as a "policy vector". If the players fail to agree unanimously on an outcome, then all players receive their "disagreement payoffs".[2] The number of negotiating rounds is finite and is known to all players. In each round, one player is randomly chosen to propose a policy vector. The probability with which a given player is chosen is referred to as that player's "access probability". This probability, which represents the strength of the player's institutional role in the bargaining process, is one determinant of the strength of its bargaining position. If all players vote to accept the player's proposal, the game ends. If some player rejects it, then they begin another round of bargaining. If the proposals are rejected in every bargaining round, then players receive their disagreement payoffs and the negotiation process is terminated.

The model is solved by backward induction. Intuitively, since each player knows every other player's access probability and objective function, a player can adjust its proposal to ensure that all players prefer to accept the proposal rather than advance to the next bargaining round. By repeatedly applying this principle, it is shown that if the number of bargaining rounds is sufficiently large, then essentially the same policy vector will be proposed in the first bargaining round, regardless of which player is randomly chosen to make the proposal.

Figure 7.1, adapted from Adams et al. 1996, illustrates the backward induction solution procedure for the Rausser–Simon model, for an example in which three players bargain over three issues. Three policymakers are negotiating over how to allocate a fixed, total national budget between three spending categories: military, education, and health. The policymakers are named M, E, and H, after the expenditure categories. Each one, if it had total control over the negotiation process, would choose to spend 100 percent of the available budget on its own expenditure category. These allocations are referred to as the policymakers' "ideal points". For example, M's ideal point is to devote the entire budget to military expenditure. In Figure 7.1, all possible expenditure allocations are represented by the unit simplex: the closer a point is to a vertex of the simplex, the greater is the fraction of expenditure devoted to the corresponding category. The utility that a player derives from an expenditure allocation in the simplex decreases quadratically with the Pythagorean distance between the allocation and the player's ideal point. (In other words, each indifference curve for player *i* is a circle centered at *i*'s ideal point.) The consequences of failing to reach an agreement are assumed to be so dire that each policymaker would prefer to accept any budget division than

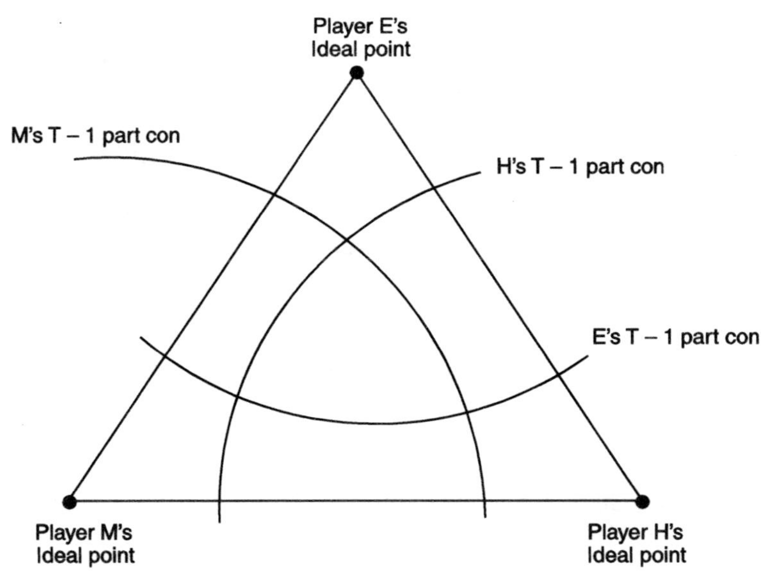

Figure 7.1 Participation constraints in round T – 1.

to incur these consequences. Given this assumption, in the final round of bargaining, T, each player will propose its ideal point, since both opponents will agree to it.

Now consider the problem facing proposer i in the penultimate round (T – 1). Since i knows the probability a_j with which each player j will be selected to make a proposal in round T, as well as the proposal, X^j_T, that j will make in this event, i can calculate all players' expected utilities conditional on disagreement in round T – 1. These expected utilities are represented by the indifference curves drawn in Figure 7.1. In order for i's round T – 1 proposal, X^i_{T-1}, to be accepted, each player $j \neq i$ must obtain at least as much utility from this proposal as its expected utility from rejecting it and advancing to round T. In the analysis below, this requirement is referred to as player j's "round T – 1 participation constraint". Formally, player i's proposal will be accepted in round T – 1 if, for all j, $U^j\ (X^i_{T-1}) \geq EU^j = \sum_{k=1}^{N} a_k\ U^j\ (X^k_T)$. The set of proposals that will be accepted by all three players in period T – 1 is represented in Figure 7.1 by the area bounded by the three indifference curves. If selected to make the round T – 1 proposal, each player will select the budget allocation in this set that yields it the highest utility, that is, the one that is closest to that player's ideal point.

By the same logic, since each player knows what the others will do in round T – 1, a player in round T – 2 can calculate what each player's expected utility will be if negotiations proceed to round T – 1. Necessarily, each player in round T – 1 receives with certainty at least its expected utility from entering round T, and strictly more than this with positive probability. Hence, for each player, the indifference curve associated with the player's expected utility conditional on entering round T – 1 is strictly closer to the player's ideal point than the curve associated with the player's expected utility conditional on entering round T; it follows that the set of proposals that will be accepted in round T – 2 is strictly smaller than the set for round T – 1. Iterating backwards to round 1, the set of acceptable proposals continues to shrink. If T is sufficiently large, the available set in the first round will be, essentially, a single point P*. Thus, in the unique equilibrium of the bargaining game, whichever player is selected in the first round of bargaining will make a proposal essentially equal to P*, and it will be accepted.

The solution P* depends on the access probabilities of the three players. Figure 7.1 is drawn assuming equal access probabilities. If one player obtains additional access at the expense of the other two players, the solution will shift closer to that player's ideal point. For example, if player H's access increased, perhaps through a successful campaign that increased the number of player H's supporters in a representative assembly, the new negotiated solution would assign more resources to healthcare than the original one.

7.4 Model implementation

To implement the Rausser–Simon model for the case of the Adour negotiations, the following functions and relationships must be specified: the hydrology and budget constraints that the outcome of the negotiation must not violate; players' objective functions as the interests of stakeholders with government-mandated access to the negotiation process; and the issue space regarding how stakeholders can negotiate over dam sizes, irrigation quotas, and quota prices.[3] There is a hydrology constraint at the measurement point separating each subbasin, and one where the negotiation area ends, at the Adour's intersection with the Midouze. Each hydrology constraint is a minimum flow requirement. To meet the hydrology constraint for a subbasin, the sum of water flowing into the subbasin, plus water released from dams in the basin, minus water use within the subbasin, must be at least equal to the residual flow requirement. Two possible cases regarding the hydrology constraints are considered: normal years, in which the hydrology constraints are not binding on water users, and drought years, in which they are. In addition to the hydrology constraints, there is a budget constraint, stipulating that the administrator's revenue from quota sales must weakly exceed operating costs.

Seven players are defined, who represent the stakeholders included in the negotiation process by the French central government: three farmers, an environmentalist, a downstream user, a water manager, and a taxpayer. Each player is risk averse, meaning that it prefers a certain income to a risky income with the same expected value. The farmers represent agricultural water users in each subbasin. The environmentalist represents citizens and environmental groups who are concerned primarily with the negative aesthetic impacts of the potential dams on rural landscapes, although they also care about the impact of reduced flows on water quality and aquatic life. The downstream user represents citizens and businesses concerned with the impacts of reduced water flows downstream from the study area, as well as citizens and environmental groups who are concerned primarily with the impacts of reduced water flows on water quality and aquatic life, although they also care about the negative aesthetic impacts of the potential dams on rural landscapes. The water manager's goal is to maximize revenue from the sale of water quotas. Since the maximum revenue obtainable is a regulated rate of return on operating costs, and since costs increase with the scale of the system administered, the water manager's goal is to maximize this scale. The taxpayer represents nonlocal, nonagricultural interests who are primarily concerned with mitigating the burden of dam construction on nonlocal taxpayers and with maximizing the benefit of residual flows for nonagricultural users, both local and nonlocal.

The administrator's budget constraint imposes certain relationships between the variables in the model. When this constraint binds, the exposition can be simplified by reducing the dimensionality of the problem, and representing players' objective functions in terms only of prices and quotas. Farmers prefer higher quotas and lower prices; to induce other parties—especially the downstream user—to agree to higher quotas, they can propose larger dams, which increase total water flows enough that both quotas and residual flows can be increased. However, since the administrator's operating costs increase with dam sizes, farmers' quota prices must increase also, in order to satisfy the administrator's budget constraint. In this sense, farmers face an upward-sloping effective supply schedule for quotas. Under drought conditions, this schedule becomes significantly steeper. Both the environmentalist and the downstream user value residual flows (the difference between total flows and total quotas) and hence prefer lower quotas to higher ones. Both these players recognize that associated with dam expansion there are pros—residual flows increase—and cons—the quality of the landscape is degraded. The environmentalist is relatively more concerned about the landscape than the downstream user. Up to a point, both players are willing to agree to higher quotas at higher prices; through the budget constraint, the higher prices can be mapped to expanded dams. For the taxpayer, a preference for lower dam construction expenditures and increased residual water flows also implies a preference for lower quotas. The water manager prefers higher quotas and higher prices, as both are associated with an increase in the scale of the system managed.

The analysis considers two alternative specifications of the issue space. These specifications are differentiated by the extent to which negotiated quotas can differ across farmers. The more restrictive regime requires that players propose a common quota level per hectare that applies to all three subbasins. This regime is referred to as CQ (common quota). The second regime allows the quota rate per hectare to vary across farmers, and is referred to as IQ (independent quotas). In both cases, farmers negotiate in addition a common price per unit quota. Only a limited degree of independence is allowed in the IQ regime, since, in the absence of any restriction, each farmer would attempt to appropriate all the available water for its own subbasin, leaving the other two subbasins with no water at all. For similar reasons, an issue space where both the negotiated prices and the negotiated dam scales can differ across basins is not considered. Intuitively, such a broad issue space would likely lead to a deadlocked negotiation in real life, as each farmer would seek to capture the entire benefit of increased water supplies while offloading all of the associated costs onto other farmers.

7.5 Analysis and results

The analysis focuses upon the implications of certain structural options available to the policymaker responsible for designing the negotiation process. It addresses two specific questions regarding the effects of the structure of the bargaining process on its outcome: Will a stakeholder ever benefit from a reduction in that stakeholder's own access to the bargaining process in favor of another stakeholder with relatively similar interests? Will a group of stakeholders with relatively similar interests ever benefit from being assigned a single negotiator, or common spokesperson, to represent their interests? For the second question, this study investigates whether or not the outcome depends on the space of issues over which the negotiation takes place. The analysis is conducted using comparative statics. The structural parameters that are varied are under the control of the policymaker designing the negotiation process. Under this approach, the parameter of interest is increased or decreased many times by a small amount each time, and the outcomes are compared. This procedure allows isolation of the effects of the parameter in question on player payoffs and on policy variables, which in this case include prices, quotas, and dam capacities.

7.5.1 Effects of reduced access on player welfare

The first question is considered in a simplified version of the model that contains only three players: the upper subbasin farmer, the middle subbasin farmer, and the downstream user. For this subsection, these players will be referred to, respectively, as U, M, and D. The simplified model considers only one potential dam: dam 1. Attention is restricted to the CQ regime, so that the two farmers must pay a common price for a common quota. In order to evaluate the effects of transferring access between players a series of scenarios are compared, where U's access is steadily reduced, and M's access is correspondingly increased. Intuitively, one would expect that reducing (increasing) a player's role in the negotiations would reduce (increase) how much the player benefits in the negotiation outcome. In this particular case, however, both farmers benefit from shifting access to the middle subbasin farmer. The simple intuition that reducing U's access will reduce that player's welfare holds only for the final round of negotiations. Because U prefers its own proposal to M's, U's expected utility declines in this round when access is reduced in favor of the other farmer. Because of the offsetting indirect effects discussed below, however, U benefits in earlier rounds from M's increased access.

Penultimate (T – 1) round effects. Recall that at any given quota level, the marginal value of irrigation water per hectare is higher for the middle subbasin farmer than for the other two subbasins. For the purpose of this exercise, it is assumed that other differences between the two farmers are relatively insignificant. Because M has a higher willingness to pay for water, the intensity of conflict between D and M is greater than it is between D and U. As a result, M's constraint is binding on D in round T – 1, while U's is slack. (Specifically, D offers fewer quotas in round T – 1 than the level that is expected to be offered in round T, but at a price sufficiently lower that M is just willing to accept the offer; U, whose willingness to pay for quotas is lower than M's, is less negatively impacted than M by the quota reduction, and so strictly prefers D's offer to the prospect of continuing to round T.) Now as M's access increases, its round T – 1 participation constraint tightens, so that D's offer in round T – 1 must become more and more palatable to M. There is a corresponding slackening in U's round T – 1 participation constraint, but D obtains no offsetting benefit from this, since this constraint was not initially binding. To summarize, the additional effect in round T – 1 of the shift in access in favor of M is that D makes a proposal in this round that is worse for D and better not only for farmer M but also for farmer U.

T – 2 round effects. The effect of the access shift in round T – 2 is illustrated in Figure 7.2. The upper panel displays players' T – 2 participation constraints and the proposals that each player will make in round T – 2, given those constraints. Notice that as in round T – 1, M's constraint is binding on D but U's constraint is slack. Note also that D's participation constraint is binding on both farmers in this round, but that M's proposed price and quota are both higher than U's, reflecting M's higher willingness to pay. The lower panel of Figure 7.2 depicts how things change as M gains access at the expense of U. The new participation constraints are labeled and indicated by dashed lines. For purposes of comparison, the original constraints are included as well, as unlabeled solid lines. The direct effect of shifting access from U to M is evident from the relative positions of the old and new constraints for the two players: U's constraint has slackened slightly because of reduced access, while M's constraint has tightened. As in round T – 1, however, only M's constraint is binding on D, and as this constraint tightens, D is obliged to adjust its proposal by increasing the quota and reducing the price. This benefits both farmers. In addition, because of the effect on D described in the previous paragraph, D's round T – 2 participation constraint is slacker than initially, enabling both M and U to make proposals that they both strictly prefer to their original ones.

The net impact of all these effects in round T – 2 is to increase U's expected utility, relative to the benchmark case, conditional on entering this round. Moreover, as one proceeds backwards up the game tree, the positive effects accumulate, so that in the first round, the shift in the equilibrium outcome resulting from the access shifts ultimately benefits U. Essentially, the result is driven by the positive externality that M's increased access generates for U, by weakening D's bargaining position. In this case, the similarity in the farmers' interests, relative to the downstream user's interests, outweighs the differences between them.

Implications of the access result for negotiation design. This result identifies some possible problems with the design of the Adour negotiation, and has implications for negotiation process design more broadly. Differences in player access and interests interact to determine player welfare under the negotiation outcome. In this case, the more "moderate" farmer, U, benefited from an increase in the access of the more "extreme" farmer, M. When designing negotiation processes, policymakers should be aware of the possibility that this scenario may hold for their moderates on either side of an issue, who may prefer to let extremists dominate the process. In an actual negotiation, however, this tactic may increase the likelihood of the negotiation failing. In this modeling of the Adour negotiation, the calibration of the model was based

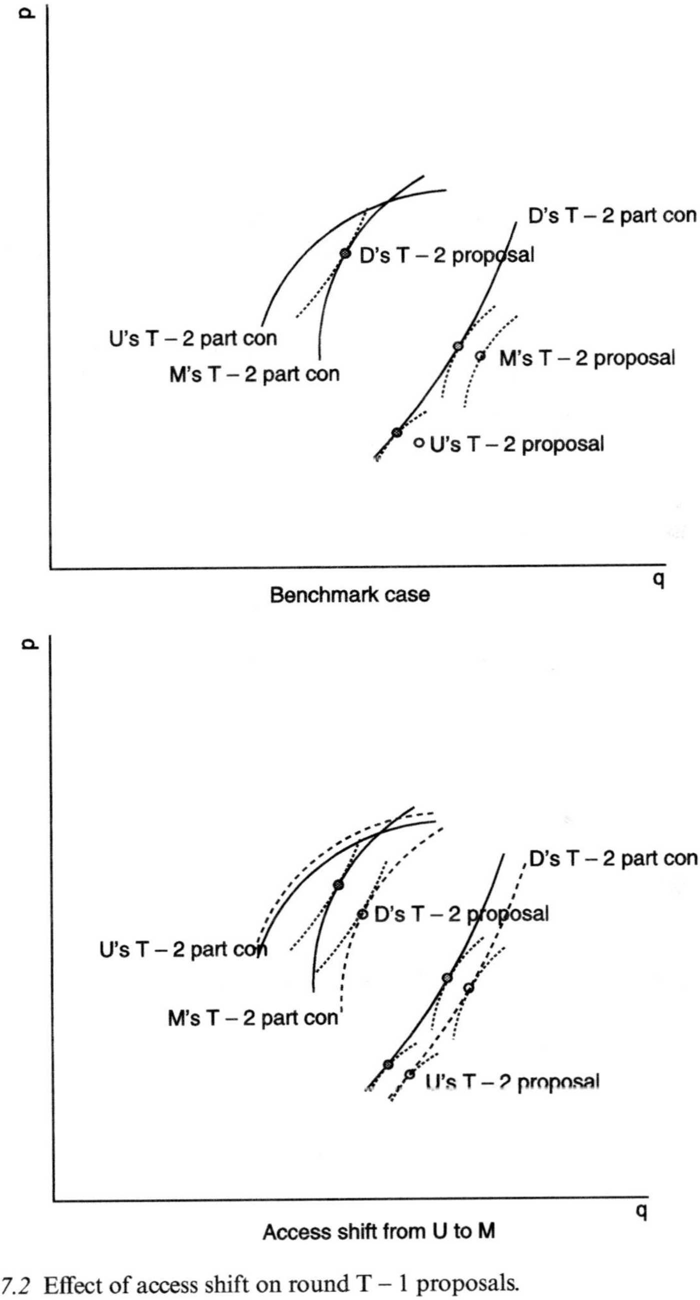

Figure 7.2 Effect of access shift on round T – 1 proposals.

on the premise that even for the most diametrically opposed stakeholders, accommodation was preferable to disagreement. Given the outcome of the negotiations, it appears that this assessment may have underestimated the intensity of the actual conflict.

7.5.2 *Effects of a common spokesperson on player welfare*

In order to consider the second question, the complete, seven-player version of the model is analyzed, and a spokesperson is introduced representing the three farmer stakeholder groups. The spokesperson has no personal interests in the negotiation process, but is charged with the task of maximizing a simple weighted average of the three farmers' objective functions. The impact of the spokesperson's participation is evaluated by transferring access from the farmers to the spokesperson, holding constant both total farmer access (including the spokesperson) and the relative access of the three farmers. Intuitively, evaluation of the spokesperson's impact addresses a negotiation design problem relating to which stakeholders should be at the bargaining table: Will defining a broad stakeholder group enable its members to present a united front, and increase their welfare in the negotiated outcome, or will it inhibit its ability to negotiate effectively, because increasing the welfare of some group members necessarily implies reducing the welfare of others? This analysis indicates that defining a broad stakeholder group represented by a single spokesperson will increase the overall welfare of group members, although some group members may be made worse off.

Issue space. In addition to selecting the configuration of interests at the bargaining table, the negotiation designer must also specify the issue space. In order to examine the interaction between these two design dimensions, the effects of the spokesperson in both the CQ and IQ regimes are considered. Under the CQ regime, the spokesperson's role is necessarily limited, because the farmers do not have sufficient flexibility to benefit themselves at the expense of the other farmers by adjusting individual quotas. In contrast, under the IQ regime each farmer has an incentive to increase its own quota, at the expense of the other two farmers.

As noted above in section 7.4, individual farmers and the spokesperson face an upward-sloping effective supply schedule for quotas. In order to increase quotas, they must "purchase" them from the nonagricultural players. Because of the downstream user's strategic location, that player's preferences shape the effective supply schedule facing farmers. Because the downstream user primarily cares about residual river flows, there is a range of dam capacities that it is willing to accept provided that a sufficiently large share of the stored water will be devoted to increasing residual flows, rather than to agricultural uses. As dam capacities increase, the downstream user demands that an increasingly large share of marginal capacity be devoted to residual flows, which increases the marginal cost per unit quota for farmers. (Recall that prices must increase with dam capacity to satisfy the administrator's budget constraint.)

Effect of the spokesperson under the common quota regime. Under the CQ regime, the effect of the spokesperson's participation on farmer welfare depends on hydrological conditions. The spokesperson increases farmer welfare in a normal year, but has no effect in a drought year. Intuitively, in a drought year the hydrology constraints limit the ability of the spokesperson and individual farmers to construct welfare-enhancing proposals. If the constraints bind sufficiently tightly, then the lowest quota on the political supply schedule that will satisfy these constraints is higher (and more expensive) than the optimal level for any of the farmers, even the middle subbasin farmer, who has the highest willingness to pay for quotas. Therefore, differences in farmer preferences do not emerge during the negotiations, because all farmers (and the spokesperson) must propose the same superoptimal quota in order to meet the hydrology constraints.

In contrast, under normal conditions each farmer maximizes individual utility by selecting the price-quota pair from the political supply schedule that is tangent to its induced indifference map in price-quota space. Because the value of water in production varies by subbasin, farmers' optimal points will vary as well. The spokesperson's effect on farmer utility hinges on risk aversion. Because the spokesperson's proposal maximizes average farmer utility, the utility that each farmer obtains from the spokesperson's proposal exceeds that farmer's expected utility from the three proposals that maximize the farmers' individual utilities. The variance in utilities associated with the farmer group's proposals declines as the spokesperson's access increases.

The above result, however, is by no means a general one. The nonfarmer players are also risk averse. If they were sufficiently more risk averse than the farmers, they would benefit more from the reduction in the variance of the farmers' proposals than the farmers themselves do, which would increase their bargaining strength relative to the farmers'. These results suggest that if the issue space for a negotiation is defined in such a way that parties with similar but nonidentical interests have very little scope to differentiate their proposals, then replacing them by a single negotiating player has an indeterminate effect on their welfare, and on the welfare of other players.

Effect of the spokesperson under the independent quota regime. Under the IQ regime, by contrast, the spokesperson has a determinate, and much more significant, effect on the farmers' welfare. Moreover, the spokesperson's role does not depend on hydrological conditions, improving the average farmer's welfare in both drought and normal conditions, although this average benefit comes at the expense of the upper subbasin farmer. The potential for improvement arises because farmers in this regime, negotiating as individuals, create two kinds of inefficiencies. The first source of inefficiency is that when farmers negotiate individually, the resulting allocation of quotas does not fully reflect the different environmental and financial costs of providing quota to different subbasins. One unit of quota assigned to the upper subbasin farmer reduces water flows at every residual flow monitoring point. One unit of quota assigned to the lower subbasin farmer only reduces water flows at the final residual flow monitoring point at Audon, which separates the lower subbasin from downstream. Similarly, an increase in dam capacity needed to provide the upper or middle subbasin farmer with an additional unit of quota must come from dams 1 or 3, while all three dams can be used to provide quota to the lower subbasin farmer. Thus, it is cheapest to provide additional quota to the lower subbasin farmer. The middle subbasin farmer has the highest marginal value of quota. Hence, the spokesperson raises average farmer utility by allocating additional quota to these farmers and reducing the quota allocated to the upper subbasin farmer, relative to the average of the proposals made by the three individual farmers.

The second source of inefficiency is that each farmer has an incentive to engage in "beggar-thy-neighbor" behavior, by assigning itself a relatively large quota at the expense of the other two farmers' shares. However, since all farmers behave this way, they end up on average buying more quotas than they would choose to buy if they had made their decisions based on the correct expected cost-benefit computation. Focusing on the difference between the spokesperson's and the average of the three farmers' proposals in the final round provides some intuition for this problem, and for how the spokesperson mitigates it. As noted above, each farmer faces an effective supply curve in this space. There is, however, an important difference between the curve a farmer faces ex post—that is, after that farmer has been selected to make a proposal—and ex ante—that is, at the start of the bargaining round. To simplify the exposition of the difference, assume for the moment that all farmers are identical in all respects, with identical induced preferences in price-quota space, identical technological considerations, and equal access. Once a proposer has been selected, that player can, and will, allocate to itself a disproportionate share of the total quotas available at any given price, and the smallest admissible level to each of the other farmers. Since all farmers behave identically, the ex ante quota level that a farmer can expect to realize, conditional on some member of the farmer group being chosen as a proposer, will be less than the level that that farmer will propose ex post if selected. It follows that ex post each farmer will propose a higher common price for quotas than the price that would have been optimal if it had selected the proposal optimally from the set of possibilities available ex ante.

This point is illustrated in Figure 7.3: the ex post supply curve is drawn as a broken line, on the assumption that the negotiation rules prohibit a farmer from allocating more than two thirds of all quotas to its own subbasin. Given this rule, each farmer will propose a common price of p, and a quota vector comprising $2q$ for its own subbasin, and $q/2$ for each of the other subbasins. Since each farmer is equally likely to be selected as the proposer, however, the quota that each farmer can expect to receive ex ante, conditional on one of the three farmers being selected, is only q. But as the indifference curve indicates, each farmer would ex ante prefer to propose a lower price in exchange for a lower expected quota, and was only willing to purchase $2q$ when able to seize a disproportionate fraction of the total available quota

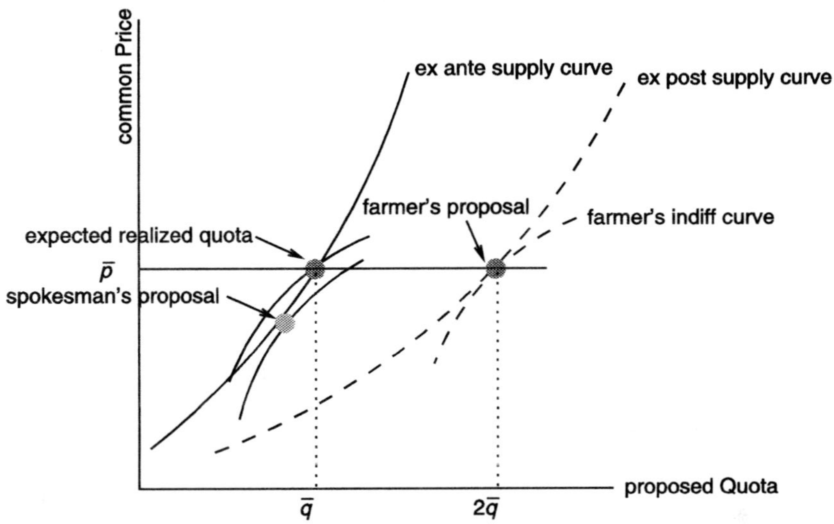

Figure 7.3 Beggar-thy-neighbor behavior by farmers.

supply from neighboring farmers. Since the spokesperson's task is to maximize aggregate farmer utility, it will allocate quotas equally, and hence is not subject to any disjunction between ex ante and ex post. The spokesperson proposes a lower quota for each subbasin and a lower common price, thus increasing farmers' ex ante utility.

Implications of the common spokesperson results for negotiation design. The spokesperson results illustrate how critical the definition of the issue space is for the negotiation outcome. First, differences in outcomes between drought and normal conditions suggest that perhaps the issue space should be specifically designed to incorporate relevant physical constraints. Second, results differ depending on the extent to which farmers' interests are forcibly aligned by requiring a common quota across subbasins, rather than individual quotas allowed to vary by subbasin. Under a common quota, there is no guarantee that farmers will benefit from the introduction of a common spokesperson. Under individual quotas, farmers as a group will benefit, although not every farmer will benefit individually. Intuitively, the difference between the two cases is that under the common quota there is very little scope for the spokesperson to improve farmer welfare by rationalizing the farmers' proposals in order to maximize their welfare, given the interests of the other players. More broadly, this suggests that from a policymaker's perspective there may be some substitutability between the definition of the players and the definition of the issue space when designing a negotiation, but that this substitutability may not exist from a player's perspective.

7.6 Conclusion and policy implications

In most bargaining models, the effectiveness with which participants can pursue their interests during a negotiation is considered to be a function of their access to the negotiation process and how well they will fare if the negotiations break down. This analysis has illustrated that in multi-issue, multiplayer negotiation situations, there are other important determinants of bargaining performance. This chapter has focused on considerations relating to the design of the negotiation structure, and has illustrated how these design factors can have important effects on the outcomes realized by stakeholders. The effect of access to the bargaining process on the outcome for a given stakeholder depends on the opportunity cost of the stakeholder's access. In considering the impact of reassigning access from one farmer to another, it has been shown that if the recipient of access had a more favorable strategic location, this transfer could benefit not only the recipient but the donor as well. In the example used, the middle subbasin farmer's higher willingness to pay for quotas enabled that farmer to extract greater concessions from the downstream user than the upper subbasin farmer could have extracted, and this benefited the upper subbasin farmer. Thus, in the context of the model, the value of access can only be assessed within the context of the composition of the group of negotiators and their preferences.

The study considered in a variety of contexts the effect of assigning a spokesperson to a set of stakeholders with similar interests, rather than allowing each of them independent access to the negotiation process. Whether or not assigning a spokesperson benefited farmers significantly depended on the issue space and the set of constraints facing the negotiators. When farmers were limited to proposing a common price and common quota (CQ regime), the spokesperson had no effect on farmer welfare under drought conditions, because all farmers and the spokesperson were constrained to offer the same proposal in order to meet the hydrology constraints. Under normal conditions, the spokesperson improved welfare slightly by reducing the riskiness of expected utility for the farmers. In contrast, when farmers could propose different quotas for each subbasin (the IQ regime), under both normal and drought conditions the spokesperson increased average farmer welfare significantly, although the upper subbasin farmer was left worse off.

The deadlocking of the Adour negotiations has contributed to a worsening of the water shortage problem. In recent years, there has been increased concern regarding the effects of reduced water flows on fish stocks in the river (Cuende and Marty 2005). Government analysts have recommended that irrigation technologies that conserve water be adopted, and have urged the rapid development of new water resources, such as the dams that were the subject of the deadlocked negotiations (Nau et al. 2005). The deadlocked negotiations have been recognized as contributing to current problems. Nau et al. (2005) urge the development of a new governance structure for the

Adour, characterized by greater cooperation between the water management agency and other government institutions.

The present analysis suggests some structural factors that may have contributed to the deadlocking of the Adour negotiations, and provides broader lessons regarding the design of water allocation negotiation processes. First, the definition of the interest groups, the stakeholders chosen to represent those groups, and their access to the negotiation process will affect how well the interest groups will fare in the negotiation. In the case of the Adour, elected farmer representatives participated in the negotiation process. Their electoral bases, defined by administrative area, did not correspond to hydrological areas, such as subbasins. Because of the differences in the environmental costs and economic benefits of increasing irrigation quotas through building dams across the subbasins, it became more complex for representatives to identify clearly the interests they were to advocate.

More broadly, when defining an interest group and identifying stakeholders to participate in a specific negotiation, policymakers should seek to define key objectives and distinguishing characteristics of specific stakeholders, perhaps through a prenegotiation process whereby stakeholder groups contribute information regarding their own objectives and characteristics, and then evaluate the information provided by other groups in terms of its similarity to their own. For example, each group could be asked to assign access to all other participating groups, excluding itself, although such a process could be subject to strategic manipulation. One interesting possibility suggested by the access experiment is that a player may wish to yield access to another one. Perhaps in some cases it would be feasible for a policymaker to identify two broad groups of stakeholders with divergent interests, such as the agricultural and environmental water users in this case, and appoint one well-defined stakeholder from each broad group to screen other stakeholders who wish to enter the negotiation. If the two screeners do not agree, then they would both have to explain their positions to the policymaker designing the negotiation process.

A second factor that may have influenced the Adour negotiations is the difference in feasible outcomes under normal conditions and under drought conditions, paired with the existing water management system. Hydrological constraints play an important role under drought conditions. It becomes more costly for a farmer to obtain an additional quota unit, because much more dam capacity must be built in order to meet residual flow demands imposed at other measurement points. If, in contrast to the present model, drought and normal conditions are not separated in terms of prices and quotas, or cannot be aggregated in a unanimously accepted way, the negotiation may fail. The existing water management approach may have exacerbated such problems, because all upstream irrigation must be halted whenever it appears that minimum flow objectives will not be met. Given the different marginal values of water in agricultural production, the efficient solution would almost certainly suggest halting irrigation in different subbasins (or for different crops, although this was not analyzed) at different times. Given this implementation constraint, which is not modeled formally in the present study, it may have simply become impossible for farmer representatives to agree upon a way of allocating scarce water resources in times of drought that the other stakeholders would approve. If the implementation method for allocating scarce resources during a drought had been included in the issue space, perhaps the players could have agreed upon a means of doing so.

This problem suggests that in general, the issue space addressed in the negotiation process must be defined carefully, and that the policymaker designing the structure must consider how the outcome of the negotiation process will interact with pre-existing policies and regulations. For example, in negotiations regarding surface water allocation, policies regarding groundwater extraction may affect the strategic location of some stakeholders, as well as their welfare if the negotiations fail. One unanticipated consequence of not incorporating groundwater use into a negotiation regarding surface water could be the overdraft of groundwater resources by some users, which could lead to negative environmental effects.

This chapter has illustrated the implementation of the Rausser–Simon multilateral bargaining model for a specific negotiation process. The issues discussed provide a good indication of the kinds of challenges that might arise when implementing the model in other contexts. First, in water allocation negotiations, a great deal of technical information regarding hydrology and water use is required, including information regarding the value of water to users. Obtaining this information can be quite difficult, but the value of modeling the negotiation process in its absence can be quite limited, as was demonstrated here by the sensitivity of some results to the difference between normal conditions, when hydrology constraints were not binding, and drought conditions, when they were. Second, the definition of the players in the negotiation process can be challenging. In the present application farmers were modeled by subbasin, based on differences in hydrology and the value of water in agricultural production. In the actual negotiations, these differences were not the basis for the identification of stakeholders, which may have contributed to the deadlock. More broadly, it may be challenging to define players, and their objective functions, in a given context. The formally identified players may not correspond to the actual interests at stake. Because the issues faced by analysts seeking to implement the Rausser–Simon bargaining model are also faced by policymakers designing negotiation processes, perhaps the best way to address them, and maximize the value of the analysis to policymakers, is to seek to implement the model during the period when the negotiation process in question is actually being designed. This will enable the model to be used to identify potential solutions to complex problems that may not be readily apparent without the formal identification of interest groups, their objective functions, and the issue space.

Notes

1. Detailed information on these subjects is available in Cemagref 1994; Faÿsse 1998; Faÿsse and Morardet 1999; Gleyses and Morardet 1997; Simon et al. 2007; Thoyer et al. 2001; and Thoyer et al. 2004.
2. The Rausser–Simon bargaining model requires that negotiation proposals be accepted unanimously, rather than simply by a majority of the participants. The reason for this stringent requirement is technical. Under majority rule, the model typically exhibits "cycling" between periods: players who are going to be excluded from the majority coalitions in round T will generally have slacker participation constraints in that round than the players who are included; in round T – 1, therefore, they will tend to be *included* in coalitions because they are easier to please. Thus, coalition membership changes from period to period, with the result that it is extremely difficult to guarantee convergence to a unique solution.
3. The interested reader is referred to Simon et al. 2007 for the complete technical specification of the model.

References

Adams, G., Rausser, G.C., and Simon, L.K. (1996) "Modeling multilateral negotiations: an application to California water policy", *Journal of Economic Behavior and Organization*, 30: 97–111.

Barreteau, O., Garin, P., Dumontier, P.A., and Arami, G. (2003) "Agent based facilitation of water allocation: case study in the Drôme River Valley", *Group Decision and Negotiations*, 12: 441–61.

Carraro, C., Marchiori, C., and Sgobbi, A. (2005) "Applications of negotiation theory to water issues", Working Paper 656.06, Fondazione Eni Enrico Mattei.

Cemagref (1994) "Schema directeur de gestion des étiages de l'Adour", Working Paper, Montpellier.

Cuende, F.X. and Marty, S. (2005) *Indicang: Rapport d'étape, bassin de l'Adour*, Institution Adour/Migradour.

Dinar, A., Ratner, A., and Yaron, D. (1992) "Evaluating cooperative game theory in water resources", *Theory and Decision*, 32: 1–20.

Dinar, A., Farolfi, S., Patrone, P., and Rowntree, K. (2008) "To negotiate or to game theorize: evaluating water allocation mechanisms in the Kat basin, South Africa", this volume, Chapter 5.

Faÿsse, N. (1998) "L'eau entre Etat et usager dans le bassin de l'Adour: de la gestion réglementaire à la gestion concertée", unpublished thesis, DEA-Sciences de l'eau dans l'environnement continental, University Montpellier II, ENGREF, Cemagref.

Faÿsse, N. and Morardet, S. (1999) "La mise en place d'une gestion négociée de l'eau en France: l'exemple de la gestion des étiages sur le bassin de l'Adour", paper presented at the 17th International Congress on Irrigation and Drainage, Granada, September 1999.

Frisvold, G.B. and Emerick, K.J. (2008) "Rural–urban water transfers with applications to the US–Mexico border region", this volume, Chapter 8.

Gleyses, G. and Morardet, S. (1997) *Barrage de Gardères-Eslourenties: évaluation économique de l'agriculture irriguée*, Cemagref, Division Irrigation, Institution interdépartementale pour l'aménagement hydraulique du bassin de l'Adour, Compagnie d'Aménagement des Coteaux de Gascogne, Montpellier.

Nau, F., Payen, D., Ollivier, J.-Y., and Roux, A. (2005) *Bilan du plan de gestion des étiages de l'Adour et du plan de crise interdépartemental: rapport définitif*, Affaire IGE n 03/060, rapport de l'Inspection Générale de l'Environnement, Ministère de l'Ecologie et du Développement Durable.

Parrachino, I., Dinar, A., and Patrone, P. (2006) "Cooperative game theory and its application to natural, environmental, and water resource issues: 3. Application to water resources", Policy Research Working Paper 4074, World Bank, Washington, D.C.

Rausser, G.C. and Simon, L.K. (1999) "A non-cooperative model of collective decision-making: a multilateral bargaining approach", Working Paper number 620, Department of Agricultural and Resource Economics, University of California, Berkeley.

Simon, L.K., Goodhue, R.E., Rausser, G.C., Thoyer, S., Morardet, S., and Rio, P. (2007) "Structure and power in multilateral negotiations: an application to French water policy", Monograph No. 47, Giannini Foundation of Agricultural Economics, University of California.

Thiesse, E., Loucks, D., and Stedinger, J. (1998) "Computer assisted negotiation of water resource conflicts", *Group Decision and Negotiations*, 7: 109–29.

Thoyer, S., Morardet, S., Rio, P., and Goodhue, R.E. (2004) "Comparaison des procédures de décentralisation et de négociation de la gestion de l'eau en France et en Californie", *Natures Sciences Sociétés*, 12: 7–17.

Thoyer, S., Morardet, S., Rio, P., Simon, L., Goodhue, R., and Rausser, G. (2001) "A bargaining model to simulate negotiations among water users", *Journal of Artificial Societies and Social Simulation*, 4. Available online: www.soc.surrey.ac.uk/JASSS/4/2/6.html (accessed January 3, 2007).

8 Rural–urban water transfers with applications to the US–Mexico border region

George B. Frisvold and Kyle J. Emerick

This chapter examines large-scale rural–urban water transfers between irrigation districts and municipal providers as a bargaining game, drawing on empirical examples from the US–Mexico border region. While irrigation technology subsidy programs have been instituted to facilitate transfers and conserve water, increased irrigation efficiency can have unintended effects that actually discourage transfers and conservation. Numerical simulations of the bargaining game, based on the largest rural–urban water transfer agreement in US history, illustrate the usefulness of a game-theoretic approach to assessing transfers. While the transfer agreement provides significant benefits to contracting parties, policy questions concerning negative externalities of transfers remain unresolved.

8.1 Introduction

The population in the US–Mexico border region has grown twelvefold since 1945. In 1980, about 4 million people lived within 100 kilometers of the border. By 2000, this population exceeded 11 million and is projected to exceed 19 million by 2030 (Peach and Williams 2000). Rapid growth in urban water demand has accompanied this population growth. Yet water supplies are limited in this arid environment, with little scope for increasing future supplies. Most major rivers have already been dammed for water supply, the Colorado River is overallocated, and the scope for new dam projects is limited by economic and environmental constraints.

Growing urban water demand is being, and will continue to be, met by transfers from agriculture. Surveying data from 12 western US states, Libecap et al. (2005) estimate that more than 3 million acre-feet (3.7 billion cubic meters) of water was transferred out of agriculture through leases and sales in 2003.[1] This is significant considering that total water withdrawal by public water providers was 17.4 million acre-feet (21.5 billion cubic meters) in 2000 (Hutson et al. 2004). Table 8.1 shows water lease and sales activity in the four US border states from 1990 to 2000. While agriculture-to-agriculture leases are common, permanent sales transferred water out of agriculture in 97 percent of transactions. Agriculture remains the dominant water user in the

Table 8.1 Water sales and leases in US border states 1990–2000

State	*Number of sales*	*Number of leases*	*Mean acre-feet per sale*[a]	*Mean acre-feet per lease*[a]	*Number of agricultural buyers*	*Number of agricultural leasees*
Arizona	34	71	3,148	85,168	0	14
California	31	299	16,222	19,301	4	104
New Mexico	36	25	574	14,050	0	4
Texas	31	131	2,002	2,572	0	40

a. 1 acre-foot ≈ 1,233.5 cubic meters.

Source: Adams et al. 2004.

region. Irrigation accounts for 80 percent of the water withdrawals in Arizona, 79 percent in California, and 88 percent in New Mexico (Hutson et al. 2004). Small reductions in agricultural water use translate into large percentage increases in urban water use.

While rural–urban water transfers vary in size, this chapter focuses on large-scale agreements between irrigation districts and municipal water providers. Such agreements can reallocate significant amounts of water from agricultural to urban use. For example, the Metropolitan Water District of Southern California has entered into a number of bilateral agreements with different irrigation districts, with transfers of 50,000–100,000 acre-feet (61.7 to 123.3 million cubic meters) of water per agreement per year. The Quantification Settlement Agreement reached between the Imperial Irrigation District and the San Diego County Water Authority will transfer up to 200,000 acre-feet (246.7 million cubic meters) of water per year after 10 years with the option of increasing this amount to 303,000 acre-feet (373.7 million cubic meters) per year in later years (Imperial Irrigation District 2004). Transfers of this size are much larger than the historical average (Table 8.1).

Given growing demand for water in the border region, large-scale transfer agreements may become increasingly common on both sides of the border. Some smaller-scale rural–urban transfers have already occurred in Mexico (Rosegrant and Schleyer 1996). Tijuana is considering construction of an aqueduct to transfer agricultural water from the Mexicali Valley to the coast. At one point, there was serious discussion of a binational aqueduct that would transport Colorado River water to San Diego, Tijuana, and Ensenada (Michel 2002; Mumme and Lybecker 2006). The North American Development Bank has also established a Water Conservation Investment Fund to ease transfers of water used in Mexico to the United States in order to meet treaty obligations. The Central Arizona Groundwater Replenishment District is looking to meet over 200,000 acre-feet (246.7 million cubic meters) in future demand from leases from Indian tribes and from irrigation districts along the Colorado River (Jacobs and Megdal 2004).

This chapter examines rural–urban negotiations over large-scale water transfers as a bilateral game between an irrigation district and municipal water provider. Large municipal wholesalers of water control large segments of urban markets, while individual irrigation districts with secure and senior water rights may face little competition on the supply side. Hence, a game-theoretic approach is a useful way to characterize these negotiations and transactions.

The problem is similar to a classic bilateral monopoly problem with some important institutional differences that create deviations from a textbook bilateral monopoly solution. First, the municipal water provider is not a profit-maximizing buyer, but a regulated public utility. The provider maximizes consumer benefits minus charges for cost recovery of acquiring new water supplies. The provider may also engage in two-part pricing, so that consumers pay a fixed rate plus a per unit fee for water. This per unit fee need not equal the marginal cost of acquiring water. Such two-part pricing for urban water is common in the United States and much of the developing world (Dinar and Subramanian 1998).

Second, bilateral water transfer agreements have considerable scope to generate negative pecuniary and environmental externalities. Reductions in irrigated acreage can reduce demand for agricultural labor and other agricultural inputs, leading to localized unemployment and reduced sales by input suppliers. While many economists might argue that these pecuniary externalities should be ignored, Howe (2005) counters that reduced economic activity in already depressed areas may lead to longer-term unemployment and reductions in the value of nonmobile farm assets. Further, water transfers may erode the rural tax base needed to provide social services at just the time when such services are most needed. By reducing local water used for irrigation, transfers can reduce subsequent groundwater recharge and surface water return flows. This, in turn, can take water away from other local water users or reduce water that maintains riparian habitats. Negotiated terms of large-scale transfers will thus be sensitive to laws governing mitigation of these external costs. Western states rely on state agencies or the courts to redress third-party damages and compensation to affected parties may be required to approve a transfer (Howe 2005). Transboundary externalities are an additional complication, affecting international relations between the United States and Mexico.

In an analysis of urban demand for farmers' water rights, Merrett (2003) points out that water markets are not the active, perfectly competitive markets portrayed in simple supply and demand graphs of basic neoclassical economic textbooks, and observes that "the absolutely predominant form of transaction is the bilateral deal" (p. 326). He argues that because markets are thin, there is significant scope for exercise of monopsony power by buyers or monopoly power by sellers. Asymmetries in technical capacity and bargaining power are likely to be important (particularly if sellers are small-scale producers in developing countries). Also important are negative effects of transfers on third parties. He concludes that "the market equilibrium approach is rarely applicable" (p. 319) and "theory of the neoclassical type does not well represent social processes" (p. 326) of water transfers. Merrett does concede, however, that a "rural monopoly–urban monopsony approach" might be useful and that "concepts of a farmer's minimum release price and an urban actor's maximum bid price are still appropriate, as is the account of how these prices are shaped" (p. 326).

This chapter illustrates that, rather than being outside the realm of neoclassical economics, the real-life complexities Merrett introduces—bilateral monopoly, asymmetric capacity and bargaining power, and negative externalities of transfers—can be addressed using a game-theoretic approach. The minimum release price and maximum bid price, important in Merrett's discussion, are central to determining the game-theoretic bargaining solution.

This chapter characterizes the nature of efficient contracts and the bargaining process. It also considers the impacts of agricultural policy, technology, and environmental change. For example, agricultural price support payments can discourage transfers. National and bilateral agencies have instituted technology adoption subsidies, financing improved irrigation efficiency to encourage reallocation of irrigation water to environmental and urban uses. Increasing irrigation efficiency, however, does not necessarily encourage transfers, and can actually discourage them. Increasing irrigation efficiency means less water needs to be applied to achieve a given yield of a crop. But greater efficiency also reduces the cost of effective water (water actually used by crops). It may thus encourage expanded production or shifts to more water-intensive crops, both of which could increase demand for water for irrigation and thus discourage transfers.

In other cases, improving efficiency increases transfers and overall consumptive use. Here, rather than truly conserving water, there is only a reduction in return flows. Reduced return flows can also have unanticipated negative implications for downstream users or the environment.

The rest of this chapter proceeds as follows. Sections 8.2 and 8.3 introduce simple models of rural supply of water for transfers and urban demand for water transfers. Section 8.4 combines these models to derive the set of efficient water transfer contracts. Section 8.5 then characterizes negotiations over water transfers as a sequential offer, noncooperative bargaining game. The analysis makes use of the fact that the two-player Nash bargaining game closely approximates the sequential offer game (as long as time between offers is short enough or player discount rates are high enough). Section 8.6 presents comparative static results, showing how changes in agricultural prices, input costs, water endowments, irrigation efficiency, mitigation requirements, and drought parameters affect the volume of water transferred. Section 8.7 examines the effect of interactions between irrigation technology and mitigation requirements on third-party impacts of water transfers. A main finding is that third-party damages of water transfers are highly sensitive to changes in irrigation technology; to limit these damages, mitigation requirements will need to be regularly monitored and recalibrated. Section 8.8 uses numerical simulations based on the Quantification Settlement Agreement involving the Imperial Irrigation District and the San Diego County Water Authority to illustrate this point. While the transfer agreement provides significant benefits to contracting parties, policy questions concerning transboundary externalities remain unresolved. The conclusion summarizes policy implications and discusses the applicability of the approach to other regions and situations.

8.2 Rural supply of water

The source of water for rural–urban transfers is an irrigation district that receives a fixed allotment of water, X. The district can enter into a contract to transfer x acre-feet of its allotment to a municipal water provider for a fixed price of w per acre-foot of water.

If an agreement is reached, the district may be required to allocate a portion of its water allotment, m, to mitigate damages to other water users or to the environment resulting from the transfer. Irrigation water not taken up by the crop or lost to evaporation may become available to other water users as return flows or groundwater recharge. Data from Shiklomanov (2000) suggest that globally, 70 percent of irrigation withdrawals are consumed and 30 percent are recycled. Return flows may also replenish riparian habitat and be critical for environmental protection. By sending water out of the basin, a transfer agreement may reduce return flows or recharge. This can lead to third-party damages to other water users, damages to the environment, or both.

The district's problem is to allocate water between rural–urban transfers and irrigation to maximize overall district profits, π, subject to any mitigation requirements. To simplify analysis, one can think of the district as made up of many homogeneous irrigators. This abstracts from differences between irrigators, intradistrict bargaining, and the distribution of gains between irrigators (Rosen and Sexton (1993) provide an interesting analysis of these issues). Following Caswell and Zilberman (1986), agricultural production q is a function of effective water e used by the crop, $q = f(e)$, where $f'(e) > 0$ and $f''(e) < 0$. Effective water $e = h[X - (x + m)]$, where X is the district's water allotment, x is the quantity of water transferred, m is water devoted to mitigation, and h measures irrigation efficiency. The efficiency parameter h measures the share of applied water that is used by the crop.

If there is no sale to the municipal provider, the district's entire allocation of water X is used for irrigation and no water is required for mitigation. The cost of production is a constant c per acre-foot of water used for irrigation. Quantity rationing of irrigation district water is also assumed. In the United States and Mexico, districts often receive water at subsidized rates. Because of low water prices, irrigation demand for water is frequently quantity rationed rather than price rationed (Moore 1991; Moore and Dinar 1995).

8.3 Urban demand for water

A municipal water provider supplies an urban market. The provider can supply water at a low cost P^0 but faces a capacity constraint, C (Figure 8.1). If the urban water demand curve is D^0 then demand can be met at price P^0. If demand increases to D^1, however, the municipal provider would have to either raise price or impose quantity rationing. Politically, these may be unattractive options. For example, in June 1976 a "reform" coalition of four city council members in Tucson, Arizona, voted to increase water rates in response to water demand growing beyond existing capacity. Public response to water rate increases was swift and negative. By November, three of the council members were removed in a special recall election, while the fourth resigned prior to the election (Gelt et al. 1998). To avoid rationing or price increases, the provider can seek out additional supplies by leasing water or purchasing water rights from agricultural sources. In Figure 8.2, urban total willingness to pay for agricultural water is represented by the area under D^1 to the right of the capacity constraint, C. Net consumer benefits from consuming water up to the provider's current capacity, C, is v. The provider can expand urban consumption to $C + x$ through transfers, where $B(x)$ is total willingness to pay for x units of transferred water, given that C units are already being consumed.

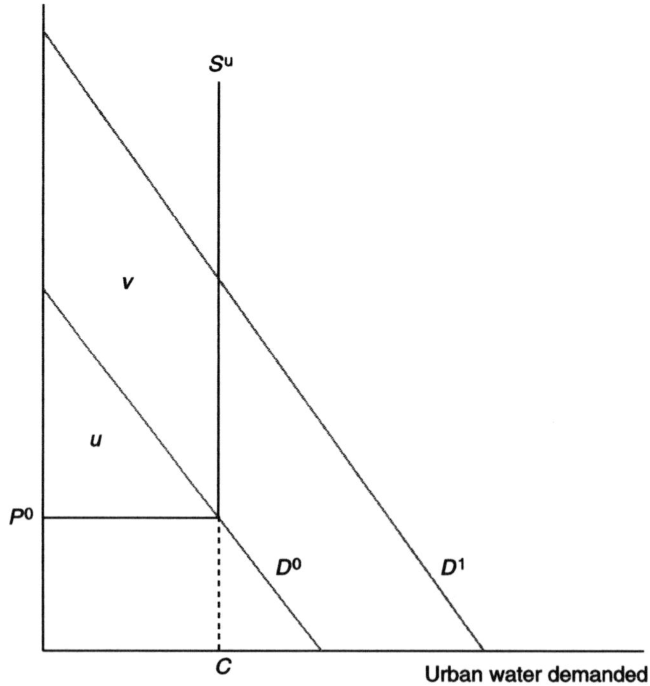

Figure 8.1 Urban water demand.

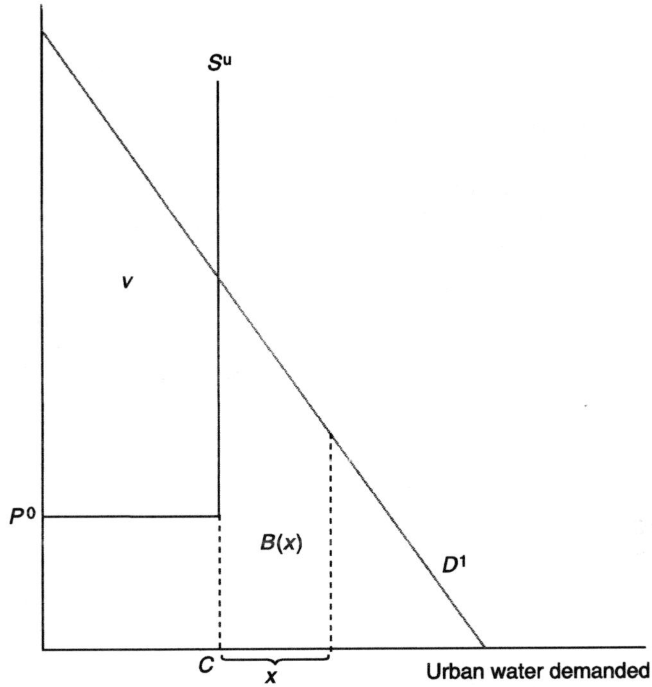

Figure 8.2 Urban willingness to pay for transferred water.

The provider is a regulated public utility whose objective is to maximize consumer benefits of new water supplies acquired from agriculture, $B(x)$, subject to a break-even constraint. The provider must cover the cost of acquiring water by charging fees to urban customers. Total fees collected must equal wx, the amount paid to the irrigation district. Fees can be collected from urban users through a combination of fixed payments and per unit charges. Urban water consumers in the United States and many parts of the developing world are typically charged this type of two-part fee for water, with marginal payments lower than the marginal cost of supplying water (Dinar and Subramanian 1998). The municipal water provider's objective is to negotiate a quantity and price of transferred water to maximize $V(x, w) = v + (B(x) - wx)$. If no transfer agreement is reached, the provider's payoff $V(x = 0) = v$.

8.4 The contract curve

Rural–urban water transfer agreements can be thought of as contracts that jointly specify x (the quantity transferred) and w (the transfer price), subject to any mitigation requirement, m. The contract curve is the set of Pareto optimal contracts between the irrigation district and municipal provider. Pareto optimal contracts are $x - w$ pairs that, once agreed upon, cannot be altered to make one party better off without making the other worse off. There are no deviations from Pareto optimal contracts that can simultaneously make both parties better off.

Mathematically, the contract curve is the locus of points of tangency between the iso-payoff curves of the district and the municipal provider (Appendix 8.1 at the end of this chapter). These iso-payoff curves are like indifference curves and show which combinations (of x and w) yield a given level of return to the district and the provider. Figure 8.3 shows the iso-payoff curves along with disagreement points and the contract curve. The contract curve is the vertical line AB in $w - x$ space. Given all the exogenous variables, there is a single, efficient quantity of water, x^*, to be transferred that is determined independently of w. The vertical contract curve with a unique, optimal quantity of the good transferred is a basic result of simple bilateral monopoly models (Blair et al. 1989; Truett and Truett 1993).

The transfer price, w, determines the distribution of gains between the irrigation district and the municipal water provider. At a transfer price of w_A, the irrigation district captures all the gains from the transfer, while at w_B the municipal provider captures all the gains from trade. The transfer price w_A represents the provider's maximum willingness to pay for x^* acre-feet of water, while price w_B represents the minimum payment that the irrigation district would be willing to accept to sell x^* acre-feet of water.

The highest payoff the irrigation district can obtain is $\pi(x^*, w_A)$, while the highest payoff the municipal provider can obtain is $V(x^*, w_B) = v + (B(x^*) - w_B x^*)$. The remaining Pareto optimal points are all the combinations where $x = x^*$ and $w_A > w > w_B$. All the points on the contract curve—the set of all

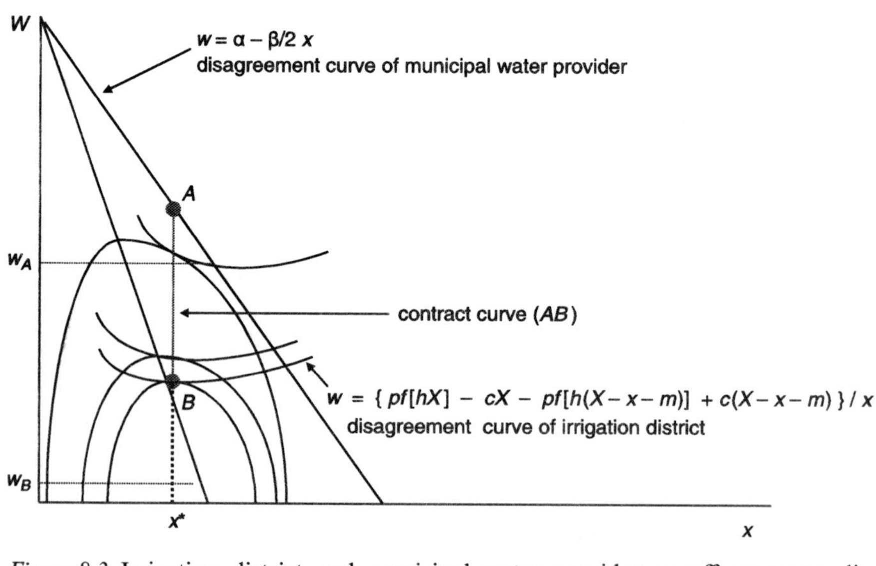

Figure 8.3 Irrigation district and municipal water provider payoff contours, disagreement curves, and the contract curve.

Pareto optimal $x - w$ combinations—have a single common, optimal quantity of water transferred, x^*.

The set of efficient (Pareto optimal) water transfer contracts have been identified, but which (x^*, $w_A \geq w \geq w_B$) combination will be chosen? Most economics textbooks are vague about how the transfer price (in particular) is determined. Here, game theory is particularly useful in describing the negotiated contract as the result of a bargaining process. In the next section, the well-known Nash solution concept (1953) is applied to determine which Pareto optimal point on the contract curve is selected (that is, the negotiated x, and w).

There are other solution concepts that require Pareto optimality as a criterion, such as the Kalai–Smorodinsky (1975), the asymmetric Kalai–Smorodinsky (Dubra 2001), or the equal loss (Chun 1988) solutions. These alternative solutions may yield different values of the transfer price w^*, but—because of the Pareto optimality requirement—will all pick some point on the contract curve AB and the same x^* as in Figure 8.3. The Nash bargaining solution has an advantage of being straightforward to implement. Also, the Nash solution can be more easily extended to multilateral bargaining than other bargaining solutions (Roth 1979; Thomson and Lensberg 1989). This would facilitate extending the present analysis to more than two players.

8.5 Nash bargaining model for water transfers

This section considers negotiations between the municipal water provider and the irrigation district as a two-person Nash bargaining game. The Nash cooperative bargaining framework is relatively straightforward to implement. The Nash solution can closely approximate noncooperative games such as strategic models of sequential offers if those games have high enough discount rates or the period between offers is sufficiently short (Binmore et al. 1986; Chae 1993; Krishna and Serrano 1996).

The bargaining process may represent a noncooperative sequential game, where two parties have the opportunity to exploit some mutually advantageous opportunity, but there is an exogenous risk that negotiations will break down and the opportunity will be lost. For example, during negotiations between the Imperial Irrigation District and the San Diego County Water Authority, federal and state agencies offered to fund some mitigation programs as part of a negotiated deal. However, when negotiations temporarily broke down, the US Department of Interior moved to reduce the Imperial Irrigation District's water allotment by 275,900 acre-feet (340.3 million cubic meters) and California's overall use of Colorado River water by 800,000 acre-feet (986.9 million cubic meters).

The Nash solution is found by choosing to x and w to maximize the Nash product, N:

$$N = [\pi(X - (x + m), w) - \pi^0]^\gamma [V(x) - V^0]^{1-\gamma} \tag{8.1}$$

The term π^0 is the payoff to the irrigation district if bargaining breaks down. In this case, the district applies all of its water allotment X to irrigation. The provider's payoff is the net benefit from existing water capacity, v, plus net gains from acquiring new water, so $V(x) = v + B(x) - wx$. If negotiations break down, the municipal provider receives no additional water from agriculture, so $V^0 = v$. The parameter γ ranges from 0 to 1 and measures the bargaining strength of the irrigation district relative to the municipal provider. Including the bargaining power parameter γ allows for asymmetries in the bargaining game. The Nash product can be rewritten as:

$$N = [\pi(X - (x + m), w) - \pi(X)]^{\gamma} [B(x) - wx]^{1-\gamma}. \tag{8.2}$$

The negotiated transfer price w^* can be written as

$$w^* = \gamma w_A + (1 - \gamma) w_B \tag{8.3}$$

where w_A is the provider's maximum willingness to pay for x^* and w_B is the minimum payment that the irrigation district would be willing to accept to sell x^* acre-feet of water. As the bargaining power of the irrigation district increases (as γ tends toward 1), the transfer price approaches the maximum willingness to pay of the municipal provider. Conversely, as the bargaining power of the municipal provider increases (γ tends toward 0), the transfer price approaches the price that is the minimum willingness to accept for x^*.

8.6 Factors affecting water transfers

In this particular model, the Nash bargaining solution is recursive. The optimal quantity of water transferred x^* from the contract curve (Figure 8.3) can be determined independently of w^*. Table 8.2 summarizes how the quantity of water transferred changes in response to economic, policy, technology, and environmental parameters (Appendix 8.1 discusses how these responses were derived). The quantity of water transferred:

- increases as urban demand increases
- increases as the costs of agricultural production increase
- decreases as agricultural output price increases
- increases as the irrigation district's water allotment increases
- decreases as the mitigation requirement increases
- may increase or decrease as irrigation efficiency increases.

Increases in urban demand encourage rural–urban transfers. In the model the increased demand is captured by two parameters. A parallel outward shift is represented by an increase in a while a pivotal outward shift is captured by a reduction in slope, β. One would thus expect demand for transferred water to

Table 8.2 Effect on water transfers of changes in market, policy, or technology parameters

Effect on water transfers of an increase in:	*on the quantity of water transferred* (x)
• urban water demand (a, $1/\beta$)	Increase
• agricultural production costs (c)	Increase
• agricultural output price or subsidies (P)	Decrease
• the irrigation district's water allotment (X)	Increase
• the mitigation requirement (m)	Decrease
• irrigation efficiency (h)	May increase or decrease

increase with urban population growth, especially in areas where urban demand begins to outstrip existing urban water supply capacity.

Given the diminishing marginal productivity of effective water, an increase in the water allotment of the district X will increase the amount of water transferred. Districts with ample water allotments are more likely to extend irrigation to marginal lands or to lower-value bulk commodity crops. Such districts are more likely to take marginal lands out of production to transfer water.

Combined, these results suggest that the impacts of drought on the quantity of water transferred can be quite complex. Drought can affect both the demand for and the supply of water. Drought can cause the regular water supplies of a city to contract and increase urban water requirements (for example for landscaping), increasing demand for transferred water (a increases, β decreases, or both). On the supply side, though, drought may reduce water available to the irrigation district (X decreases). In the western United States, water rights follow a prior appropriation doctrine, where owners of junior water rights have their water allotments reduced first in times of drought. Consequently, urban water providers may not be able to negotiate larger transfers from such junior rights holders in response to severe, prolonged drought. Senior water rights holders with secure water allotments, in contrast, would be placed in a relatively strong bargaining position relative to urban providers.

The mitigation requirement acts as a tax on water transfers, so increasing m reduces the quantity transferred. In contrast, lax (or no) mitigation requirements imply not only greater transfers, but larger negative environmental impacts or larger third-party pecuniary externalities from transfers.

Turning to agricultural policies, higher agricultural prices increase returns to irrigation and reduce transfers. A policy implication is that a movement from coupled to decoupled farm support payments would encourage larger water transfers. Decoupling farm support programs would lead to a decline in P, with lump sum transfers being made to the irrigation district. In the early 1990s, Mexico instituted market liberalization of its agricultural sector, moving from direct price supports to decoupled support payments to farmers (Rosegrant and Schleyer 1996; Hearne 2004). In the United States, there has also been greater decoupling of farm program payments. Yet, the prices of cotton, rice, and sugar beet and sugar cane are still supported through federal programs. Alfalfa and other hay also receive indirect demand stimulus through US dairy price supports. Together, these crops account for a third of irrigation water applied in US border states (USDA/NASS 2003).

Increasing the cost of irrigated production c will increase the quantity of water transferred. Irrigation water is generally subsidized in both the United States and Mexico. Given that agricultural input and output subsidies discourage transfers, a policy implication is that reallocation of water to urban uses can be achieved simply by removing current market distortions.

The comparative static results suggest that increasing irrigation efficiency (increasing h) need not lead to greater water transfers. Economists have long recognized that improving irrigation efficiency can, under certain conditions, increase agricultural demand for water (Boggess et al. 1993; Caswell and Zilberman 1986).

Mirroring earlier results of Caswell and Zilberman (1986), whether improved irrigation technology increases or decreases agricultural water use depends crucially on the elasticity of the marginal product of effective water with respect to the irrigation efficiency parameter. The elasticity of the marginal product measures how responsive the crop is to further irrigation. It is denoted as e(h), where

$$\varepsilon(h) = f''(e)(e/f'(e)). \tag{8.4}$$

The specific comparative static derivative for the effects of irrigation efficiency on quantity transferred is written as

$$dx^*/dh = -[Pf'(e)(1 + m/x)(1 + \varepsilon)]/[\beta - h^2(1 + m/x)^2 Pf''(e)]. \tag{8.5}$$

The denominator is positive, so when $|\varepsilon| > 1$ (high elasticity case), $dx^*/dh > 0$, and when $|\varepsilon| < 1$ (low elasticity case) $dx^*/dh < 0$. Caswell and Zilberman (1986) show that the low elasticity of marginal product case is more likely to occur when soil quality is poor and when initial irrigation efficiency, h, is low. In this case increasing h increases demand for applied irrigation and consequently discourages transfers.

Although beyond the scope of the current study to explore in detail, this result when combined with other results of Caswell and Zilberman (1986) suggests that increased irrigation efficiency may also discourage transfers in areas with deep wells.

This result has important implications for agricultural conservation policy. Technology adoption subsidies encouraging improvements in irrigation efficiency are provided both in the United States and Mexico. The 2002 US Farm Bill expanded the Environmental Quality Incentive Program to US$5.8 billion from 2002–07. The program provides "cost sharing" (federal subsidies) for adoption of agricultural conservation practices. Program payments were budgeted to improve irrigation efficiency on nearly 25 million acres (10.1 million hectares), primarily in the arid western United States, with projected reduction of agricultural water applications of 5.4 inches per acre (1,322 cubic meters per hectare). According to the US Department of Agriculture, "any water saved would be available for alternative uses such as by municipalities, utility generation, and wildlife habitat restoration" (USDA/NRCS 2003). Assuming 20 percent loss in storage and transmission, the program would free 8.9 million acre-feet (11 billion cubic meters) of water per year over 10 years. This is sizeable considering total water withdrawals by public suppliers, industry, and domestic self-suppliers in the 17 westernmost contiguous states were 25.1 million acre-feet (31 billion cubic meters) per year in 2000 (Hutson et al. 2004).

In 2002, in response to drought-induced disputes over Rio Grande water, the North American Development Bank initiated a Water Conservation Investment Fund, which provides grants to finance investments in improved irrigation conveyance and efficiency. The fund provides US$40 million each to the United States and Mexico to be allocated principally to irrigation districts along the US–Mexico border. Also, under the 1944 Water Treaty, Mexico is required to deliver an average of 350,000 acre-feet (431.7 million cubic meters) of water per year to the United States, mostly from the Rio Conchos basin in Chihuahua. Improved irrigation efficiency is intended to increase Mexico's ability to meet this commitment.

These ambitious projects prompt the question: will improving irrigation efficiency actually conserve agricultural water and encourage transfers? Previous studies have also considered the possibility that subsidizing irrigation technology adoption can backfire, working counter to water conservation goals (Huffaker and Whittlesey 1995, 2000, 2003; Peterson and Ding 2005).

The following simple example illustrates how this might happen. Assume an irrigation district with an irrigation efficiency of $h = 0.5$ diverts 100,000 acre-feet of water from a river, allowing 20,000 acre-feet to pass downstream. Half the diverted water, 50,000 acre-feet, is consumed by the district, while 50,000 goes downstream as return flow. Downstream users then have 70,000 acre-feet of water available. Now suppose irrigation efficiency improves to $h = 0.7$ and the district reduces diversions to 90,000 acre-feet of water. Now 30,000 acre-feet pass directly downstream, but the district consumes 63,000 acre-feet, return flows fall to 27,000 acre-feet, and total water available downstream is reduced to 57,000 acre-feet. Although the district's diversions declines, consumptive use increases and there is less water available downstream.

Without imposing more structure on the general model, it is not possible to derive unambiguous signs for comparative static results regarding w^*. An exception is the bargaining power parameter, γ. The price of transferred water w^* increases with the bargaining power of the irrigation district and decreases with the bargaining power of the municipal provider. Because x^* is determined independently of w^*, an increase in γ will unambiguously increase the payoff to the irrigation district.

Binmore et al. (1986) discussed how the size of γ will depend on the length of time it takes each player to respond to proposals and make counterproposals. Negotiators who can formulate and evaluate proposals and make counterproposals more quickly will have greater bargaining power. Some researchers have pointed out that small-scale irrigators in developing countries often lack the technical capacity or knowledge to effectively bargain with larger, urban demanders of water (Bauer 2004; Galaz 2004; Merrett 2003). To change this bargaining asymmetry, one policy intervention might be for government agencies (or nongovernmental organizations) to provide technical assistance to irrigation districts or water user associations. The game-theoretic results suggest that improved technical capacity to evaluate and respond to offers would increase the bargaining position of rural water suppliers.

8.7 Water transfers and externalities

Now consider how rural–urban water transfers affect third parties. Irrigation water not taken up by the crop or lost to evaporation may become available to other water users as return flows or groundwater recharge. Return flows may also replenish riparian habitat. Unused water available for these other uses under the Nash bargaining solution $R^* = (1 - h)(X - (x + m)) + m$, where m is the quantity of water dedicated to mitigation. Water transfers reduce R^*, while mitigation increases it. Reductions in return flow thus capture third-party damages. Other damages would include pecuniary externalities resulting from lower demand for agricultural labor and other purchased inputs in the region, along with associated multiplier effects. These pecuniary externalities would be an increasing function of expenditures on agricultural inputs.

There are (at least) two possible institutional responses to limit third-party damages. One would be to allow downstream water users or groups concerned about environmental impacts into the coalition with the irrigation district. The Imperial Irrigation District in southern California has, in addition to irrigators, representatives from the local community as voting members, while neighboring Palo Verde Irrigation District does not. In the recent water transfer to San Diego, the Imperial Irrigation District negotiated for mitigation payments to compensate for pecuniary externalities, while water transfers from the Palo Verde Irrigation District do not include such provisions. However, including more heterogeneous actors may greatly increase transaction costs (Rosen and Sexton 1993).

A regulatory body may also set the mitigation requirement m such that return flows are not lower than those with no transfers, $m = m^0 = x(1 - h)/h$, where m is strictly decreasing in h. Holding transferred quantities x constant, improved irrigation efficiency (greater h) reduces the extent of mitigation required to prevent third-party damages. Conversely, if h is low, water transfers have the potential to create significant third-party damages, absent mitigation requirements. In many developing countries (and even parts of the United States) irrigation efficiency is low. In Mexico, Rosegrant and Schleyer (1996: 266) report, "losses through the irrigation infrastructure were as high as 50–70%". Yet, when h is low, the amount of water needed for mitigation is significant, relative to the amount of water transfers. If $h < 0.5$, then $m^0 > x^*$; to prevent a reduction in return flows, more water would have to be dedicated to mitigation than would be transferred.

What might this mean for water policy? One might argue that, to limit third-party damages, one should encourage investments to increase h, prior to encouraging water markets. But a complication arises here because increasing h has a direct negative effect on return flows and indirect effects (via dx/dh), which may be positive or negative. These results, and the numerical simulations in the next section, suggest that limiting third-party damages of water transfers via technological improvements alone will be difficult and that changes in mitigation requirements m will need to be continually monitored and adjusted over time.

8.8 Empirical application

This section applies the Nash bargaining model to the recent Quantification Settlement Agreement, which transfers water from the Imperial Irrigation District to the San Diego County Water Authority. The agreement is the largest rural–urban water transfer agreement in US history. It is a multiyear agreement with payments and transfers growing over time that could reach 300,000 acre-feet (370 million cubic meters) of water per year. In 2005, 30,000 acre-feet (37 million cubic meters) were transferred at US$276 per acre-foot. The numeric simulation is calibrated to price and quantity accounting relationships observed in 2005. Choosing a single year abstracts from the multiyear nature of the agreement. However, the contract curve in each year depends only on exogenous parameter values for that year. The static model is sufficient to illustrate key comparative static results.

Under the agreement, the Imperial Irrigation District was also required to provide 15,000 acre-feet (18.5 million cubic meters) in 2005 for environmental mitigation of impacts to the Salton Sea, California's largest lake, famous for sport fishing and recreational uses. The Salton Sea National Wildlife Refuge is a critical stop on the Pacific flyway for migrating birds, including several state and federally listed endangered and threatened species. Agricultural drain water from the Imperial Valley accounts for about 75 percent of the freshwater inflow to the sea. Colorado River water transferred by the Imperial Irrigation District directly to the San Diego County Water Authority reduces agricultural drain water feeding the sea. Total return flows, R, from the Imperial Irrigation District are as follows:

$$R = (1 - h)(X - x - m) + m. \tag{8.6}$$

Consumptive use of water in the system $C = X - R$.

Table 8.3 reports parameter values and accounting relationships used in the numerical simulations. Appendix 8.2 discusses how values were developed. Given model assumptions, the minimum price that the irrigation district would accept to transfer 30,000 acre-feet of water, W_B, was estimated to be \$41.41 per acre-foot. Based on Draper et al. 2003, the marginal willingness to pay of the district for additional water was assumed to be \$24.30 at the disagreement point. The Imperial Irrigation District solicits bids from its members to participate in the land-fallowing program that frees up water for transfer and mitigation, offering a flat payment of \$60 per acre-foot to participants. At this payment rate, the Imperial Irrigation District can obtain more water than the minimum needed and has instituted a policy to ration participation. This suggests that the \$41.41 figure, lying between \$24.30 and \$60, is reasonable. The San Diego County Water Authority's estimated maximum willingness to pay for 30,000 acre-feet, W_A, was \$315.46. This rate is based on what the authority (a water wholesaler) charges smaller retail providers for treated water deliveries net of treatment and delivery costs for Imperial Irrigation District water. Given estimates of W_A, W_B, and an observed transfer price w of \$276 per acre-foot, the parameter measuring relative bargaining power γ was 0.85. This suggests asymmetric bargaining power favoring the Imperial Irrigation District.

Table 8.3 Values used in numerical simulation

Variable name	*Variable description*	*Parameter or accounting value*
X	Irrigation district water allotment	3.2 million acre-feet[a]
x	Quantity of water transferred	30,000 acre-feet
m	Quantity of water used for environmental mitigation	15,000 acre-feet
h	System irrigation efficiency	0.7
$pf(X)$	Gross revenues of irrigation district without transfers	\$1.198 billion
c	Variable cost of production	\$115 / acre-foot
w_B	Irrigation district's minimum acceptable price to transfer 30,000 acre-feet	\$41.41 / acre-foot
w_A	Municipal provider's maximum acceptable transfer price to acquire 30,000 acre-feet	\$315.46 / acre-foot
w	Negotiated transfer price	\$276 / acre-foot
a	Municipal provider's inverse demand intercept for additional sources of water	\$600
β	Municipal provider's inverse demand slope parameter	0.01897
γ	Bargaining power coefficient	0.856

a. 1 acre-foot = 1233.489238 cubic meters.

Table 8.4 reports estimates of the effects of changing irrigation efficiency on water transfers and transfer price. The system irrigation efficiency for the Imperial Irrigation District was assumed to be 70 percent ($h = 0.7$), approximating previous estimates (Rosen and Sexton 1993). Around the baseline value the simulations show that $dx/dh > 0$ and $dw/dh < 0$. In the Imperial Irrigation District case, improving irrigation efficiency appears to facilitate transfers. Table 8.4 shows, however, that if h is low, a small increase in h can reduce incentives to transfer water. Increasing h from 0.35 to 0.45 reduces incentives to transfer water. This has important implications for technology subsidy policies in the border region aimed at facilitating transfers by increasing h. In some irrigation systems in the border region, system efficiency is quite low. For example, Irrigation District 005 in Delicias, Chihuahua, Mexico, has received US$40 million of North American Development Bank funding to improve irrigation efficiency to facilitate transfers to the United States. However, the system efficiency in the district is only 37 percent (North American Development Bank 2004).

Table 8.4 also shows that transfers tend to increase overall consumptive use relative to a no-transfer baseline. This occurs because transfers come at the expense of return flows. The effect of transfers on overall consumptive use is lessened as h increases, however. The environmental mitigation provisions of the Quantification Settlement Agreement are intended to prevent reductions in return flows reaching the Salton Sea. Table 8.4 shows that, based on the empirically based numerical simulations, the combination of transfers with mitigation leads to a small decrease in consumptive use (increase in return flows). Mitigation transfers of 15,000 acre-feet (18.5 million cubic meters) appear sufficient to maintain return flows for $h > 0.66$.

Table 8.4 Impact of irrigation efficiency on water transfers, transfer price, and consumptive use

Irrigation efficiency	*Transfer price*	*Acre-feet transferred*[a]	*Change in consumptive water use resulting from transfer (in acre-feet)*[a]	
h	*w*	*x*	*with mitigation*[b]	*without mitigation*
0.35	$304.60	27,738	12,780	18,049
0.40	$309.86	27,316	10,390	16,412
0.45	$311.97	27,143	8,179	14,955
0.50	$310.95	27,219	6,110	13,639
0.55	$306.79	27,544	4,145	12,427
0.60	$299.54	28,116	2,246	11,280
0.65	$289.25	28,935	377	10,162
0.70[c]	$276.00[c]	30,000[c]	−1,500[c]	9,035
0.75	$259.86	31,310	−3,422	7,861

a. 1 acre-foot = 1233.489238 cubic meters.
b. 15,000 acre-feet of water devoted to mitigation.
c. Baseline simulation values.

Table 8.5 shows the impacts on consumptive use of simultaneously introducing transfers and increasing irrigation efficiency. Here, a lower *h* no-transfer baseline can be compared with an alternative scenario with transfers combined with a higher *h*. In the early phases of the Quantification Settlement Agreement, water conservation is to be achieved primarily by land fallowing. In later phases, however, it is expected that increasing irrigation efficiency will play a more prominent role in reducing agricultural water use. In Table 8.5, consider a movement from no trades with $h = 0.70$ to one with trades and $h = 0.75$. Here, transfers combined with a larger *h* increase consumptive use by over 156,000 acre-feet (192.4 million cubic meters). Now, consider a movement from no trades with $h = 0.65$ to one with trades and $h = 0.75$. Here, consumptive use increases by over 300,000 acre-feet (370 million cubic meters), even with 15,000 acre-feet set aside to supplement return flows. These results suggest that increasing irrigation efficiency with increased trading may well lead to greater overall consumptive use of water at the expense of return flows to the Salton Sea or subsurface flows that supply Mexico with groundwater.

The results from Tables 8.3 and 8.4 suggest that (a) current mitigation provisions appear adequate to achieve stated short-run objectives, but that (b) future mitigation requirements are highly sensitive to both base levels of *h* and changes in *h*. The numerical simulations imply that close environmental monitoring and water accounting will be necessary to recalibrate mitigation requirements to protect the Salton Sea. The Quantification Settlement Agreement includes provisions for detailed and precise water accounting as well as an adaptive management plan for environmental mitigation. The simulation results suggest these types of provisions will be necessary to meet the long-term environmental objectives of the Quantification Settlement Agreement.

A key element to increase *h* and facilitate future transfers is a plan to line the All-American Canal, which transports water from the Colorado River to the Imperial Valley. The unlined canal is built on sandy soils and loses water

Table 8.5 Impacts of transfers and irrigation efficiency on consumptive use

Irrigation efficiency	*Acre-feet transferred*[a]	*Consumptive use*[b]
h	*x*	*C*
0.65	0	2,080,000[c]
0.65	28,935	2,080,377
0.70	0	2,240,000[c]
0.70	30,000	2,238,500
0.75	0	2,400,000[c]
0.75	31,310	2,396,578

a. 1 acre-foot = 1,233.489238 cubic meters.
b. 15,000 acre-feet of water devoted to mitigation.
c. Result without transfer agreement.

to seepage. The United States plans to build a concrete-lined canal parallel to the original to capture 67,700 acre-feet (83.5 million cubic meters) of water, most of which will be sent to the San Diego County Water Authority.

Lining the canal, however, will impose external costs on Mexico. Groundwater seepage from the canal recharges the aquifer below the Mexicali Valley that supplies irrigated agriculture on the Mexican side of the border. Saille et al. (2006) estimate that canal lining would reduce water available in the most affected area of Mexicali Valley by 14 percent. Also, because water from seepage is less saline than other water in the aquifer, canal lining would also increase the concentration of dissolved salts in the aquifer. The increase in soluble salts could lead to a 9 percent reduction in agricultural production and a 13 percent increase in energy costs (Saille et al. 2006). Canal seepage also supports the Andrade Mesa wetlands in Mexico (Dibble 2005). The wetlands consist of 500 acres (202 hectares) of marsh and more extensive areas of mesquite bosque. They also provide habitat for endangered birds and migratory birds on the Pacific flyway.

Historically, the position of the United States has been that canal seepage is US surface waters allocated to it by the 1944 Water Treaty with Mexico and that environmental mitigation under the Endangered Species Act (or other US environmental laws) is not required for impacts beyond the US border. Mexico has argued that the United States is taking action that adversely affects its groundwater and that a protocol to the 1944 treaty requires that the countries formally consult with each other prior to undertaking such a project. However, a requirement for consultation is not necessarily a requirement for mitigation. The United States and Mexico have never signed a comprehensive agreement to manage transboundary groundwater resources.

In August 2006 the US Ninth Circuit Court, in response to a lawsuit brought by Mexican business groups and environmental groups from both countries, issued a temporary injunction halting the lining of the All-American Canal (Dibble 2006a). At the time of writing the case was scheduled to be heard by a three-judge panel of the Ninth Circuit in December 2006. The injunction illustrates how third parties can increase the negotiating parties' transaction costs if mitigation is not undertaken. According to the San Diego County Water Authority, litigation has delayed the project at least a year and raised costs by tens of millions of dollars (Dibble 2006b). One of the last acts of the 109th Congress was to pass legislation ordering the US secretary of the interior to begin canal lining "without delay" (Dibble 2006b). The secretary's ability to do this will depend on the Ninth Circuit's ruling. Given that the United States and Mexico share 17 groundwater basins (Hall 2004), this could also have implications for future rural–urban transfers on the border.

8.9 Conclusions and policy implications

Rural–urban water transfers in developing countries have received increasing attention from economists and other social scientists (Bauer 2004; Biswas 2006; Galaz 2004; Hearne 2004; Merrett 2003; Michel 2002; Mumme and Lybecker 2006; Riaz 2002; Rosegrant and Schleyer 1996). Demand for large-scale transfers will probably increase with the rise of megacities (large urban agglomerations) in many developing countries (Biswas 2006; Varis 2006).

The issues highlighted in the stylized theoretical and numerical models address issues pertinent to other areas. For example, Chennai, India, is facing questions of whether improved irrigation efficiency truly conserves water or merely reduces return flows available to third parties (Biswas 2006). Irrigators supplying water to Chennai have begun to address questions of collective bargaining in determining water transfers and whether and how to include third parties in negotiations (Ramalingam 2005).

A goal of this chapter has been to demonstrate how a game-theoretic analysis of rural–urban water transfers can examine not only gains from trade but also issues of bilateral monopoly, asymmetric capacity and bargaining power, and negative externalities of transfers. But what policy lessons can be drawn from this analysis?

First, agricultural output and input subsidies discourage rural–urban water transfers. One way to encourage reallocation of water to urban uses through market transfers would be to remove pre-existing distortions in agricultural output and input markets.

Second, the model results suggest that improving irrigation efficiency need not encourage water transfers. Depending on agronomic conditions, increasing efficiency can increase the demand for irrigation water, thus discouraging transfers. The numerical simulations suggest a U-shaped relationship, where improving efficiency reduces transfers in areas with very low efficiency, but encourages transfers once efficiency reaches a minimum threshold. This has important implications for Mexico and other developing countries, where baseline efficiencies are often quite low.

Third, resource management agencies on both sides of the border appear to be counting on improving irrigation efficiency to conserve water. Yet, increasing efficiency can reduce return flows and groundwater recharge and increase consumptive use of water. Numerical simulations based on the Quantitative Settlement Agreement in California suggest that introducing transfers and new irrigation technology simultaneously can generate large increases in consumptive use. This occurs because, while diversions for irrigation decline, return flows decline even more, causing consumptive use to increase. So, irrigation technology subsidies intended to encourage transfers or conserve water may not achieve these goals. This is not to discourage efforts to improve irrigation efficiency per se, but care must be taken to ensure that hydrologic and economic factors are such that improved efficiency truly conserves water.

Fourth, mitigation requirements will be needed to prevent negative environmental impacts or other third-party impacts of transfers. Huffaker and Whittlesey (2000) have noted that a number of western states have passed laws to encourage increased farm-level irrigation efficiency and to reduce diversions for irrigation. Their analysis and the analysis in this chapter suggest that, in areas where return flows are important, this can lead to increases in water consumption. They recommend that policies to encourage conservation should define water conservation in terms of reduced consumptive use, not merely reduced diversions.

Fifth, not only are mitigation requirements important, they are highly sensitive to small changes in irrigation efficiency. The numerical analysis suggests that mitigation policies will have to be dynamic and flexible to prevent third-party damages. To avoid such damages, large-scale water transfers will require regular environmental monitoring, recalibration of mitigation requirements, and adaptive environmental management.

Finally, in the US–Mexico border region, large-scale transfers are confronting (and will continue to confront) unresolved questions about how to deal with transboundary externalities of water transfers. Unresolved legal questions remain over transboundary environmental mitigation requirements and transboundary groundwater management.

Note

1. An acre-foot is a unit of volume commonly used in the United States for large-scale water resources. It is the volume of water necessary to cover one acre of surface area to a depth of one foot. One acre-foot ≈ 1,233.5 cubic meters.

References

Adams, J., Crews, D., and Cummings, R. (2004) "The sale and leasing of water rights in the western United States: an update to mid-2003", Water Policy Working Paper 2004–004.

Bauer, C.J. (2004) "Results of Chilean water markets: empirical research since 1990", *Water Resources Research*, 40: W09S06, doi:10.1029/2003WR002838.

Binmore, K., Rubinstein, A., and Wolinsky, A. (1986) "The Nash bargaining solution in economic modelling", *RAND Journal of Economics*, 17: 176–88.

Biswas, A.K. (2006) "Water management for major urban centres", *International Journal of Water Resources Development*, 22: 183–97.

Blair, R.D., Kaserman, D.L., and Romano, R.E. (1989) "A pedagogical treatment of bilateral monopoly", *Southern Economic Journal*, 55: 831–40.

Boggess, W., Zilberman, D., and Lacewell, R. (1993) "Economics of water use in agriculture", in G. Carlson, D. Zilberman, and J. Miranowski (eds) *Agricultural and Environmental Economics*, Oxford: Oxford University Press.

Burt, C. (1999) "Irrigation water balance fundamentals", paper presented at USCID Conference on Benchmarking Irrigation System Performance Using Water Measurement and Water Balances.

Caswell, M. and Zilberman, D. (1986) "The effects of well depth and land quality on the choice of irrigation technology", *American Journal of Agricultural Economics*, 68: 798–811.

Chae, S. (1993) "The n-person Nash bargaining solution with time preference", *Economics Letters*, 41: 21–4.

Chun, S. (1988) "The equal-loss principle for bargaining problems", *Economics Letters*, 26: 103–6.

Dibble, S. (2005) "Wetlands become a focus in debate over canal lining", *San Diego Union-Tribune*, June 6, 2005.

Dibble, S. (2006a) "All-American canal lining halted by court injunction", *San Diego Union-Tribune*, August 26, 2006.

Dibble, S. (2006b) "Congress OKs bill to speed lining of all-American canal", *San Diego Union-Tribune*, December 12, 2006.

Dinar, A. and Subramanian, A. (1998) "Policy implications from water pricing experiences in various countries", *Water Policy*, 1: 239–50.

Draper, A., Jenkins, M., Kirby, K., Lund, J., and Howitt, R. (2003) "Economic-engineering optimization for California water management", *Journal of Water Resource Planning and Management*, 129: 155–64.

Dubra, J. (2001) "An asymmetric Kalai–Smorodinsky solution", *Economics Letters*, 73: 131–6.

Galaz, V. (2004) "Stealing from the poor? Game theory and the politics of water markets in Chile", *Environmental Politics*, 13: 414–37.

Gelt, J., Henderson, J., Seasholes, K., Tellman, B., and Woodard, G. (1998) *Water in the Tucson Area: Seeking Sustainability*, Tucson, AZ: Water Resources Research Center, College of Agriculture, University of Arizona.

Hall, R. (2004) "Transboundary groundwater management: opportunities under international law for groundwater management in the United States–Mexico border region", *Arizona Journal of International and Comparative Law*, 21: 873–5.

Hearne, R.R. (2004) "Evolving water management institutions in Mexico", *Water Resources Research* 40: W12S04, doi:10.1029/2003WR002745.

Howe, C. (2005) "Property rights, water rights and the changing scene in western water", in C. Gopalakrishnan, C. Tortajada, and A.K. Biswas (eds) *Water Institutions: Policies, Performance and Prospects*, Berlin: Springer.

Huffaker, R. and Whittlesey, N. (1995) "Agricultural conservation legislation: will it save water?", *Choices*, 10: 24–8.

Huffaker, R. and Whittlesey, N. (2000) "The allocative efficiency and conservation potential of water laws encourage investments in on-farm irrigation technology", *Agricultural Economics*, 24: 47–60.

Huffaker, R. and Whittlesey, N. (2003) "A theoretical analysis of economic incentive policies encouraging agricultural water conservation", *International Journal of Water Resources Development*, 19: 37–53.

Hutson, S., Barber, N., Kenny, J., Linsey, K., Lumia, D., and Maupin, M. (2004) *Estimated Use of Water in the United States in 2000*, USGS Circular 1268, US Geological Survey, Reston, Virginia.

Imperial Irrigation District (2004) *Quantification Settlement Agreement Annual Implementation Report*, Imperial, California.

Jacobs, K. and Megdal, S. (2004) "Water management in the active management areas", in P. Likins (ed.) *Arizona's Water Future: Challenges and Opportunities Background Report*, Tucson, AZ: University of Arizona.

Kalai, E. and Smorodinsky, M. (1975) "Other solutions to Nash's bargaining problem", *Econometrica*, 43: 513–18.

Krishna, V. and Serrano, R. (1996) "Multilateral bargaining", *Review of Economic Studies*, 63: 61–80.

Libecap, G., Glennon, R., and Ker, A. (2005) "The implications of legal and political institutions on water transfers in the west", *Arizona Review*, 3: 3–4.

Merrett, S. (2003) "The urban market for farmers' water rights", *Irrigation and Drainage*, 52: 319–26.

Michel, S. (2002) "The geography of water transfers and urbanization in Baja and Southern California", in L. Fernandez and R. Carson (eds) *Both Sides of the Border: Transboundary Environmental Management Issues Facing Mexico and the United States*, Dordrecht: Kluwer Academic Publishers.

Moore, M. (1991) "The Bureau of Reclamation's new mandate for irrigation water conservation: purposes and policy alternatives", *Water Resources Research*, 27: 145–55.

Moore, M. and Dinar, A. (1995) "Water and land as quantity-rationed inputs in California agriculture: empirical tests and water policy implications", *Land Economics*, 71: 445–61.

Mumme, S.P. and Lybecker, D. (2006) "The All-American Canal: perspectives on the possibility of reaching a bilateral agreement", in V.S. Munguía (ed.) *The US–Mexican Border Environment: Lining the All-American Canal—Competition or Cooperation for the Water in the US–Mexican Border?* SCERP Monograph Series 13, San Diego: San Diego State University Press.

Nash, J. (1953) "Two-person cooperative games", *Econometrica*, 21: 128–40.

North American Development Bank (2004) *Water Conservation Investment Fund Fact Sheet: Irrigation District 005 Delicias, Chihuahua*, San Antonio, Texas.

Peach, J. and Williams, J. (2000) "Population and economic dynamics on the US–Mexican border: past, present and future", in P. Ganster (ed.) *The US–Mexican Border Environment*, SCERP Monograph Series, San Diego: San Diego State University Press.

Peterson, J. and Ding, Y. (2005) "Economic adjustments to groundwater depletion in the high plains: do water-saving irrigation systems save water?" *American Journal of Agricultural Economics*, 87: 147–59.

Ramalingam, K. (2005) "Chennai sucking up rural water", *India Together*, 25 April 2005.

Riaz, K. (2002) "Tackling the issue of rural–urban water transfers in the Ta'iz Region, Yemen", *Natural Resources Forum*, 26: 89–100.

Rosegrant, M.W. and Schleyer, R.G. (1996) "Establishing tradable water rights: implementation of the Mexican Water Law", *Irrigation and Drainage Systems*, 10: 263–79.

Rosen, M.D. and Sexton, R.J. (1993) "Irrigation districts and water markets: an appli-Cation of cooperative decision-making theory", *Land Economics*, 69: 39–53.

Roth, A.E. (1979) *Axiomatic Models of Bargaining*, Berlin: Springer.

Saille, G.G., López, A.L., and Navarro Urbina, J.A. (2006) "Lining the All-American Canal: its impact on aquifer water quality and crop yield in Mexicali Valley", in V.S. Munguía (ed.) *The US–Mexican Border Environment: Lining the All-American Canal—Competition or Cooperation for the Water in the US–Mexican Border?* SCERP Monograph Series 13, San Diego: San Diego State University Press.

Shiklomanov, I.A. (2000) "Appraisal and assessment of world water resources", *Water International*, 25: 11–35.

Thomson, W. and Lensberg, T. (1989) *Axiomatic Theory of Bargaining with a Variable Population*, Cambridge: Cambridge University Press.

Truett, D.B. and Truett, L.J. (1993) "Joint profit maximization, negotiation, and determinacy of price in bilateral monopoly", *Journal of Economic Education*, 24: 260–70.

USDA/NASS (United States Department of Agriculture, National Agricultural Statistics Service) (2003) *Farm and Ranch Irrigation Survey, 2003*, Washington, DC: United States Government Printing Office.

USDA/NRCS (United States Department of Agriculture, Natural Resources Conservation Service) (2003) *EQIP Benefit Cost Analysis: Final Report*, Washington, DC: Natural Resource Conservation Service.

Varis, O. (2006) "Megacities, development and water", *International Journal of Water Resources Development*, 22: 199–225.

Appendix 8.1: Mathematical derivations

Irrigation district's problem

If no water is transferred, irrigation district profits π^d are

$$\pi^d = Pf(hX) - cX, \tag{8.A1}$$

where P is the agricultural output price, X is the district's water allotment, h is the irrigation efficiency parameter, and c is the cost of production per acre-foot of water. If the district sells water to an urban municipal provider, its profits are

$$\pi = Pf(h(X - x(1 + s))) - c(X - x(1 + s)) + wx \tag{8.A2}$$

where x is the quantity of water transferred. Water required for mitigation m can be expressed as a proportion of water transferred, $m = sx$. The district's iso-payoff function in w–x space shows combinations of x and w that yield a fixed profit π^0

$$w = \frac{\pi_0 - Pf(h(X - x(1 + s))) + c(X - x(1 + s))}{x} \tag{8.A3}$$

For a given value of x, the minimum acceptable price for the district is

$$w = \frac{Pf(hX) - Pf(h(X - x(1 + s))) - cx(1 + s)}{x}. \tag{8.A4}$$

Municipal water provider's problem

Urban demand for additional water (beyond current capacity) is

$$p^m = a - \beta x, \tag{8.A5}$$

where p^m is the inverse demand for water and a and β are positive constants. The urban water provider is a regulated public utility whose objective is to maximize consumer benefits from obtaining new water subject to a cost recovery constraint. Urban water consumers must pay wx, the cost of obtaining agricultural water. The urban water provider seeks to negotiate x and w to maximize this objective function, V, where

$$V = (v + a\,x) - \left(\frac{1}{2}\beta\,x^2 + wx\right) \tag{8.A6}$$

In w–x space, the iso-payoff curve for the provider shows combinations of x and w that yield a fixed payoff V^0:

$$w = \frac{ax - \frac{1}{2}\beta x^2 - (V_0 - v)}{x}. \tag{8.A7}$$

For a given value of x, the provider's maximum willingness to pay is

$$w = a - \frac{1}{2}\beta\,x. \tag{8.A8}$$

The contract curve

The contract curve, the set of efficient x–w pairs, is obtained by equating the slopes of the district's and provider's iso-payoff curves (8.A3) and (8.A7). The efficient volume of water transferred can then be derived:

$$x^* = \frac{(1 + s)(-hPf_e + c) + a}{\beta}. \tag{8.A9}$$

The efficient volume of water traded can be determined independently of the transfer price, w. Comparative static results can then be obtained by totally differentiating (8.A9) to determine how changes in exogenous variables affect x^*. These comparative static results (such as dx/ds, dx/da, etc.) are summarized in Table 8.2.

Appendix 8.2: Simulation model parameters

Water diversions (X): obtained from US Bureau of Reclamation Colorado River Accounting and Water Use Report: Arizona, California, and Nevada (various years). BOR Boulder Canyon Operations Office, Boulder City, Nevada.

Gross farm returns ($Pf(hX)$): estimated from statistics for Imperial County from USDA, NASS 2002 Census of Agriculture and the Environmental Working Group Farm Subsidy Data Base (www.ewg.org/farm/findings.php).

Variable cost of production (c): obtained from USDA, NASS 2002 Census of Agriculture production expenditure data for Imperial County. Excludes livestock purchases, feed purchases, and land rental payments.

Irrigation production function ($f(e)$): the irrigation district production function was assumed quadratic. Parameters of the marginal value product (MVP) of water function were chosen such that the marginal willingness to pay for additional water at X was \$24.30, approximating Draper et al. (2003) and integrating the MVP curve yielded gross farm returns.

Irrigation efficiency (h): results from Rosen and Sexton (1993) suggest an h close to 0.7. Internal Imperial Irrigation District documents state on-farm efficiency is about 0.79, but Burt (1999) suggests Imperial Irrigation District system efficiency will be less than on-farm efficiency.

Water transfer parameters (x, w, m): the quantity of water transferred, price paid, and water used for Salton Sea Mitigation come from the Quantification Settlement Agreement Annual Implementation Report.

Return flows and consumptive use (R, C): data were derived based on return flows going to the Salton Sea and Colorado River. These came from US BOR Reclamation Colorado River Accounting and Water Use Report and the California Environmental Protection Agency California Regional Water Quality Control Board, Colorado River Basin Region.

Municipal demand for additional water (α, β): parameters were derived to obtain maximum willingness to pay for transferred water based on wholesale rates charged by the San Diego County Water Authority to retail municipal providers.

9 WAS-guided cooperation in water management

Coalitions and gains

Franklin M. Fisher and Annette T. Huber-Lee[1]

The water allocation system (WAS) is a tool for optimal water management and the resolution of water conflicts. Countries or parties using WAS-guided cooperation all gain in total benefits from water. Estimates of the benefits of such cooperation to Israel, Jordan, and Palestine are given, for various coalitions among them, and the relation of bilateral cooperation to trilateral cooperation is discussed and generalized. The tool is applicable anywhere in the world and not just in the Middle East.

9.1 Introduction

Starting in 1993, the Water Economics Project developed economics-based methods for optimal water management, especially for resolving water disputes. That work is extensively described in Fisher et al. 2005.[2] There it is shown that an agreement to trade in water permits (permits to use each other's water in specified amounts) sold at the shadow values generated by the water allocation system (WAS) model is never inferior and usually superior to the standard international treaty with a fixed division of water quantities among countries.

Of course, the realization that water can (and should) be treated as an economic commodity—with special properties—is not new. It goes back at least as far as the Harvard Water Program of the 1950s and 1960s (see Eckstein 1958; Maass et al. 1962; see also Hirshleifer et al. 1960). That realization has led to a large literature on water markets, to policy recommendations for water markets, particularly by the World Bank, and, in some cases, to the adoption of such recommendations (see for example Saliba and Bush 1987).

But, for reasons discussed below, actual water markets will often, indeed generally, not lead to efficient or optimal results. There are several reasons for this, but perhaps the most important is because, in allocating water, there are social benefits and costs that are not simply private benefits and costs. While some of these "externalities" are recognized in the literature (environmental issues, for example), others are not. In particular, the fact that subsidization of water for a particular use (usually agriculture) implies a policy decision that there are social benefits from such use above and beyond those realized by the direct users themselves is typically ignored. But private markets will fail to handle this, and, where such views are held by policymakers, such failure will properly make them unwilling to institute actual market mechanisms.

The present work, as will be seen, overcomes such difficulties by the use of WAS, an explicit optimizing model that allows the policymaker to impose his or her own views as to water values or, equivalently, to impose certain water policies and also to take other externalities into account. This is a form of systems analysis.

The application of systems analysis to water management is also not new (see for example Rogers and Fiering 1986 and the papers cited in endnote 3). As Rogers and Fiering state (1986: 146S):

> Systems analysis is particularly promising when scarce resources must be used effectively. Resource allocation problems are worldwide and affect the developed and the developing countries that today must make efficient use of their resources. At any given time some factor such as skilled manpower, energy, transport, material or capital is in short supply and that scarcity impedes progress. Under these conditions, governments and their planners, but particularly those in less developed countries (LDCs), are constantly faced with the need to make the best use of those resources that will be needed at further stages of progress, and with the future need to balance current needs against investment for the future. This is the domain of systems analysis.

However, they also point out (p. 150S) that:

> [M]any, if not most, conventional uses of mathematical programming to solve water resource problems pursue optimality indiscriminately. . . . Unfortunately, the stated . . . goals and measures of merit do not fully reflect the true concerns of the decision makers, and experience shows that mathematical programming, while often used in some vague way by planners, is rarely used to make the critical decisions associated with project planning.

As discussed above, the WAS tool attempts to solve this problem by permitting the user of WAS to impose his or her own values.

In addition, there are other differences between the present work and its predecessors. First, much previous work concerns the cost-benefit analysis of particular projects, whereas the present work provides a tool for the overall management of water and water projects taking into account countrywide (or regionwide) effects while using a model that is geographically quite disaggregated.

Second, perhaps the most familiar use of such models is that of minimizing cost in connection with fixed demand quantities.[3] The WAS model described below goes beyond that in more than one respect:

- WAS takes account of demand considerations and the benefits to be derived from water use rather than fixing water quantities to be delivered.
- WAS permits the user to impose social values that differ from private ones and to impose policies that the optimization must respect.
- Beyond mere optimization, WAS models can be used in conflict resolution, an area highly important in water issues, as they can be used to value disputed water, thus effectively monetizing and de-emotionalizing the dispute. Moreover, for international disputes, the water systems of each party can be analyzed separately, testing options of links to other parties, or of analyzing the combined territory of two or more parties as one. This provides estimates of the benefits of cooperation, which can then be weighed against the political issues involved in such cooperation.

Fisher et al. (2005) applied those methods to Israel, Jordan, and Palestine, which have a long history of water disputes, and examines the gains to such cooperation compared to the much lower values of plausible changes in ownership of the water in dispute. Most of the discussion concerned two-way cooperation between Israel and Palestine[4] and Israel and Jordan.[5] Cooperation among all three countries was considered, giving results on particular aspects of the water flows that would occur, but the analysis did not exhibit fully the gains from three-way cooperation (although the case in which the third country joins a cooperative agreement already in place between the other two was also discussed).

This chapter considers the benefits of different coalitions and presents a number of results thereon. Before this is done, however, a related question is considered. Suppose that there are *n* countries. Is the "grand coalition"—that is, WAS-guided cooperation among all countries involved—stable? In game-theoretic terms, is the grand coalition in the core or can there exist a subset of countries that, by cooperating among themselves, can all do better than they can in the grand coalition and therefore block the latter?

It is shown that the answer to this question is in the negative[6]—an important fact with an easy demonstration, once one understands how WAS-guided cooperation works. Since this is also essential for an understanding of the results for the three countries discussed below, that mechanism is first summarized.

9.2 Water ownership and the value of water

There are two basic questions involved in thinking about water agreements. These are:

- The question of water ownership.
- The question of water usage.

One must be careful to distinguish these questions.

All water users are effectively buyers irrespective of whether they own the water themselves or purchase water from another. An entity that owns its water resources and uses them itself incurs an opportunity cost equal to the amount of money it could otherwise have earned through selling the water. An owner will thus use a given amount of its water if and only if it values that use at least as much as the money to be gained from selling. The decision of such an owner does not differ from that of an entity that does not own its water and must consider buying needed quantities of water: the nonowner will decide to buy if and only if it values the water at least as much as the money involved in the purchase. *Ownership only determines who receives the money (or the equivalent compensation) that the water represents.*

Water ownership is thus a property right entitling the owner to the economic value of the water. Hence a dispute over water ownership can be translated into a dispute over the right to monetary compensation for the water involved, taking into account social and environmental values.

The property rights issue of water ownership and the essential issue of water usage are analytically independent. For example, resolving the question of where water should be efficiently pumped does not depend on who owns the water. While both ownership and usage issues must be properly addressed in an agreement, they can and should be analyzed separately.[7]

The fact that water ownership is a matter of money can be brought home in a different way. It is common for countries to regard water as essential to their security because water is essential for agriculture and countries wish to be self-sufficient in their food supply. This may or may not be a sensible goal, but the possibility of desalination implies the following. Every country with a seacoast can have as much water as it wants if it chooses to spend the money to do so. Hence, so far as water is concerned, every country with a seacoast can be self-sufficient in its food supply if it is willing to incur the costs of acquiring the necessary water. Disputes over water among such countries are merely disputes over costs, not over life and death.

Note that, in valuing water, one must consider the following:

- Unless water is very abundant, the value of water does not simply consist of the costs of extraction, treatment, and conveyance. In general, water value also includes a scarcity rent, reflecting the opportunity cost of its use—the value of using or selling it elsewhere.
- When one speaks of "the value of water", one is speaking of the value of molecules of H_2O. That is not the only value of importance, however. There may be religious or historical values placed on particular water sources. Further, water in certain uses (agriculture, for example), may be considered as more valuable to society than is reflected in the private valuations of the users. And water in sources thought to be

secure may be considered more valuable than water in less secure sources. In the WAS model, the user is permitted to take such things into account by constraining the model to reflect such considerations. In that sense, this analysis is not confined to narrowly defined economic considerations.

9.3 Why actual free water markets will not work

In the case of most scarce resources, free markets can be used to secure efficient allocations. This does not always work, however; the important propositions about the efficiency of free markets require the following conditions:

- The markets involved must be competitive, consisting only of very many, very small buyers and sellers.
- All social benefits and costs associated with the resource must coincide with private benefits and costs, respectively, so that they will be taken into account in the profit-and-loss calculus of market participants.

Neither of these conditions is generally satisfied when it comes to water, for two reasons. First, water markets will not generally be competitive, with many small sellers and buyers. Second (and perhaps more important), water in certain uses—for example agricultural or environmental uses—is often considered to have social value in addition to the private value placed on it by its users. The common use of subsidies for agricultural water, for example, implies that the subsidizing government believes that water used by agriculture is more valuable than the farmers themselves consider it to be.

This does not mean, however, that economic analysis has no role to play in water management or the design of water agreements. One can build a model of the water economy of a country or region that explicitly optimizes the benefits to be obtained from water, taking into account the issues mentioned above.[8] Its solution, in effect, provides an answer in which the optimal nature of markets is restored and serves as a tool to guide policymakers.[9]

Such a tool does not itself make water policy. Rather it enables the user to express his or her priorities and then shows how to implement them while maximizing the net benefits to be obtained from the available water. While such a model can be used to examine the costs and benefits of different policies, it is not a substitute for, but an aid to, the policymaker.

It would be a mistake to suppose that such a tool only takes economic considerations (narrowly conceived) into account. The tool leaves room for the user to express social values and policies through the provision of low (or high) prices for water in certain uses, the reservation of water for certain purposes, and the assessment of penalties for environmental damage. These are, in fact, the ways that social values are usually expressed in the real world.

9.4 The water allocation system (WAS) tool

As already indicated, the tool is called WAS for "water allocation system" (see Appendix at the end of this chapter for a detailed mathematical description). As here discussed, it is a single-year, annual model, although the conditions of the year can be varied and different situations evaluated.[10] The objective function maximizes net water benefits for a given region, subject to constraints. These include:

- The capacity of various pieces of infrastructure (water treatment plants, desalination plants, and conveyance lines or canals)
- The constraints represented by social policies arising from environmental or other factors, including constraints restricting the water taken from any natural source to be no greater than the annual renewable flow available from that source (although that maximum can be adjusted according to climatic conditions)
- The price policies (such as subsidization of water for agriculture) that are prescribed by the model user
- Most important of all, the constraint that the quantity of water consumed in a particular location cannot exceed the quantity produced there plus the quantity imported into less the quantity exported from that location.

The country or region to be studied is divided into districts. Within each district, demand curves for water are defined for household, industrial, and agricultural use of water. Extraction from each water source is limited to the annual renewable amount. Allowance is made for treatment and reuse of wastewater and for interdistrict conveyance. This procedure is followed using actual data for a recent year and projections for future years.

Environmental issues are handled in several ways. Water extraction is restricted to annual renewable amounts; an effluent charge can be imposed; the use of treated wastewater can be restricted; and water can be set aside for environmental (or other) purposes. Other environmental restrictions can also be introduced.

The WAS tool permits experimentation with different assumptions as to future infrastructure. For example, the user can install wastewater treatment plants, expand or install conveyance systems, and create seawater desalination plants.

Finally, the user specifies policies toward water. Such policies can include specifying particular price structures for particular users; reserving water for certain uses; and imposing ecological or environmental restrictions.

Given the choices made by the user, the model allocates the available water so as to maximize total net benefits from water. These are defined as the total amount that consumers are willing to pay for the amount of water provided less the cost of providing it.[11]

9.5 Shadow values and scarcity rents

It is an important theorem that, under very general conditions, when an objective function is maximized under constraints, the solution also generates a set of nonnegative numbers, usually called "shadow prices", but here called "shadow values" to emphasize that these are not necessarily the prices to be charged to water users. Such shadow values (which are the LaGrange multipliers corresponding to the various constraints) have the property that they show the amount by which the value of the thing being maximized would increase if the corresponding constraints were to be relaxed a little.

In the case of the WAS model, the shadow value associated with a particular constraint shows the extent by which the net benefits from water would increase if that constraint were loosened by 1 unit. For example, where a pipeline is limited in capacity, the associated shadow value shows the amount by which benefits would increase per unit of pipeline capacity if that capacity were slightly increased. This is the amount that those benefiting would just be willing to pay for more capacity.

The central shadow values in the WAS model, however, are those of water itself. The shadow value of water at a given location corresponds to the constraint that the quantity of water consumed in that location cannot exceed the quantity produced there plus the quantity imported less the quantity exported. That shadow value is thus the amount by which the benefits to water users (in the system as a whole) would increase were there an additional cubic meter per year available free *at that location*. It is also the price that the buyers at that location who value additional water the most would just be willing to pay to obtain an additional cubic meter per year, given the net-benefit maximizing water flows of the model solution.[12]

Experience shows that the following points about shadow values cannot be overemphasized:

- Shadow values are not necessarily the prices that water consumers are charged. That would be true in a purely private, free-market system. But in the WAS model, as in reality, the prices charged to some or all consumers can (and often will) be a matter of social or national policy. When such policy-driven prices are charged, the shadow values of water will reflect the net benefits of additional water given the policies adopted.
- Related to this is the fact that shadow values are *outputs* of the model solution, not inputs specified a priori. They depend on the policies and values put in by the user of the model.

It is important to note that the shadow value of water in a given location does not generally equal the direct cost of providing it there. Consider a limited water source whose pumping costs are zero. If demand for water from that source is sufficiently high, the shadow value of that water will not be zero; benefits to water users would be increased if the capacity of the source were greater. Equivalently, buyers will be willing to pay a nonzero price for water in short supply, even though its direct costs are zero.

A proper view of costs accommodates this phenomenon. When demand at the source exceeds capacity, it is not costless to provide a particular user with an additional unit of water. That water can only be provided by depriving some other user of the benefits of the water; that loss of benefits represents an opportunity cost. In other words, scarce resources have positive values and positive prices even if their direct cost of production is zero. Such a positive value—the shadow value of the water *in situ*—is called a "scarcity rent".

Where direct costs are zero, the shadow value of the resource involved consists entirely of scarcity rent. More generally, the scarcity rent of water at a particular location equals the shadow value at that location less the direct marginal cost of providing the water there.[13] Just as in a competitive market, a positive scarcity rent is a signal that more water from that source would be beneficial were it available.

Water shadow values and, accordingly, water scarcity rents depend upon the infrastructure assumed to be in place.

When water is efficiently allocated, as in the solution of the WAS model, the following relationships must hold. Equivalently, if they do not hold, then water is not being efficiently allocated. (All values are per unit of water.)

- The shadow value of water used in any location equals the direct marginal cost plus the scarcity rent. For water *in situ*, the shadow value is the scarcity rent.
- Water will be produced at a given location only if the shadow value of water at that location exceeds the marginal cost of production. Equivalently, water will only be produced from sources whose scarcity rents are nonnegative.
- If water can be transported from location *a* to location *b*, then the shadow value of water at *b* can never exceed the shadow value at *a* by more than the cost of such transportation. Water will actually be transported from *a* to *b* only if the shadow value at *b* exactly equals the shadow value at *a* plus the transportation cost. Equivalently, if water is transported from *a* to *b*, then the scarcity rent of that water will be the same in both locations.

This situation is illustrated in Figure 9.1, where water in a lake (*L*) is conveyed to locations *a*, *b*, and *c*. It is assumed that the only direct costs are conveyance costs. The marginal conveyance cost from the lake to *a* is denoted t_{La}; similarly, the marginal conveyance cost from *a* to *b* is denoted t_{ab}; and that from *b* to *c* is denoted t_{bc}. The shadow values at the four locations are denoted P_L, P_a, P_b, and P_c, respectively.

To see that the equations in Figure 9.1 must hold, begin by assuming that $P_a > P_L + t_{La}$ and that there is extra conveyance capacity from *L* to *a* at the optimal solution. Then transferring one more cubic meter of water from *L* to

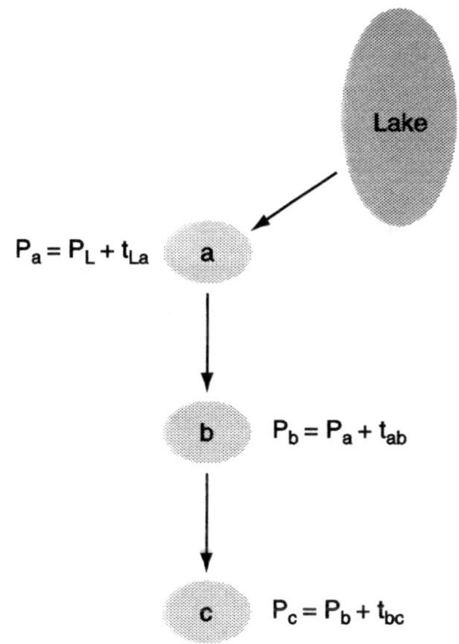

Figure 9.1 Efficient water allocation and water values.

a would have the following effects. First, since there would be 1 cubic meter less at *L* net benefits would decline by P_L, the shadow value of water at *L*. (That is what shadow values measure.) Second, since conveyance costs of t_{La} would be incurred, there would be a further decline in net benefits of that amount. Finally, however, an additional cubic meter at *a* would produce an increase in net benefits of P_a, the shadow value of water at *a*. Since, by assumption, $P_a > P_L + t_{La}$, the proposed transfer would increase net benefits; hence, we cannot be at an optimum.

Similarly, assume that $P_a < P_L + t_{La}$. Then too much water has been transferred from *L* to *a*, and transferring one less cubic meter would increase net benefits. Hence, again, we cannot be at an optimum.

It follows that, at an optimum, $P_a = P_L + t_{La}$, and a similar demonstration holds for conveyance between any two points.

Now, the first part of the demonstration just given requires the assumption that conveyance capacity is adequate to carry an additional cubic meter of water from *L* to *a*. Even were this not true, however, it would remain true that, in a generalized sense, $P_a = P_L + t_{La}$ at an optimum. Suppose that, with the conveyance system operating at capacity, it would increase net benefits if an additional cubic meter of water could be transferred from *L* to *a*. In this case, the capacity of the conveyance system would itself have a positive shadow value measuring the additional benefit that would occur if that capacity were increased by 1 cubic meter. If one includes that shadow value in t_{La} (adding it to the operating costs), then the relation, $P_a = P_L + t_{La}$ is restored.

Note that shadow values play a guiding role in the same way that actual market prices do in competitive markets. An activity that is profitable at the margin when evaluated at shadow values is one that should be increased. An activity that loses money at the margin when so evaluated is one that should be decreased. In the solution to the net-benefit maximizing problem, any activity that is used has such shadow marginal profits zero, and, indeed, shadow profits are maximized in the solution.

That shadow values generalize the role of market prices can also be seen from the inference that, where there are only private values involved, at each location, the shadow value of water is the price at which buyers of water would be just willing to buy and sellers of water just willing to sell an additional unit of water.

Of course, where social values do not coincide with private ones, this need not hold. In particular, the shadow value of water at a given location is the price at which the *user* of the model would just be willing to buy or sell an additional unit of water there. That payment is calculated in terms of net benefits measured according to the user's own standards and values.

This immediately implies how the water in question should be valued. Water *in situ* should be valued at its scarcity rent. That value is the price at which additional water is valued at any location at which it is used, less the direct costs involved in conveying it there.

Note that the propositions about profitable and unprofitable activities involve water being so valued. Those propositions take full account of the fact that using or processing water in one activity can reduce the amount of water available for other activities. The shadow values accompanying the maximizing solution include such opportunity costs, taking into account systemwide effects. (This is particularly important in using WAS for cost-benefit analysis.)

One should not be confused by the use of marginal valuation in all this (the value of an additional unit of water). The fact that people would be willing to pay much larger amounts for the quantity of water necessary for human life is important. It is taken into account in the optimizing model by assigning correspondingly large benefits to the first relatively small quantities of water allocated. But the fact that the benefits derived from the first units are greater than the marginal value does not distinguish water from any other economic good. It merely reflects the fact that water would be (even) more valuable if it were scarcer.

It is the scarcity of water and not merely its importance for existence that gives it its value. Where water is not scarce, it is not valuable.

9.6 Cost-benefit analysis of infrastructure and the treatment of capital costs

Before proceeding, it is important to understand how WAS can be used in the cost-benefit analysis of proposed infrastructure projects and how it handles capital costs (which can be quite substantial).

Consider the discussion of the lake and the conveyance line (Figure 9.1). Suppose that there were no existing conveyance line to carry water from the lake to city *a*. Suppose further that, if the WAS model were run without such a conveyance line, the resulting shadow values would be such that $P_a < P_L + t_{La}$, where t_{La} is the per-cubic-meter conveyance *operating* cost that would be incurred were such a conveyance line in place. Such a result would show that the conveyance line in question should not be built, because it would not be used even if it were. On the other hand, if the inequality were reversed, so that $P_a > P_L + t_{La}$, then that conveyance line might well be worth building—but whether it should be built would depend on the capital costs involved.

There are two ways of incorporating capital costs into the analysis of WAS results. One option is to impute an appropriate charge for capital costs to each cubic meter of water processed by the proposed facility. The second option is more direct. The WAS model is run with and without the proposed infrastructure.[14] This generates an estimate of the annual increase in benefits that would result from having such infrastructure in place. Given the estimated life of the projected infrastructure, the exercise can be repeated for the expected conditions of the various years of that life. Then, choosing a discount rate, the present value of such benefits is calculated and compared with the capital costs.

In MYWAS, the multiyear version of WAS (discussed briefly in section 9.10), the treatment of capital costs is more straightforward. They are treated as cash outlays incurred at various times when the project in question is begun or expanded. MYWAS then takes them directly into account in the calculation of the present value of total net benefits being maximized.

9.7 Conflict resolution: negotiations and the gains from trade in water permits

By using the WAS model to value water in dispute, water disputes can be monetized, and this may be of some assistance in resolving them. Consider bilateral negotiations between two countries, A and B. Each of the two countries can use its WAS tool to investigate the consequences to it (and, if data permit, to the other) of each proposed water allocation. This should help in deciding on what terms to settle, possibly trading off water for other, non-water concessions. Indeed, if, at a particular proposed allocation, A would value additional water more highly than B, then both countries could benefit by having A get more water and B getting other things that it values more. (Note that this does not mean that the richer country gets more water. That only happens if it is to the poorer country's benefit to agree.)[15]

Of course, the positions of the parties will be expressed in terms of ownership rights and international law, often using different principles to justify their respective claims. The use of the methods here described in no way limits such positions. Indeed, the point is not that the model can be used to help decide how allocations of property rights should be made. Rather the point is that water can be traded off for nonwater concessions, with the trade-offs measured by WAS.

In addition to monetizing water disputes, WAS can facilitate water negotiations by permitting each party, using its own WAS model, to evaluate the effects on it of different proposed water arrangements. As exemplified below, this can show that the trade-offs just discussed need not be large.

Water on the Golan Heights (see Figure 9.2) is sometimes said to be a major problem in negotiations between Israel and Syria, because the Banias River, which rises on the mountains of the Golan, is one of the three principal sources of the Jordan River. This question is evaluated by running the Israeli WAS model with different amounts of water.

In 2010, the loss of an amount of water roughly equivalent to the entire flow of the Banias River (125 million cubic meters annually) would be worth no more than $5 million per year to Israel in a year of normal water supply and less than $40 million per year in the event of a reduction of 30 percent in naturally occurring water sources.[16] These results take into account Israeli fixed-price policies toward agriculture.

Note that it is not suggested that giving up so large an amount of water is an appropriate negotiating outcome, but water is not an issue that should hold up a peace agreement. These are trivial sums compared to the Israeli gross domestic product of approximately $100 billion per year or to the cost of fighter planes.

Similarly, a few years ago, Lebanon announced plans to pump water from the Hasbani River—another source of the Jordan. Israel called this a *casus belli* and international efforts to resolve the dispute were undertaken. But whatever one thinks about Lebanon's right to take such an action, it should be understood that the results for the Banias apply equally well to the Hasbani. The effects on Israel would be fairly trivial.[17] Water, it would seem, is not worth war.

Monetization of water disputes, however, is neither the only nor, perhaps, the most powerful way in which the use of WAS can promote agreement. Indeed, WAS can assist in guiding water cooperation in such a way that all parties gain.

The simple allocation of water quantities, after which each party then uses what it "owns", is not an optimal design for a water agreement. Suppose that property rights issues have been resolved. Since the question of water ownership and the question of water usage are analytically independent, it will generally not be the case that it is optimal for each party simply to use its own water.

Instead, consider a system of trade in water permits—short-term licenses to use each other's water. The purchase and sale of such permits would be in quantities and at prices (shadow values) given by an agreed-on version of the WAS model run jointly for the two (or more) countries together. (The fact that such trades would take place at WAS-produced prices would prevent monopolistic exploitation.) There would be mutual advantages from such a

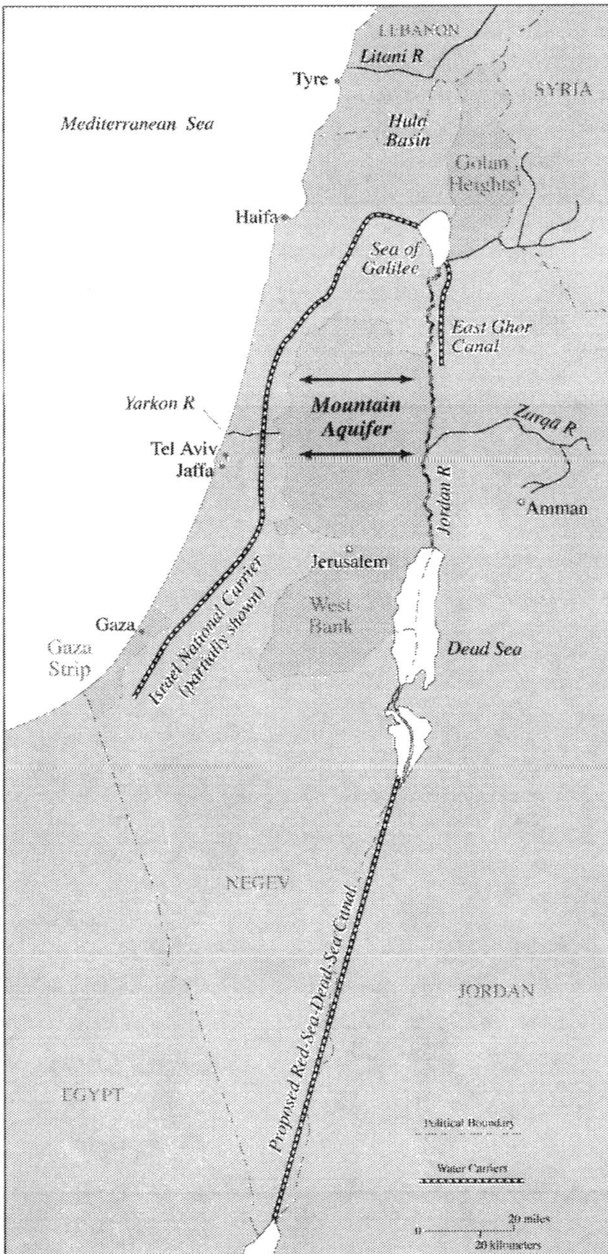

Figure 9.2 Partial regional map with water systems.

Note: This map shows the name of the large lake on the Jordan River as "Sea of Galilee". That lake is called "the Kinneret" by Israel and "Lake Tiberias" by Jordan and Palestine. The map is adapted from Wolf (1994: 27).

system, and the economic gains would be a natural source of funding for water-related infrastructure.

Both parties would gain from such a voluntary trade. The seller would receive money it values more than the water given up (otherwise, it would not agree); the buyer would receive water it values more than the money paid (otherwise, it would not pay it). While one party might gain more than the other, such a trade would not be a zero-sum game but a win–win opportunity.

Such gains are illustrated for Israel, Jordan, and Palestine below. They generally exceed the value of changes in ownership that reflect reasonable differences in negotiating positions. First, however, the stability of the grand coalition is considered.

9.8 Stability of the grand coalition

Consider a group of n countries all in the same river basin or otherwise connected as to water. Let G denote the grand coalition—where, by "coalition" is meant a WAS-guided cooperation arrangement as described above. It will be shown that G is stable—that is, that there exists no nonempty proper subset, S, of the countries such that a coalition of the members of S, using their own water resources, can lead to all such members being better off than they would be in G. (Of course, this applies to single-member subsets where a single country optimizes the use of its own water.)

> *Proposition:* The grand coalition is stable.
> *Proof:* The proof begins with:
> *Lemma:* The proposition is true for $n = 2$.
> *Proof of lemma:* Let the two countries be denoted by X and Y respectively. Suppose that the joint WAS solution calls for a transfer of water from some district of X to some district of Y. (For simplicity of notation, assume that each country has only one district.) Then it must be the case that, without such transfer, $P_X < P_Y + t^*_{XY}$, where P_X denotes the (pretransfer) shadow value of water at X, P_Y denotes the (pretransfer) shadow value of water at Y, and t^*_{XY} denotes the per-cubic-meter conveyance cost—this time including the appropriate capital charge.[18] Then, after the transfer from X to Y, the shadow value at X will be higher than P_X, and the shadow value at Y will be lower than P_Y, the respective pretransfer shadow values. But then both X and Y must have total benefits higher than they would be without the transfer—X because it has received money that it values more than the water given up, and Y because it has received water that it values more than the money it has paid.[19]
> *Proof of proposition:* Now, consider any nonempty proper subset, S, whose members form a coalition. Let T be the set of all countries that are not in S. Consider a coalition formed by the members of T. Then the water resources of the members of S are being used optimally as though

S were a single country, and similarly for T. But the grand coalition is nothing but a coalition formed by S and T. This is a bilateral agreement. Hence, by the lemma, both S and T must be at least as well off as they were alone. It follows that it cannot be true that the members of S (or the members of T) can all be better off than they would be in G.

Some comments need adding. First, note that the special water valuations of each of the countries are respected in any WAS-guided arrangement in which they participate.

Second, because the use of WAS shadow values, given those valuations, essentially restores the properties of a competitive market, the result just given is not surprising. It essentially mirrors the well-known result that the competitive equilibrium is always in the core of the economy.

Third, as already remarked in note 6, the result does not imply that every country does better in the grand coalition than it would in other coalitions. For example, in the case of Israel, Jordan, and Palestine, for plausible (but not inevitable) distributions of water ownership, Israel is a net seller and the other two countries are net buyers of water under WAS-guided arrangements.[20] It is therefore not surprising that Israel does better in the grand coalition than in bilateral ones, while the other two countries do (slightly) better in bilateral coalitions with Israel than in the grand coalition. Note, however, that, as should be expected from the proposition under discussion, if side payments are allowed, then all three countries can be made better off when a third one joins an existing bilateral coalition.[21]

9.9 Gains from WAS-guided cooperation: case studies involving Israel, Jordan, and Palestine

Results are now presented for Israel, Jordan, and Palestine, illustrating the gains from cooperation—and especially those from participation in the grand coalition. Attention is concentrated on predictions for 2010, those for 2020 being qualitatively similar.[22]

For perspective on the scarcity of water in these three countries, a great proportion is arid to semi-arid, receiving as little as 50 to 250 millimeters of rainfall per year—drier than Phoenix, Arizona. The estimated total renewable water supply for the region is approximately 2,400 million cubic meters per year, while water use averages 3,000 million cubic meters—an amount that is clearly unsustainable without some level of intervention.

The study concentrates on two sources of water that are the subjects of conflicting claims. These are the Jordan River and the so-called Mountain Aquifer (see Figure 9.2). Both of these are (very roughly) of equal size, each yielding about 650 million cubic meters a year. The Jordan River is claimed by all three countries, while the Mountain Aquifer is claimed only by Israel and Palestine. Since the gains from cooperation depend on assumptions about water ownership, results are obtained for selected varying assumptions about such ownership. *It must be emphasized that such assumptions are not meant as a political statement.* They are illustrative only.

For the Jordan River, ownership cases are examined as follows:

- Israel 92 percent; Jordan 8 percent; Palestine 0 percent (This is approximately the existing situation.)
- Israel 66 percent; Jordan 17 percent; Palestine 17 percent
- Israel 33 percent; Jordan 33 percent; Palestine 34 percent.

For the Mountain Aquifer, ownership cases are examined varying from Israel 80 percent–Palestine 20 percent (close to the existing situation) to Israel 20 percent–Palestine 80 percent by shifts of 20 percent at a time.[23]

First, results on two-way cooperation are presented. These differ from those in Fisher et al. 2005 primarily because of the expanded set of ownership assumptions.[24] It is assumed that, both for Israel and for Jordan, fixed-price policies are in place. For both countries, this means subsidies for agriculture and, for Israel, higher fixed prices for the other sectors. The Palestinian water price in each district is assumed to equal the corresponding shadow value.

9.9.1 Israel–Palestine bilateral cooperation

First, bilateral Israeli–Palestinian cooperation is considered. Figure 9.3 shows the gains from such cooperation as a function of the different ownership assumptions. (Note that cooperation here is not merely cooperation on the Mountain Aquifer but full cooperation in water.) Effectively, this and the following two figures represent slices of a multidimensional diagram. (In Figures 9.3, 9.4, and 9.5 Israel is represented by black, Jordan by white, and Palestine by white with diagonal stripes.)

Look first at Figure 9.3(i)—the case of an 80:20 Israel–Palestinian division of the Mountain Aquifer. As should be expected, in case A of Jordan River ownership, where Israel has most of the river and Palestine has none, it is Palestine that benefits most from cooperation—far more than relatively water-rich Israel. Further, the same is true for the other cases of Jordan River ownership. But an interesting phenomenon appears. As expected, Palestine gains more from cooperation in case A, where it owns no Jordan River water, than it does in case B, where it has 17 percent of the river. Moving on to case C, however, where Palestine has 34 percent of the river, the gains to Palestine once again increase, being very nearly as high ($171 million dollars per year) as in case A ($172 million dollars per year), even though Palestine has considerably more water in case C than in case A.

The reason for this is not hard to find. Palestine has considerably more water in case C than in case A, but Israel has considerably less. Both buyer and seller gain from WAS-guided cooperation, and, in case C, Palestine gains by selling water to Israel, despite its low share of the Mountain Aquifer.

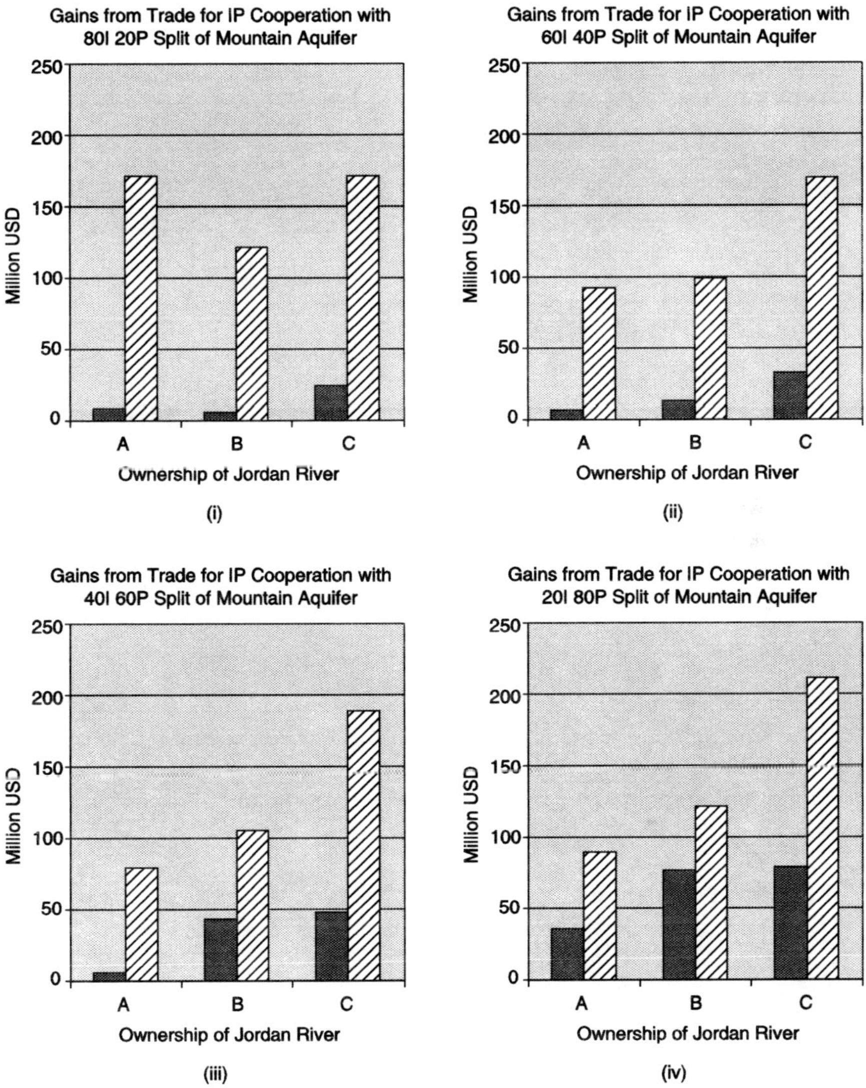

Figure 9.3 Gains from bilateral cooperation between Israel and Palestine in 2010.

Correspondingly, Israel, whose gains as a seller decrease slightly from case A to case B, has increased gains as a buyer in case C.

This phenomenon becomes even more pronounced in the other panels of Figure 9.3, in which Palestine has increased shares of the Mountain Aquifer. In each of those panels, Palestinian's gain from cooperation increases as it owns more and more Jordan River water; that is because it gains as a *seller*. Correspondingly, Israel gains as a buyer.

Moreover, the gains from cooperation are typically higher—and often much higher—than the gains from a shift in ownership of the Mountain Aquifer of 10 percent of that aquifer's water. This is true even if it is assumed that there is no cooperation when shifting ownership; it is even more true if cooperation is assumed, where the value of such a shift (to either party) ranges from about $5.5 million per year in case A to about $17 million per year in case C. The corresponding numbers for the noncooperation case range from $3 million to $36 million per year but are not the same for the two countries. And, where such values are highest, so are the gains from cooperation.

9.9.2 Israel–Jordan bilateral cooperation

Figure 9.4 shows the gains from bilateral cooperation between Israel and Jordan for 2010. Here the gains are generally lower than in the Israel–Palestine case. The interesting phenomena are as follows.

Where Israel has the lion's share of the Mountain Aquifer (Figures 9.4(i) and 9.4(ii)) and also 92 percent of the Jordan River (case A), there are no gains from cooperation at all. Neither Israel nor Jordan gains from trading Jordan River water—despite Jordan's small share thereof. Staying with the same shares of the Mountain Aquifer and moving to case B for the Jordan River, small gains do appear[25] and the gains are quite a bit larger in case C. In Figures 9.4(iii) and 9.4(iv), where Israel has 40 percent and 20 percent of the Mountain Aquifer, respectively, there are larger gains for Jordan, the increase being most noticeable in case B (although the gains remain largest in case C).

What is going on here is that, as Jordan owns more and more of the river, it pays both parties for Jordan to transfer water *to* Israel by *selling* water permits. This is even true when Israel owns 92 percent and Jordan only 8 percent of the river,[26] but is more pronounced in cases B and C as Israel's share of the river goes down and Jordan's goes up. This reflects the finding in Fisher et al. 2005 that Jordan has a major problem of conveyance infrastructure in using Jordan River water.

It is also interesting to note that Israel's gains from such purchases in case C are greatest when it owns 80 percent of the Mountain Aquifer (Figure 9.4(i)) and hence presumably needs the Jordan River water less than it does when it has less Mountain Aquifer water. The explanation is that Figure 9.4(i) shows a case in which Israel has sufficient water to make the shadow value of the Jordan River water lower than in the other cases. Hence the price that Israel pays for such water is also lower, and this benefits Israel (while correspondingly reducing Jordan's gains as a seller).

9.9.3 Jordan–Palestine bilateral cooperation

The case of bilateral cooperation between Jordan and Palestine can be handled quickly (and requires no figure to illustrate it). Here the only

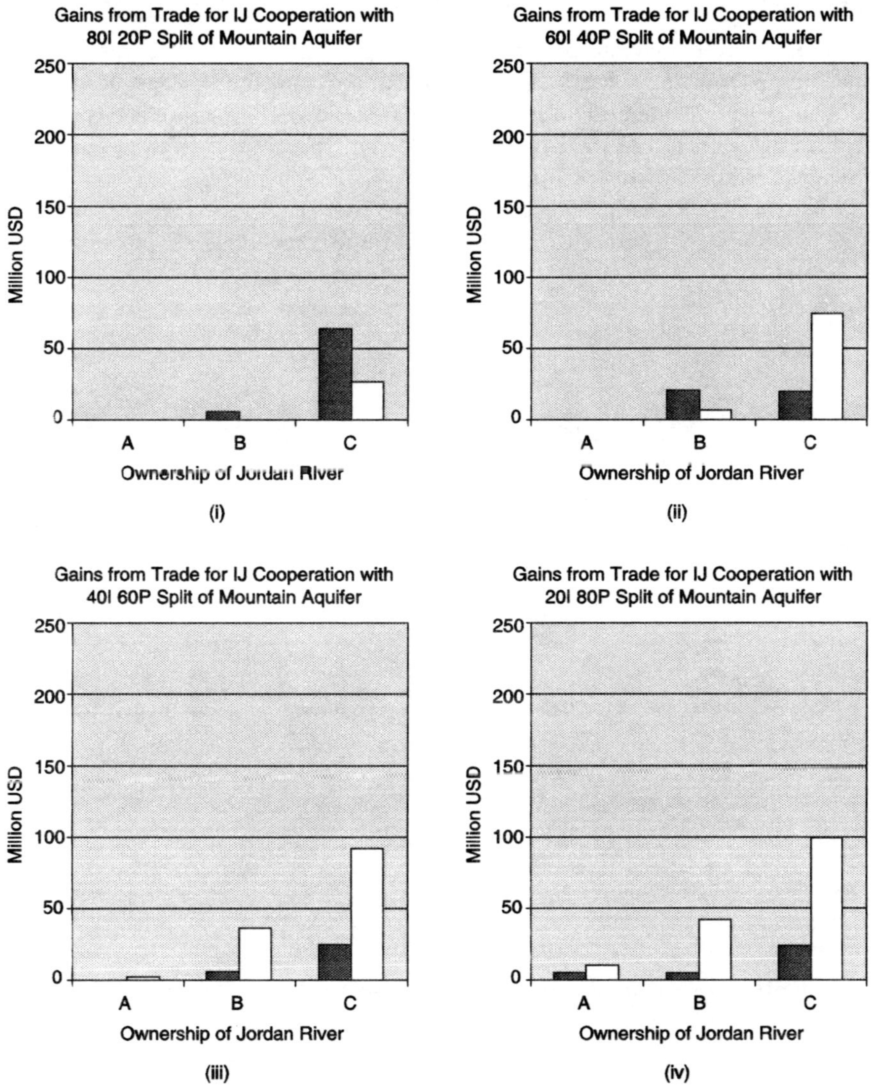

Figure 9.4 Gains from bilateral cooperation between Israel and Jordan in 2010.

possible beneficial trades are those that involve Jordan River water—and both countries have relatively little water to trade. As a result, the only case in which gains from cooperation are not totally negligible is that where Palestine owns only 20 percent of the Mountain Aquifer and none of the Jordan River (case A). There, even though Jordan only owns 8 percent of the river, it gains \$10 million per year by selling a permit to Palestine to use river water, and Palestine gains \$12 million per year by buying it.

9.9.4 Trilateral cooperation (grand coalition)

Figure 9.5 gives the results for trilateral cooperation. Here what stands out is that, in general, the largest gains are Palestinian and the smallest Jordanian. Not surprisingly, generally the less water Israel owns, the more it has to gain from cooperation. On the other hand, Jordan gains more from cooperation the more water it owns—selling permits to Israel.

Like Jordan, Palestine presents a picture that is somewhat more mixed. It also tends to benefit more from cooperation the larger its share of the Jordan,

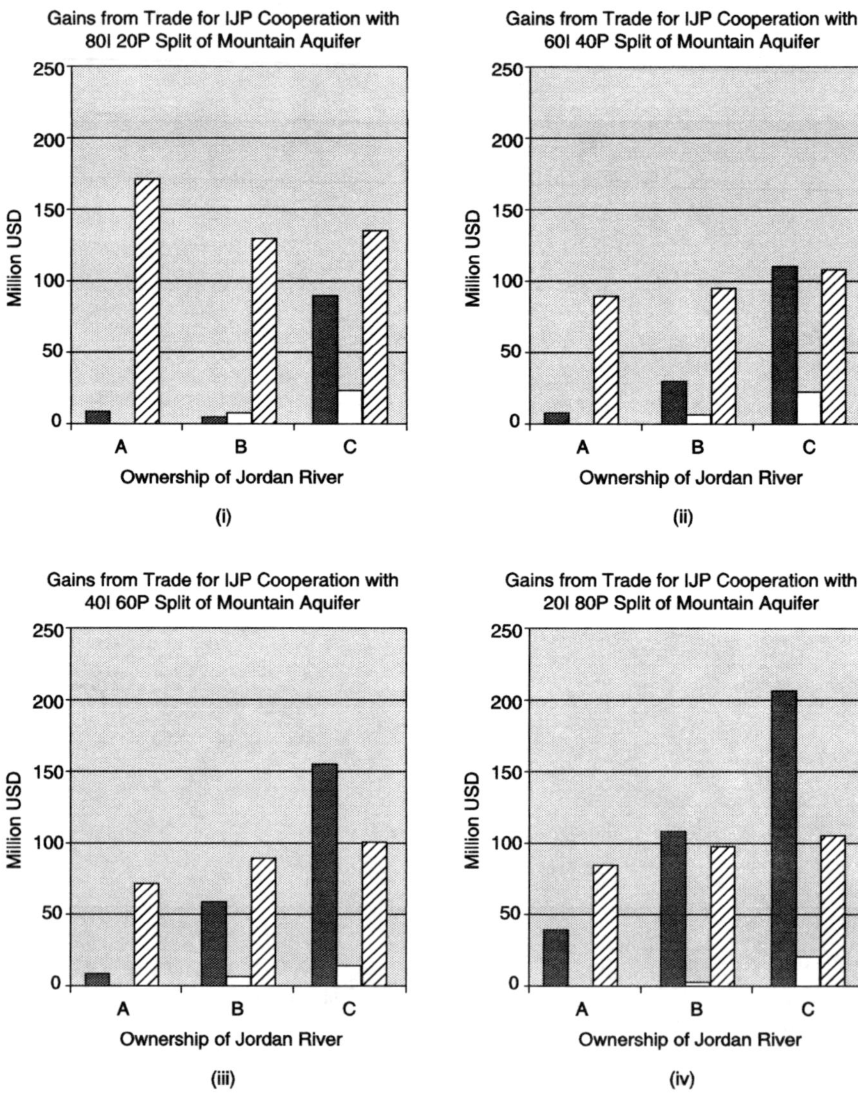

Figure 9.5 Gains from trilateral cooperation (grand coalition) in 2010.

selling permits to use river water to Israel. On the other hand (Figure 9.5(i)), when Palestine owns relatively little Mountain Aquifer water, it also benefits as a buyer.

9.10 Conclusion and policy implications

So far as Israel, Jordan, and Palestine are concerned, the most important general conclusion from all these cases should be clear. WAS-guided cooperation in water would benefit all parties—Israel and Palestine the most. The gains from such cooperation exceed those that can be had from obtaining reasonable amounts of ownership shifts.

Moreover, such cooperation would benefit the parties in other ways. For one thing, runs of the WAS model strongly suggest that Israel and Palestine would both benefit from the construction of an effluent treatment plant in Gaza with the output sold to Israel for use in agriculture in the Negev desert, which would provide needed investment for Gaza and cheaper water for Israel. This means that Israel has a positive economic incentive to assist with the construction of such a plant. That would be a positive gesture of cooperation, and one that does not impinge on the really difficult core issues that separate the two parties.

Further, the usual quantity-oriented water treaty may be adequate for a time, but as populations, economies, and climate change it can become dated and place strains on a peace arrangement. This certainly can happen in the region under discussion. WAS-guided cooperation provides a flexible method for adjusting water allocations with all parties benefiting from the adjustments.

Is such cooperation likely in the region? It is not easy to tell. On the one hand, for the last several years, the Israeli–Palestinian situation has been dominated by conflict over nonwater issues. On the other hand, that conflict has not prevented all cooperation in water. Even during the al-Aqsa Intifada, which began in September 2000, the Israeli and Palestinian water authorities met regularly to discuss water issues—particularly because of the substantial drought conditions then prevailing. Further, there were no attacks on the water system of either party.

Israel and Jordan signed a water (quantity) treaty in 1994, and have generally lived up to its terms, even though there have been various difficulties.

Nevertheless, cooperation has been limited. Israel's attitude towards providing more water for the Palestinians has generally been that it will assist the Palestinians in finding new sources of water, but that it will not change the allocations of existing water resources. At one point, Israel even suggested providing, for the southern West Bank, water that would be pumped uphill a considerable distance from a desalination plant in Ashkelon (Israel). That would be a very inefficient way of providing water to that area—much less efficient than keeping the desalinated water in Israel and permitting the Palestinians to pump more water from the Mountain Aquifer than they have heretofore been permitted to do. That suggests that symbolism plays a considerable role in these matters.[27]

Nevertheless, there are some hopeful signs. Partly because of the efforts of the Water Economics Project and others, the Israeli government appears to have come to understand that water ownership issues can be considered a matter of money and that, in particular, desalination provides a way out of the problem. Large desalination facilities have been constructed with more planned, and there is discussion of the fact that there may now be enough water to avoid the water conflict problem. Even though such a solution is inferior to WAS-guided cooperation, and even though such cooperation would still be useful when desalination becomes a necessity some years from now, solving the ownership issue with the aid of desalinated water is better than not solving it at all.

The prospects for the use of WAS and even WAS-guided cooperation are certainly not dead, however. The Palestinian and Lebanese water authorities have both expressed interest in using WAS at least for their own domestic management purposes, and there has been some activity at the Israeli Water Commission in this regard. Further, the Lebanese have indicated an interest in the use of WAS to prepare for negotiations with Lebanon's neighbors, and, at least before the Israel–Hezbollah war of 2006, that interest included negotiations with Israel.

For Lebanon, as for all countries either in the Middle East or elsewhere, the use of WAS can serve two purposes. First, WAS provides a powerful tool for infrastructure planning. Second, it provides a way of resolving water conflicts, changing water negotiations from a zero-sum game to a win–win situation, whether the conflict is between competing uses domestically, or between countries.

As suggested earlier, the usefulness and power of WAS is being enhanced by the construction of a multiyear version (MYWAS). In that version, one can specify a menu of possible infrastructure projects, together with their costs and useful lives. MYWAS will then give results as to which (if any) of the specified projects should be built, in what order, at what time, and to what capacity. Those results can be examined for their dependence on predictions as to such things as population growth, climatic conditions, and technological advances.

Further, MYWAS can be easily used to examine issues of storage (in aquifers or reservoirs). And it can give guidance on aquifer management—advising on overpumping in dry years and replenishment in wet ones. Finally, MYWAS can be used to study the effects of climatic uncertainty.

WAS and MYWAS have recently (mid-1996) been adopted by the Energy, Water, and Environment Community—a high-profile program on environmental cooperation in the Middle East coordinated academically at the London School of Economics and Political Science and politically sponsored by Prague Forum 2000, a leading European vehicle of track two (informal, nongovernmental) diplomacy. Efforts are currently under way to feature WAS and MYWAS in a major round of awareness raising on water optimization in the Middle East, seeking long-term political sponsorship from sympathetic European Union States and institutions.

While WAS has been first developed for the Middle East (in particular, for Israel, Jordan, and Palestine), its methods and use are not restricted to that region. Domestically, the WAS (or MYWAS) model provides a powerful tool for water management, especially for infrastructure planning and water policy with respect to allocation, pricing, and subsidies. Thus WAS is relevant for all countries, even those with well-developed infrastructure.

Beyond that, however, there are disputes over water ownership and allocations all over the globe. These are not only international disputes: sectoral and subnational disputes are increasing as well. WAS methodology provides a way of resolving those disputes to the advantage of all parties.

Notes

1. We are indebted to Ariel Dinar for raising the questions that prompted this chapter and for urging us to write it.
2. See also our expository article (Fisher and Huber-Lee 2006).
3. For examples that go beyond mere cost minimization, see Brown and McGuire 1967; Dandy et al. 1984; and McCarl 1999. The model that appears most similar to WAS is the CALVIN model, an optimizing water model for California developed at the University of California, Davis (see for example Newlin et al. 2002; Jenkins et al. 2003; and Jenkins et al. 2004).
4. We use the term "Palestine" and refer to Palestine as a "country" both for convenience and out of respect for our Palestinian colleagues.
5. Two-way cooperation between Palestine and Jordan was not investigated. This is considered in subsection 9.9.3.
6. Note that this result does not imply that every country in the grand coalition does better than it would in some partial coalition. This problem is discussed in Fisher et al. 2005: 216–17.
7. This is an application of the well-known Coase Theorem of economics. See Coase 1960.
8. The pioneering version of such a model (although one that does not explicitly perform maximization of net benefits) is that of Eckstein et al. 1994.
9. There is a large literature and much discussion about *actual* free water markets, but it is crucial to understand that what is proposed here is *not* such a market. Later in the chapter a system of water trading among (in this case) countries is recommended, but even such trading is not at freely bargained prices but rather at prices and quantities prescribed by joint operation of an optimizing model. Hence, the water market literature is not discussed explicitly.
10. A multiyear version has now largely been developed and is briefly discussed in section 9.10.
11. The total amount that consumers are willing to pay for an amount of water, Q*, is measured by the area up to Q* under their aggregate demand curve for water. Note that "willingness to pay" includes ability to pay. The provision of water to consumers that are very poor is taken to be a matter for government policy embodied in the pricing decisions made by the user of WAS.
12. If the user of the model—for example the government of a country—would value additional water in a particular location more than would private buyers, then the shadow value reflects that valuation.

13. If this calculation gives a negative figure, then the scarcity rent is zero, and water is not scarce at the given location.
14. A similar method can be used to analyze different proposed water policies.
15. If trading off ownership rights considered sovereign is unacceptable, the parties can agree to trade short-term permits to use each other's water. See later in this section.
16. All money values are in 1995 US dollars.
17. Of course, the question naturally arises as to what the effects on Syria and Lebanon, respectively, would be in these two situations. Without a WAS model for those two countries, that question cannot be answered. Both countries would surely profit from such a model.
18. The simplest case is to assume that the conditions of the year being studied will continue for the life of the project. More realistic cases only complicate the discussion without changing the nature of the proof.
19. Note that this is true regardless of how the transfer costs, t^*_{XY}, are allocated. *Y* ultimately pays those costs—either directly to *X* or by building and operating the conveyance system itself—but the surplus benefits generated are large enough to cover such payments and still leave *Y* better off than it would have been without the transfer.
20. This does not mean that sales go in only one direction. It is typically found that the direction of sales differs at different locations.
21. See Fisher et al. 2005: 216–17.
22. At this time, it is appropriate for us to correct a minor error in Fisher et al. 2005 pointed out to us by Yoav Kislev, whom we thank (not desiring to shoot the messenger). In that work, all prices are meant to be at the city gate, so to speak. But, when dealing with Israeli fixed-price policies for households and industry, we mistakenly set the fixed price for urban and industry at the prices charged at the user tap. We also calibrated the household and industry demand curves using those too-high prices. Correction of this slip has almost no effect on our important qualitative statements (indeed, it strengthens the finding that desalination is not yet an efficient technology except in times of extreme drought), but it does affect numerical results. We have corrected the mistake in the results discussed in this section.
23. The Mountain Aquifer in fact consists of several subaquifers. No attempt is made to divide ownership except in the arbitrary manner described in the text.
24. In Fisher et al. 2005, the gains from cooperation for each of the two water sources are examined separately while assuming that Israel had most or all of the other source.
25. In the 80:20 split of the Mountain Aquifer, the gain to Jordan in case B rounds off to zero.
26. In the 40:60 split of the Mountain Aquifer, Jordan gains $1 million a year—too small to show up in Figure 9.3(iii).
27. Incidentally, a country that withholds a particular water source from a WAS-guided agreement hurts both itself and its cooperating partners. WAS can be used to evaluate such effects. See Fisher et al. 2005: 217–18.
28. The notation used is restricted to this Appendix.
29. Note that the first term of the objective function is the integral of the *inverse* demand function:

$$P_{id} = B_{id} \times (QD_{id} + QFRY_{id})^{ALPHA_{id}}.$$

References

Brown, G. and McGuire, C. (1967) "A socially optimum pricing policy for a public water agency", *Water Resources Research*, 3: 33–43.

Coase, R. (1960) "The problem of social cost", *Journal of Law and Economics*, 3: 1–44.

Dandy, G., McBean, E., and Hutchinson, B. (1984) "A model for constrained optimum water pricing and capacity expansion", *Water Resources Research*, 20: 511–20.

Eckstein, O. (1958) *Water-Resource Development: The Economics of Project Evaluation*, Cambridge, MA: Harvard University Press.

Eckstein, Z., Zackay, D., Nachtom, Y., and Fishelson, G. (1994) "The allocation of water resources between Israel, the West Bank and Gaza: an economic analysis", *Economic Quarterly*, 41: 331–69 (Hebrew).

Fisher, F.M. and Huber-Lee, A. (2006) "Economics, water management, and conflict resolution in the Middle East and beyond", *Environment*, 48(3): 26–41.

Fisher, F.M., Huber-Lee, A., Arlosoroff, S., Eckstein, Z., Haddadin, M., Hamati, S.G., Jarrar, A., Jayyousi, A., Shamir, U., and Wesseling, H., with a special contribution by Salman, A.Z. and Al-Karablieh, E.K. (2005) *Liquid Assets: An Economic Approach for Water Management and Conflict Resolution in the Middle East and Beyond*, Washington: RFF Press.

Hirshleifer, J., DeHaven, J.C., and Milliman, J.W. (1960) *Water Supply: Economics, Technology, and Policy*, Chicago: University of Chicago Press.

Jenkins, M.W., Lund, J.R., and Howitt, R.E. (2003) "Using economic loss functions to value urban water scarcity in California", *Journal of the American Water Works Association*, 95: 58–70.

Jenkins, M.W., Lund, J.R., Howitt, R.E., Draper, A.J., Msangi, S.M., Tanaka, S.T., Ritzema, R.S., and Marques, G.F. (2004) "Optimization of California's water supply system: results and insights", *Journal of Water Resources Planning and Management*, 130: 271–80.

Maass, A., Hufschmidt, M.M., Dorfman, R., Thomas, H.A., and Marglin, S.A. (1962) *Design of Water Resource Systems*, Cambridge, MA: Harvard University Press.

McCarl, B. (1999) "Limiting pumping from the Edwards Aquifer: an economic investigation of proposals, water markets and spring flow guarantees", *Water Resources Research*, 35: 257–68.

Newlin, B.D., Jenkins, M.W., Lund, J.R., and Howitt, R.E. (2002) "Southern California water markets: potential and limitations", *Journal of Water Resources Planning and Management*, 128(1–2): 21–32.

Rogers, P.P. and Fiering, M.B. (1986) "Use of systems analysis in water management", *Water Resources Research*, 22: 146S–158S.

Saliba, B.C. and Bush, D.B. (1987) *Water Markets in Theory and Practice: Market Transfers, Water Values, and Public Policy*, Boulder, CO, and London: Westview Press.

Wolf, A. (1994) "A hydropolitical history of the Nile, Jordan, and Euphrates River basins", in A.S. Biswas (ed.) *International Waters of the Middle East from Euphrates–Tigris to Nile*, Oxford: Oxford University Press.

Appendix: the mathematics of the WAS model

The WAS model is written in the generalized algebraic modeling system (GAMS) language. As the objective function is not linear, the MINOS nonlinear operating system is used to solve the WAS model. The model is presented below in the standard form for optimization, namely, the objective function followed by the constraints. In mathematical terms,[28] the model is as follows:[29]

$$Max \quad Z = \sum_i \sum_d \left(\frac{B_{id} \times (QD_{id} + QFRY_{id})^{ALPHA_{id}+1}}{ALPHA_{id} + 1} \right) - \sum_i \sum_s (QS_{is} \times CS_{is})$$

$$- \sum_i \sum_j (QTR_{ij} \times CTR_{ij}) - \sum_i \sum_j (QRY_{ij} \times CR_{ij})$$

$$- \sum_i \sum_j (QTRY_{ij} \times CTRY_{ij}) - \sum_i \sum_j [CE_{id} \times (QD_{id} + QFRY_{id})]$$

Subject to:

$$\sum_d QD_{id} = \sum_s QS_{is} + \sum_j QTR_{ji} - \sum_j QTR_{ij} \quad \forall\, i$$

$$QRY_{id} = PR_{id} \times (QD_{id} + QFRY_{id}) \quad \forall\, i,d$$

$$\sum_d QFRY_{id} = \sum_d QRY_{id} + \sum_j QTRY_{ji} - \sum_j QTRY_{ij} \quad \forall\, i$$

$$(QD_{id} + QFRY_{id}) \geq \left(\frac{PMAX}{B_{id}} \right)^{1/ALPHA_{id}} \quad \forall\, i,d$$

With the following bounds:

$$QS_{is} \leq QSMAX_{is} \quad \forall\, i,s$$

$$PR_{id} \leq PRMAX_{id} \quad \forall\, i,d$$

All variables positive.
Where:

Indices

i = district (Israel: I1,I3 . . . I15; Jordan: J1 . . . J8; Palestinian Authority: P1 . . . P10; Golan: GOL; Jerusalem: JER)

d = demand type (urban, industrial, or agricultural)
s = supply source or steps (S1 . . . S5)

Parameters

ALPHA_{id}	Exponent of inverse demand function for demand d in district i
B_{id}	Coefficient of inverse demand curve for demand d in district i
CE_{id}	Unit environmental cost of water discharged by demand sector d in district i ($\$/\text{m}^3$)
CR_{id}	Unit recycling cost of water supplied from demand sector d in district i ($\$/\text{m}^3$)
CS_{is}	Unit cost of water supplied from supply step s in district i ($\$/\text{m}^3$)
CTR_{id}	Unit cost of water transported by demand sector d in district i ($\$/\text{m}^3$)
CTRY_{id}	Unit cost of recycled water transported by demand sector d in district i ($\$/\text{m}^3$)
PMAX_{id}	Maximum price of water from demand sector d in district i
PRMAX_{id}	Maximum percent of water from demand sector d that can be recycled in district i
QSMAX_{is}	Maximum amount of water from supply step s in district i (Mm^3)
P_{id}	Shadow value of water for demand sector d in district i (computed) in \$

Variables

Z	Net benefit in from water in million \$
QS_{is}	Quantity supplied by source s in district i in Mm^3
QD_{id}	Quantity demanded by sector d in district i in Mm^3
QTR_{ij}	Quantity of freshwater transported from district i to j in Mm^3
QTRY_{ij}	Quantity of recycled water transported from district i to j in Mm^3
QRY_{id}	Quantity of water recycled from use d in district i in Mm^3
QFRY_{id}	Quantity of recycled water supplied to use d in district i in Mm^3
PR_{id}	Percentage of water recycled from sector d in district i

Note: Mm^3 = million cubic meters.

Figures 9.6A and 9.6B give an illustration of continuity for freshwater and recycled water, respectively, as given in the first two constraint equations listed above.

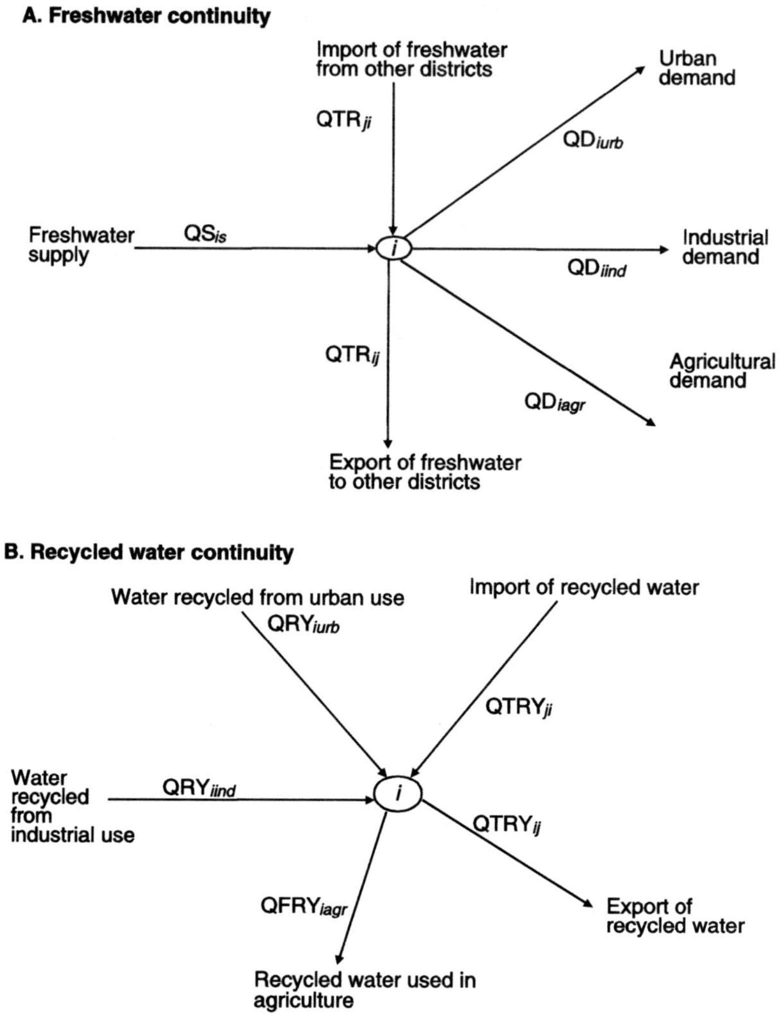

Figure 9.6 Continuity of flows at node *i*.

10 Experimental insights into the efficiency of alternative water management institutions

Aurora García-Gallego, Nikolaos Georgantzís, and Praveen Kujal[1]

An increasing number of local governments all over the world are adopting regional hydrological plans aimed at integrated management of their water resources. The effect of such integrated management plans on the final quality of water is a central but underinvestigated issue in the ongoing debate concerning the desirability of centralization of the decisionmaking process and the use of market mechanisms to allocate water among its users. Based on the results obtained from a series of experimental sessions, this chapter presents insights into the efficiency and quality-enhancing effects of centralized public management using block rate bidding in a water market with uniform market-clearing prices.

10.1 Introduction

Water is a basic input to many human activities. Its allocation among different users has traditionally depended on centralized, publicly controlled management mechanisms that significantly underestimate the true value of water to most of its users. The increasing threat of drought and water shortages, even in the most developed areas of the world, has forced many governments to adopt innovative ways of managing their water resources. Among the numerous institutional innovations, privatization of infrastructure[2] and private management[3] of water resources have attracted the attention of researchers and policymakers. This chapter focuses on the performance of private and public water management institutions, taking the underlying infrastructure as given.

Several countries have already implemented measures of water utility privatization. In the European Union, the Water Framework Directive (European Commission 2004) explicitly refers to the need for improved management to help ensure the quality and sustainability of water resources.[4] A central point of the European Union directive is recognition of the need for groundwater body identification (European Commission 2005). A groundwater body is defined as a set of aquifers whose management is undertaken through a unified decisionmaking process. Clear examples of such unified management of aquifers can be found in Finland and Norway, which are confronted with a huge number of groundwater bodies because of their specific geological characteristics. Finland has identified about 3,700 groundwater bodies. In Norway the grouping of individual aquifers (8,000–11,000) into less than 1,000 groundwater bodies is based on the analogous principle and has the aims of reducing administrative costs and enhancing water management efficiency.

Whereas the management of water resources within the geographic and administrative limits of a river basin has been analyzed by economists in the framework of common pool extraction, the role of aquifer integration has not been studied so far. In fact, despite the clear message of the Water Framework Directive and the implementation of integrated aquifer management by the aforementioned Nordic countries, the joint management of different water resources has found little if any administrative support in the European Union, especially in countries where the tradition often defines water as a regional or national property.

Despite numerous technical and institutional barriers, water transfers across regions with different endowments and the use of market mechanisms are two innovative (and controversial) strategies that are increasingly being adopted by water managers all over the world. A successful example of jointly managing water resources from different sources in order to satisfy different types of users can be found in California.[5] Apart from a centralized plan of action, a market mechanism has been established in California, the volume of which in 2001 was well above 10 times its initial volume in 1985, although it still accounts for only 3 percent of the state's water use. Even small suppliers participate in this market, contributing with their excess water endowments, or even with the entire resource they own, idling their crops to sell water.

The debate concerning the desirability of the aforementioned management strategies has often focused on the issue of water quality arising from different water allocation mechanisms. Concerning this, two issues are of special interest to this study: the heterogeneity of users' needs, and the crucial effects of decentralized actions on water quality.

Both the California Water Plan update of 2005 and the European Union Water Framework Directive discuss these two points. Concerning the first point, the California Water Plan update[6] states that:

> High quality water sources can be used for drinking and industrial purposes that benefit from higher quality water, and lesser quality water can be adequate for some uses, such as riparian streams with plant materials benefiting fish.
>
> (Department of Water Resources 2005, vol. 2, ch. 12)

With respect to the second point, which is a central issue dealt with in the present chapter, the California Water Plan update states:

> Many water management actions, such as conjunctive use, water banking,

water use efficiency, and water transfers, intentionally or unintentionally, result in one type of water quality traded for, or blended with, another.
(Department of Water Resources 2005, vol. 2, ch. 12)

The importance of mixing multiple water qualities is further highlighted by the increasing use of recycled water. Regarding the role of recycled water in providing water of sufficient quality to the users, the California Water Plan update mentions that:

> Further, some new water supplies like recycled water can be treated to a wide range of purities that can be matched to different uses. The use of other water sources, again, like recycled water, can serve as a new source of water that substitutes for uses not requiring potable water quality.
> (Department of Water Resources 2005, vol. 2, ch. 12)

Needless to say, as in the aforementioned example of the two Nordic countries, defining the scope of aquifer integration and identifying the right type of agent who should be responsible for the decisionmaking process are two central questions to be answered from a number of different viewpoints: cultural, political, social, and economic. This chapter focuses on the economic aspects of the decisionmaking process.

An ideal solution to the problem described above would require a heavily computerized and, necessarily, centralized engineering approach to the management of adjacent aquifer systems. From a theoretical point of view, this approach is feasible as long as the specification of the underlying demand model and the statistical properties of the stochastic elements on the supply side are known to the decisionmaker. However, this is a rather unrealistic desideratum, especially because policymakers, even when assisted by complex computerized decisionmaking algorithms, cannot have but approximations of the underlying demand model. Thus, even if the stochastic elements affecting the supply of water were ruled out, the decisionmaking process should not be expected to be the result of applying explicit optimization rules. In such cases, adaptive decisionmaking[7] based on trial-and-error strategies can be seen as the real-world counterpart of explicit optimization assumed by optimal control theory. However, in the presence of such an adaptive decisionmaking process, there is no clear-cut prediction concerning the performance of different water management institutions. For example, the obvious advantages from centralized public management may be offset by decision errors arising from the increased complexity of the problem faced by the decisionmaker.

Decentralized management, and even the use of privately participated water markets, might offer an appealing alternative with much weaker requirements concerning the rationality of decisionmakers. In fact, a decentralized solution would reduce the complexity of the partial decisionmaking problems faced by private profit-driven agents. But, again, there is no readily available prediction on the degree to which such a reduction in the complexity of the problem faced by each decisionmaking entity would offset the social welfare loss arising from the suboptimality induced by the decentralization of incentives. Furthermore, even in the case of resources that are fully controlled by the public sector, allocations are implemented using price mechanisms, which are the only feasible way of eliciting an individual user's need for an extra unit of water. Thus, a further nontrivial problem that arises concerns the interaction between each one of the aforementioned management alternatives and the price mechanisms usually adopted in order to allocate different quantities of water to different users. An underinvestigated question arising in this context is which institution can best allocate water to different users, accounting for each institution's effect on the quality resulting from a mixture of different water qualities.

This chapter experimentally investigates the case in which water quality is endogenously determined as the combination of water extracted from two sources of different qualities. Without replicating any real-world market, insights are gained into the advantages and disadvantages of alternative water management institutions, considering two different scenarios of centralization—private and public—and two different levels of decentralized market competition.

Although not addressing the issue of water quality, a number of authors have studied various aspects of water markets and water management. Moench (1992) showed that competitive water withdrawal can lead to overexploitation. This was previously argued by Gordon (1954), who showed that complete rent dissipation may occur from the exploitation of an open-access resource. This result was confirmed and generalized by the findings of Mason and Phillips (1997). Walker et al. (1990), Walker and Gardner (1992), and Gardner et al. (1997) also focused on the issue of water resource exploitation in a common pool framework. In a more recent study, Murphy et al. (2000) used laboratory experiments to design "smart" computer-assisted markets for water inspired by computer-assisted market institutions developed by McCabe et al. (1989, 1991). A smart market allows decentralized agents to submit messages to a computer dispatch center. The center then computes prices and allocations by applying an optimization algorithm that maximizes the possible gains from exchange. Using California as a case study, Murphy et al. (2000) tested alternative institutional arrangements for a computer-assisted spot market. They showed that the smart uniform price double auction yields highly efficient outcomes in a thin market characterized by a limited set of trading opportunities. Although their study is the only one that is potentially applicable to integrated aquifer management, their framework accommodates multiple locations but it does not consider any quality dimension.

The studies mentioned above clearly show that experiments can be a useful test bed to study alternative property right mechanisms (Walker et al. 1990; Walker and Gardner 1992), or to undertake a direct test of an existing market mechanism (Murphy et al. 2000). Experiments can replicate important characteristics of existing markets and test them in a laboratory setting at minimal cost to the regulator.

The results discussed in this chapter are based on an overview of experimental results reported in three different manuscripts. They reflect different phases of a broader research project undertaken at the experimental laboratories of the Universities of Valencia[8] and Castellón[9] in Spain. In the first paper, Georgantzís et al. (2004) present an experimental design with decentralized action on the demand side. They highlight the difficulties encountered when the allocation of water to its users takes place in a fully decentralized and possibly inefficient way. The second and third papers present an improved market design where, once water prices are posted by private or public suppliers, water allocation is implemented by a smart market-clearing mechanism guaranteeing maximal surplus on the demand side. In the second paper, García-Gallego et al. (2005) study the same management institutions as in Georgantzís et al. 2004, with the exception that the demand side is simulated to achieve maximal consumer surplus. In the third paper, García-Gallego et al. (2006) consider an alternative market mechanism that focuses on market competition versus coordination. These experiments replicate the private duopoly case, implementing a pre-play coordination mechanism inspired by games with confirmed strategies. The results indicate that centralized management of the resources by a public entity leads to the highest levels of economic efficiency, lowest prices for users, and the highest quality-to-price ratio.

The experimental markets whose results are reported here represent an aquifer system with extraction from two different sources of water. The terminology adopted is specific to the case of potentially integrated aquifers that, adopting the European Union Water Framework Directive definition, may form a unified groundwater body. However, the framework adopted is applicable to a broader spectrum of real-world cases including, for instance, the case of mixing different types of recycled water. Such situations would lead to endogenous water quality supplied to different types of users. Furthermore, a costly depuration procedure is introduced, which is activated at the consumer's location when quality falls below the potable standard. To simplify the analysis, it is assumed that the quality of farm water has no such restriction.

Four different market structures (private monopoly, duopoly, coordinated duopoly, and public monopoly) and their effects on market outcomes are studied, paying special attention to the resulting stock depletion, quantity supplied to the users, and the quality-to-price ratio. The four management scenarios are chosen to compare, on the one hand, private to public centralized management of the integrated aquifer system and, on the other hand, a more competitive to a less competitive decentralized market structure. The results reported below indicate that a social welfare-driven monopolist offers the highest quality-to-price ratio and exploits the resource at a faster rate than a private monopoly or duopoly. However, the most stable stocks correspond to the private management scenarios (monopoly and duopoly), although the corresponding stock levels are inefficiently high.

The chapter is structured as follows. Section 10.2 discusses the features of the aquifer system. Section 10.3 presents the experimental design. Section 10.4 discusses the results. Section 10.5 concludes.

10.2 Underlying theoretical framework

In this section the theoretical framework is discussed, focusing on the intuition underlying the formal assumptions of the model. There are two renewable stocks, of high quality (S_H) and low quality (S_L), from which water may be extracted. Quality of water in an aquifer may be lower because of marine intrusion, or because of infiltration of fertilizer from agriculture. The two qualities are assumed to be constant over time. However, any intermediate quality may be supplied to the consumers as a result of mixing water from the two sources. Note that, usually, quality choice is studied in the context of product differentiation models. In these models, product quality is directly and consciously chosen in a way that takes into account the competition-reducing effect of product differentiation. Thus, quality choice determines the fierceness of price competition. In the present framework, the causality is reversed. Final product quality is the result of price-setting behavior. Thus, strategic interaction determines the final quality, which is unique and not directly controlled by any single agent. Because of this, and the dynamic nature of this framework, there are no theoretical predictions that are immediately applicable to the case studied here.

For the sake of simplicity, the recharge to the respective basin is assumed to be deterministic and constant. This assumption does not imply that the extra uncertainty arising from fluctuating supply conditions observed in the real world is a problem of negligible importance. However, it has been ruled out from the analysis in order to reduce the complexity of the microworld managed by the subjects and isolate the dynamic and complex elements of the water market studied from the nonstrategic uncertainty on the supply side. Thus, the aim of the experiments reported here is to illustrate possible advantages and disadvantages of the alternative institutions in isolation from the aforementioned types of supply-side uncertainty. The inflow to the basins is assumed to cease when the storage capacity of the aquifer is reached. That is, once the maximum storable stock is reached, extra water inflow is lost. The return flow of consumed water is assumed to be negligible. Thus, changes in the stocks exclusively arise from extraction and recharge.[10] Extracting more is assumed to be more costly and the marginal cost increases. Also, it is easier to extract from a stock in which the water surface is at a higher level.

Mixing the two qualities (Q_H and Q_L) results in water whose quality is given by the weighted average (Q_M) of both types of water. Quality of potable water should weakly exceed the constant minimum quality standard Q_{min}, where

$Q_H > Q_{min} > Q_L$. Mixed water of quality Q_M may, or may not, satisfy the minimum quality standard. This depends on the quantities and the qualities that are mixed. The quality of water supplied to households may be improved to achieve the potable quality standard at a cost. This cost is an increasing function of the difference between the quality before and after purification. Moreover, a given improvement of a lower quality is less costly than the same improvement performed on a higher quality. Resource flow between the sources and the consumers is coordinated by a pair of knots, which centralize the mixing process at the consumer's location. The distribution scheme described above is presented in Figure 10.1.

Suppose that the behavior of the consumers can be aggregated under one of two types: households (h) or farmers (F). Consumers differ in their respective preferences regarding the quality of water. Both types prefer a higher quality and quantity of water to a lower one. Households consume water whose quality weakly exceeds a minimum standard. If mixed quality does not satisfy this condition, it will be subject to purification. The purification procedure is assumed to be costly enough that it is not profitable to improve quality above the minimum standard. Hence, the quality consumed by households is the maximum between the minimum and the mixed quality.

The assumptions presented here concerning consumer utility are qualitatively similar to those in Williams et al. 1986 on multiple commodities that are interdependent in consumption. Two features, which are rather specific to the dynamics of water, are added to the structure: first, buyers are restricted to purchase up to a certain amount of each type of water;[11] second, a constant inflow (recharge) in each period maintains the stock of water in the basins of each producer. In fact, following a standard formulation of similar

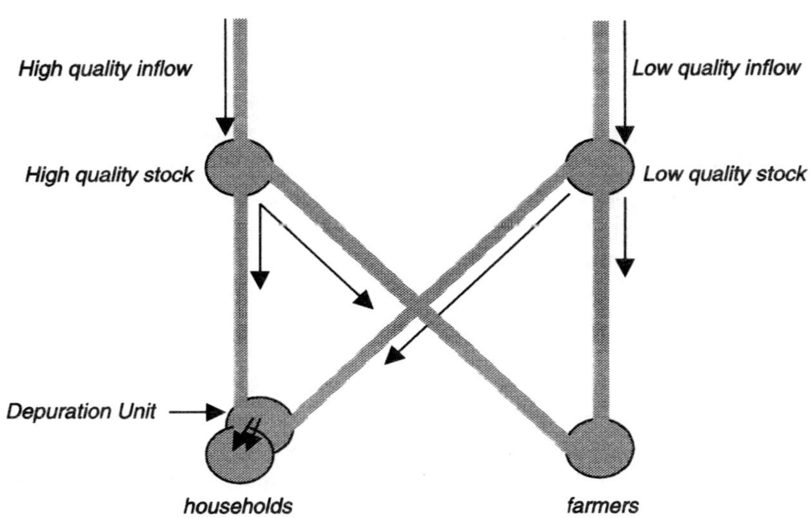

Figure 10.1 Pipeline network.

groundwater extraction problems, a lower stock implies a higher extraction cost. Thus, each period's marginal cost and past levels of extraction are positively correlated.[12]

10.3 Water resource management: from theory to the lab[13]

From an experimentalist's or a behavioral economist's point of view, the management of a complex system raises a large number of stimulating questions that can be addressed within many scientific disciplines and studied from many different points of view. Psychologists, for almost half a century, have used simulated complex "microworlds" to identify patterns of human learning and assess the performance of behavior in the presence of complex tasks. An economist's, or a management scientist's, interest in complex systems relates not only to efficiency-related issues, but also to the behavioral aspects of market functioning. These cannot be reliably described by static models based on standard simplifying assumptions aimed at analytical tractability. In static simple contexts, rational decisionmaking is often observationally equivalent to the limited behavior of adaptive subjects using trial-and-error learning algorithms. However, as complexity increases, explicit optimization becomes a less powerful descriptive model of human behavior, yielding predictions that significantly deviate from trial-and-error processes. Such deviations arise from different factors, among which are strategic interaction, multitask and multiperiod interdependence, complexity of the mathematical functions (relating actions to feedback), and imperfect information. Rather than the exception, complexity arising from the aforementioned factors is the rule in most real-world markets.

A central question motivating the focus and methodology adopted here is whether some well-known results on static market functioning are applicable to complex dynamic markets. Textbook industrial organization theory recommends competition almost as the panacea to market imperfections. This very influential view has inspired the most widespread type of state intervention in the market, namely the establishment of competition authorities. While enhanced competition is a relatively straightforward process in simple static markets, complex market setups call for policies based on a more informed view of market management by imperfect human decisionmakers.[14] For example, in imperfect information settings, a multiproduct monopolist who is learning from trial-and-error algorithms may fail to behave differently from a duopoly selling the same products (García-Gallego and Georgantzís 2001). Thus, the standard result concerning efficiency losses arising from monopoly may not be as globally applicable as is usually thought. Such learning failures seem to be more likely the more we deviate from the standard static symmetric setup.[15] The addition of dynamics and nonlinearities in the functions dictating the feedback from the underlying cost and demand models render most textbook wisdom inapplicable to some real-world markets. A further obstacle in understanding the functioning of complex dynamic markets stems from the fact that human actions exhibit a far richer range of behavioral patterns than any analytically tractable model can afford or any simulated market can include. Over the last half a century, economic experiments with human subjects have increasingly proved to be a reliable source of information tackling all (or at least most) of the aforementioned problems. Far from attempting an exhaustive review of the related literature in economics, management, and psychology, some related issues and their corresponding stylized facts are reviewed as they are reported in the literature.

The experimental design focuses on how different market structures affect water resource management. The analysis focuses on the case in which property rights to a given groundwater source are exclusively granted to a single decisionmaking unit. Implicitly, the model highlights the vertical nature of these markets, in which coordination is required between water extraction, supply, and purification.[16] The quality of water reaching the final consumer is not directly controlled by any one of the agents in the market. Instead, it is the result of the agents' bidding behavior controlling the resources. Their only strategic variables are the minimum price bids at which they are willing to sell each unit of each type of water to the consumer.

The model described in the previous section is implemented with the following values for the parameters:[17]

- Recharge ($a_i = 3$, $i \in \{H,L\}$)
- Initial and maximum stocks: $(S_H,S_L) = (20,20)$
- Water qualities: $(Q_H,Q_L) = (5,1)$
- Minimum quality standard demanded by the household: $Q_{min} - 3$.

In the steady state of the social optimum a stock size of $(S_H,S_L) = (4.84,5.01)$ is obtained associated with the prices $(p_H, p_L) = (102,86)$. However, in order to simplify subjects' perceived feedback conditional on their strategies, the design allows only for discrete quantities and prices. Thus, if subjects adopt the discrete version of this steady state equilibrium stock policy, they must aim at stabilizing stocks at $(S_H,S_L) = (5,5)$. Obviously, this is achieved when the 3-unit inflow of each water quality equals the quantity sold in each period. Assuming that this is so, then, maximizing total surplus the quantities, $(K_{Hh},K_{Lh}) = (2.55,0)$ and $(K_{HF},K_{LF}) = (0.45,3)$, are automatically assigned by the server, respectively, for household (h) and agriculture (F) consumption. To avoid subjects making uncontrolled guesses concerning the end of the session, a deterministic end game horizon was used (a total of 50 periods), which was known by subjects from the beginning of the experiment.[18]

Subjects knew the type of water they were managing.[19] Furthermore, they were conscious of the generic preference of consumers for the high-quality good over the other. Moreover, they knew that their products were demand substitutes (though not perfect) and that their extraction cost structures were identical. Subjects received a table with their unit costs depending on the stock size. A simulator (made available to them on their decision screen) informed them of the hypothetical costs and gains they would make if they sold all the units of each product for which they were currently submitting post bids. They knew that the actual number of units they sold would be known only after they had posted their period bids and that the (automated) demand's reaction to these bids was announced to them on the feedback screen.

Four different market structures are studied and compared in terms of economic efficiency. These are designed to address the effect of two different factors that are usually considered to be competing solutions to the problem of efficiency losses arising from monopolistic market power: competition, and centralization of decisions by a public monopolist. In order to address these issues, four experimental conditions are implemented. Three of them are labeled as treatments, while the fourth condition is implemented as a variation to one of the three basic treatments. They are as follows.

- *Treatment 1 (private monopoly)*. The monopoly has joint ownership of both sources and optimal (simulated) coordination of consumer behavior is assumed. Each subject posts price offers for both water qualities.[20] The bundle of high- and low-quality water that produces the highest consumer rent is allocated in the economy. The corresponding offers of the subjects then establish the clearing prices. Monetary rewards are proportional to accumulated total profits over the whole session.
- *Treatment 2 (noncooperative duopoly)*. Under the noncooperative duopoly each firm-subject offers one of the two types of water and independently decides on prices. This is a noncooperative decisionmaking scenario. As in all treatments reported here, there is optimal simulated coordination in the downstream part of the market. Each water resource is managed by a different subject sitting in front of a separate PC[21] and communication between competitors is not permitted. Duopolies are formed randomly at the beginning of the session and then matching remains fixed over the 50 periods of the experiment ("partners" protocol). Although a "strangers" protocol, forming a different subject pair (duopoly) in each period, would be an interesting extension, the resulting noisy feedback would require allowing for longer experimental sessions.[22] An alternative duopolistic setup, coordinated duopoly, was run using the same experimental market as in treatment 2. A single session was run using this design as a collusion-facilitating variant of the duopoly treatment. In the coordinated duopoly session, each resource is managed by a different subject, but the interface used is that of the private monopoly with two subjects sitting in front of each of the PCs. Communication and agreement on the timing of decision submission and the possibility of iterated inspection of the competitor's strategy before jointly pressing the "OK" button render this setup highly collusive. However, individual incentives remain uncoordinated and no side payments were feasible. Therefore, as the results show, this structure lies in many senses between treatments 1 and 2.

- *Treatment 3 (public monopoly)*. Here subjects decide on both water qualities. As in all treatments reported here, there is optimal simulated coordination in the downstream part of the market. In this treatment subjects act as public monopolists caring for total social welfare rather than private profits. Subject rewards were proportional to total social welfare (also referred to as social utility).

The experiment was run using the software Hydromanagement.[23] The results reported here are based on data obtained from experimental sessions run during 2004 in the Laboratori d'Economia Experimental (Universitat Jaume I, Castellón, Spain). Three 20-subject sessions were run for each of treatments 1 and 3 (20 monopolies per session) and five 20-subject sessions (10 duopolies per session) were run for treatment 2. Finally, a single 40-subject (20 duopolies) session was run for the coordinated duopoly treatment. A number of pilot sessions (not presented here) were run at the beginning of the whole project. Subjects were recruited following standard protocols of the Laboratori d'Economia Experimental and they were assigned only once to a single session. Sessions from different treatments were run in a random order in order for undesirable social learning to be avoided. Each session lasted an average of 80 minutes.

Demand is simulated in all treatments according to the assumption of utility maximization contingent to the block rate bids made by suppliers in each period. The simulated consumer represents a population of farmers and households with different preferences for water quantity and quality. Quality is the result of mixture of the two qualities of water sold in the market. If quality falls below a certain potable threshold, an extra cost is borne by the consumer (reducing consumer surplus) for the water to be depurated up to the threshold level. The resulting feedback from the demand model is difficult to interpret by the subjects or linearize in any sense. Unit extraction costs increase in steps as the level of stock of each water quality decreases. Therefore, in each period costs depend on extraction in previous periods.

Given each bidding schedule posted by a seller, consumer surplus maximization determines the quantity of each quality consumed. This is a rather complex problem, especially because of the uncoordinated action of sellers in the duopoly scenario, but also because of the dynamic nature of the market and the horizontal externality effect of each water type on the other.

A history window on each subject's screen displays all past outcomes regarding own decisions on quantities, payoffs, and market prices. In duopoly markets, each subject also receives the clearing price at which the other water quality was sold. In each period, subjects are asked to submit their respective reservation prices (offer bids) for each unit of product (from the 1st unit to the 5th, the maximum quantity each one of them could trade). Subjects were told that offer bids had to weakly exceed the extraction cost of the corresponding unit, and offers of subsequent units would have to be nondecreasing. Once offer bids were submitted, behavior on the demand side was simulated by a program that automatically calculates the optimal consumption of each water quality by consumer type for which total consumer surplus is maximized. All units of the same product were sold at the same market price, which was determined from the intersection of product-specific supply and demand schedules resulting from sellers' and buyers' behavior described above.

10.4 Results

In order to make the discussion of the results easier to follow, let us first recall some intuitive implications of the benchmark solution obtained in the previous section. The steady state equilibrium stock level for the discrete strategy space version of the model implemented here is 5 for both water qualities. Once this level is reached, the optimal extraction rate is dictated by the rate of the inflow. That is, in each period a manager should aim at selling 3 units of each water type (the same number as enter the tank because of the natural rate of inflow). Also, the socially optimal path leading to this steady state requires the fastest possible rate of convergence to it. Under efficient management all these predictions are fulfilled over the maximum number of periods possible. Of course, behavioral and strategic considerations are absent from the socially optimal solution. For example, competing private profit-driven subjects should be expected to restrict present extraction in order to guarantee lower extraction costs and a competitive advantage against rival managers in future periods. There might even be idiosyncratic factors such as risk aversion, or excessive fear of overexploitation, that are relevant in the observed behavior but have not been accounted for in the benchmark model. Other indicators that are usually not present in other studies on complex resource markets are price levels for each water quality, and price-weighted average quality sold to the consumers.

Results concerning market-clearing prices, quantities sold, stocks, and quality/price ratios are discussed using descriptive statistics and plots of period averages in each treatment.[24] Table 10.1 (see page 222) provides descriptive statistics concerning aggregate treatment averages obtained in the last 30 periods of each experimental session.

Some graphs that combine comparable magnitudes across treatments will now be discussed. Given that a fixed matching protocol was used, data obtained from the same subject acting in the same market in different periods cannot be treated as independent observations. Consequently, the evolution of averages over the 50 periods of a session is a reliable source of information from which useful insights can be gained.

In all cases, average stocks (Figures 10.2 and 10.3) are above the steady state optimum levels (5 units). Comparison across treatments indicates that the average stock of both water qualities is the lowest in the public monopoly treatment. Note that a sustained decreasing tendency is only observed in the public monopoly case (although without reaching the benchmark

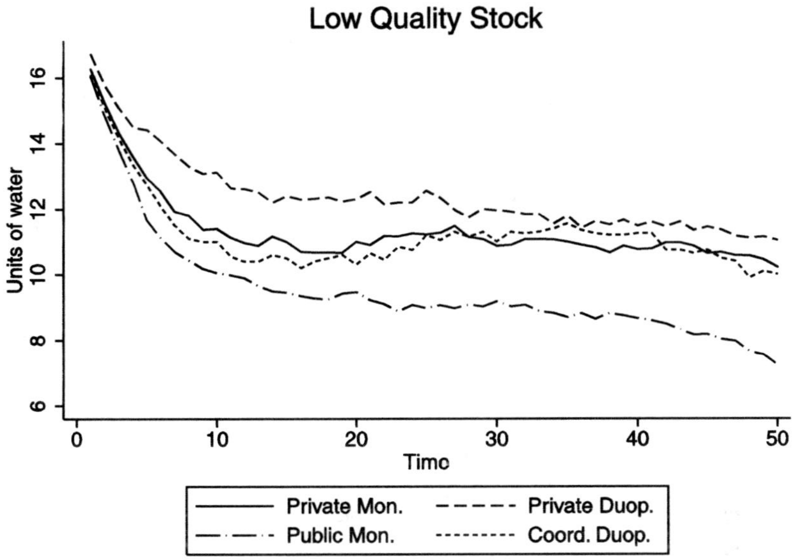

Figure 10.2 Evolution of average stock levels (low-quality water).

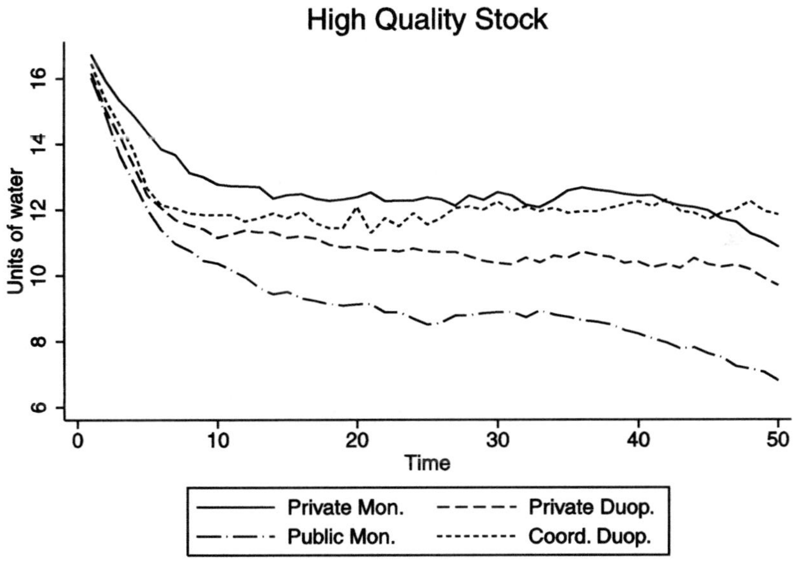

Figure 10.3 Evolution of average stock levels (high-quality water).

optimum). Other treatments stabilize their corresponding average stock levels from early periods onwards to over 10 units, which is well above the optimum.

In general, competitive environments and social welfare-oriented decisionmaking yield lower stock levels for both water qualities. Both monopolies (especially the public one) present persistently decreasing stocks, while a rather stable pattern over time is observed in duopolies.

As is seen in Table 10.1 and Figures 10.4 and 10.5, prices are highest in the private monopoly case and lowest in the public monopoly case. The differences in average qualities are reflected for all treatments (but to different degrees).

Overall, price averages per treatment indicate that the public monopoly reflects significantly the effects of different qualities on prices (51.05 of the low quality versus 67.14 of the high quality), while the coordinated duopoly produces an insignificant difference (98.63 versus 96.67). The other two treatments produce intermediate differences (75.71 versus 69.82 in the duopoly and 108.76 versus 101.85 in the private monopoly case). Data yield the expected result that relatively more competitive environments and social welfare maximization yield lower prices. Also note that public monopolies reach a stable average level throughout the second half of the session.

Table 10.2 presents quantity averages per treatment in 5-period intervals. Even in the initial 5 periods, subjects achieve the objective dictated by the hydrological equilibrium of the system. Namely, in all treatments and for

Table 10.1 Descriptive statistics (treatments 1, 2, 3 and coordinated duopoly aggregates and last 30 period averages and standard deviations)

	Low-quality stock				*High-quality stock*			
	T1	*T2*	*CD*	*T3*	*T1*	*T2*	*CD*	*T3*
Average	11.4	12.4	11.2	9.5	12.7	11.1	12.2	9.3
Std. Dev.	1.2	1.2	1.2	1.7	1.1	1.2	0.9	1.9
Av. 20–50	10.9	11.7	10.9	8.7	12.2	10.5	12.0	8.3
St. D. 20–50	0.3	0.4	0.4	0.5	0.4	0.3	0.2	0.7

	Low-quality quantity				*High-quality quantity*			
	T1	*T2*	*CD*	*T3*	*T1*	*T2*	*CD*	*T3*
Average	3.0	3.0	3.0	3.1	2.9	3.0	2.9	3.1
Std. Dev.	0.3	0.2	0.3	0.3	0.2	0.3	0.3	0.3
Av. 20–50	2.9	2.9	2.9	3.0	2.8	2.9	2.8	3.0
St. D. 20–50	0.1	0.2	0.2	0.1	0.2	0.1	0.3	0.1

	Low-quality price				*High-quality price*				*Average quality/price*			
	T1	*T2*	*CD*	*T3*	*T1*	*T2*	*CD*	*T3*	*T1*	*T2*	*CD*	*T3*
Average	101.9	69.8	96.7	51.1	108.8	75.7	98.6	67.1	0.04	0.04	0.04	0.11
St. Dev.	11.9	5.2	12.8	9.1	12.0	8.1	11.7	12.3	0.01	0.01	0.01	0.05
Av. 20–50	108.4	72.2	102.9	52.0	113.6	80.0	103.8	69.8	0.03	0.04	0.03	0.10
St. D. 20–50	6.5	3.6	4.4	8.7	9.5	5.3	4.3	12.7	0.00	0.00	0.00	0.03

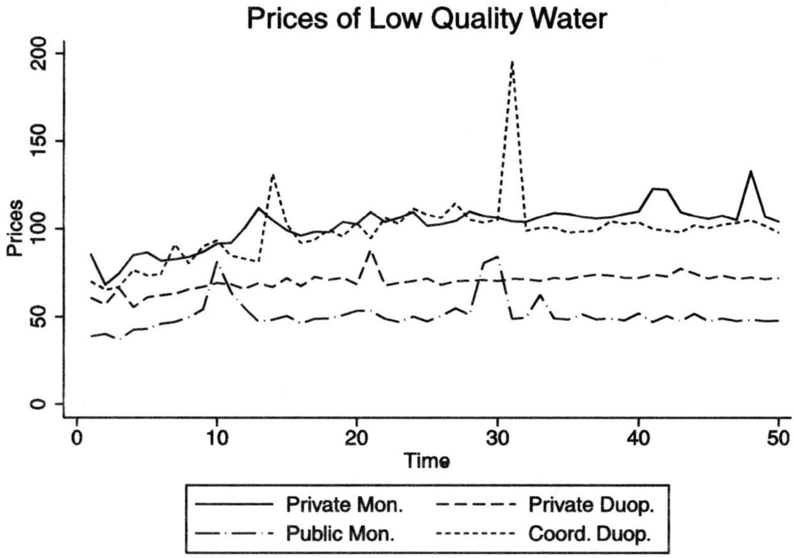

Figure 10.4 Average market-clearing prices (low-quality water).

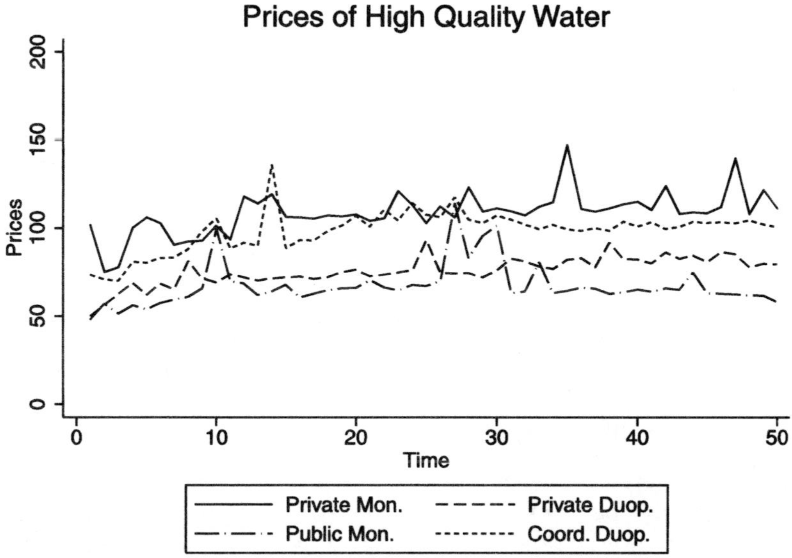

Figure 10.5 Average market-clearing prices (high-quality water).

most 5-period intervals considered, quantity averages are not significantly different from 3, implying levels of water consumption that are equal to the constant per period recharge of the aquifers. Learning this rule has taken subjects less than 5 periods, which confirms their decisionmaking capacity and understanding of the functioning of the market.

Table 10.2 Average quantity sold for high- and low-quality water, by five-period intervals

	High-quality quantity				*Low-quality quantity*			
Periods	*T1*	*T2*	*CD*	*T3*	*T1*	*T2*	*CD*	*T3*
5–10	3.04	3.08	2.95	3.20	3.22	3.00	3.19	3.22
11–15	2.83	2.88	2.78	3.11	2.94	2.96	2.91	3.02
16–20	2.76	2.95	2.75	3.03	2.93	2.81	2.84	2.97
21–25	2.80	2.91	2.88	3.08	2.89	2.82	2.67	3.03
26–30	2.76	2.92	2.66	2.89	2.95	3.00	2.91	2.90
31–35	2.76	2.84	2.82	2.97	2.89	2.94	2.84	3.05
36–40	2.83	2.87	2.76	3.07	2.88	2.98	2.99	2.97
41–45	2.87	2.91	2.91	3.09	2.90	2.90	3.01	3.08
46–50	3.05	3.00	2.78	3.15	2.95	3.02	3.02	3.17

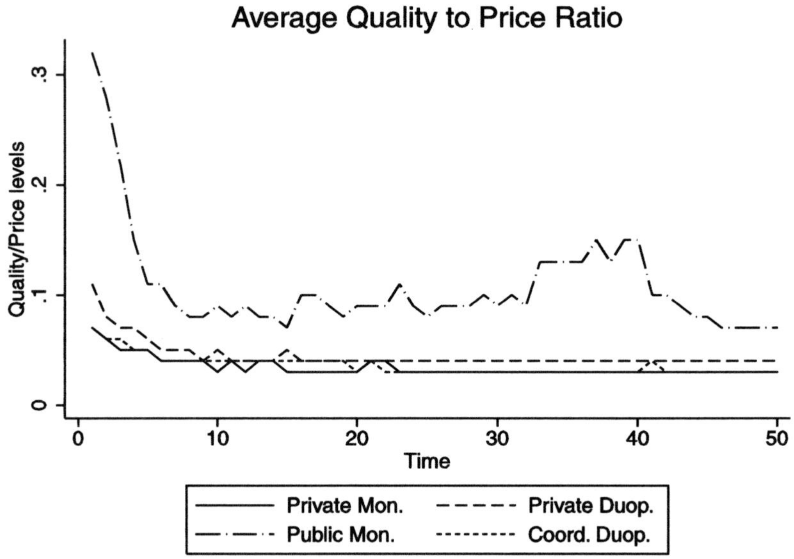

Figure 10.6 Average quality/average price ratio for the four treatments.

The more interesting aspect of the design, namely the quality of water offered, will now be considered. Define the quality/price ratio obtained as a division of average quality by the average price paid in each period. Figure 10.6 presents quality/price index averages defined as $\bar{Q}/\bar{P}$, where $\bar{Q} = \dfrac{K_H \cdot Q_H + K_L \cdot Q_L}{K_H + K_L}$ and $\bar{P} = \dfrac{K_H \cdot P_H + K_L \cdot P_L}{K_H + K_L}$. It can be easily seen that the public monopoly provides the highest quality of all the treatments. It is worth noting that the average quality offered by the public monopoly follows a relatively volatile pattern compared to the (average) qualities offered by other treatments. This difference can be understood by recalling that the public monopolist receives feedback that partly, but directly, depends on the consumer's social welfare. Specifically, the consumer's loss associated with the cost of depuration (in case quality falls below the potable threshold) negatively affects the feedback received by subjects on the success of their strategies. Volatility is thus the result of a continuous effort by subjects (representing the social planner) to cope with the additional objective of maintaining quality above a certain level in order not to trigger the inefficient (understood as costly) depuration procedure.

Noncooperative duopoly stocks are closer to those of the public monopoly while the collusive duopoly treatment leaves stock levels closer to those obtained under the private monopoly scenario.[25] As expected, under a duopoly resource depletion is faster than under tacit or explicit monopolization. This result resembles a similar finding that is recurring in most common pool resource experiments (see Walker et al. 1990; Walker and Gardner 1992). It should be noted that in this setting there is no competition at the extraction stage. Instead, this setup assumes extraction from separate pools and competition in the distribution stage. The findings reported here could be used to support the view that disaggregation increases the rate at which a resource is depleted. This is confirmed by comparison of treatments 1 and 2, in which subjects' incentives are related to private profitability. Price levels present a similar but weaker confirmation of the same principle applied to price competition.

Only the public monopoly solution gets close to the socially optimal steady state stocks. Full convergence, however, is not achieved at the end of 50 periods. The aspect in which management by a public monopolist dominates all other configurations is the quality/price ratio. Achieving a high quality/price ratio implies balanced exploitation of the two resources by the public monopolist using exactly the same instruments as those used by subjects in the other treatments. Of course, the public monopoly has the right incentives to do so, as it aims at maximization of total social welfare. Some volatility is obtained regarding the quality/price ratio as the public monopolists' task requires a global view of the performance achieved with both policy instruments available (bids for the two water qualities). This contrasts with the smoother patterns of the other treatments in which profit maximization is the only objective pursued.

In profit-driven treatments subjects post significantly higher bids aimed at lowering present consumption. This is done in order to maintain a higher stock and, thus, lower unit extraction costs in the future. This is a basic difference between pure profit-driven, and social welfare-oriented, decisionmaking. In the former, subjects seem to care too much for maintaining a high stock in order to guarantee low extraction costs and competitiveness (in treatment 2) in future periods. On the other hand, the social planners' treatment induces higher levels of extraction, without letting overexploitation emerge.

This finding, however, should not be understood as an argument against the use of market mechanisms altogether. On the contrary, it must be emphasized that in this setup social planners achieve their results by learning to set the appropriate prices for each type of water. This also implies the need for distinguishing between incentives and market-clearing mechanisms. The behavior of the social planners in treatment 3 is an example of how market-clearing prices can be used as a strategy aimed at social welfare maximization.

10.5 Conclusion and policy implications

Most of the debate concerning the use of market mechanisms as a means of allocating water among different types of users has focused on the appropriate definition of property rights. Several researchers have argued that the market mechanism may entail unfair outcomes because of some agents' greater economic or political power over other less strategic or weaker players.[26]

Although the definition of property rights is of great importance in decentralized water management, the present study looks at a different and certainly underinvestigated issue that should be taken into account when assessing the desirability of alternative water management institutions. The central question addressed here is whether adaptively learning from past actions negatively interferes with the usual benefits from centralized and, thus, informationally demanding public management. If this conjecture were true, decentralization and privatization of water resource management might mitigate the complexity of the problem faced by each decisionmaker, offsetting the well-known welfare losses because of incompatibility between private and social incentives. Thus, a further argument would be provided to support decentralized private management of the resources.

Contrary to the aforementioned conjecture, the answer provided by the experiments reported here is that the decisionmakers' incentives dominate possible efficiency losses because of an increased complexity of the underlying system managed by each type of agent. This main result establishes a straightforward ranking of management institutions in terms of their corresponding levels of social welfare.

Therefore, despite the fact that centralized public management has to deal with a more complex problem whose feedback depends on a broader set of signals than private profit alone, the public monopoly treatment of the experimental setup yields the most efficient outcomes both in terms of social welfare and in terms of the final quality of water supplied to the users. Interestingly, centralization of the decisionmaking process by a private profit-driven entity yields the worst outcomes, suggesting that centralization alone is not enough and it should always be accompanied by the right incentives.

Otherwise, if the public monopoly solution is impossible to implement because of, say, economic, institutional, legal, or historical reasons, market mechanisms should be implemented under the usual rule of thumb, favoring competition against monopolization of the market by a private agent. The coordinated duopoly treatment was shown to move the market away from the competitive outcome and close to the monopoly solution. Therefore, another well-known rule of thumb for market regulation is also shown to hold in the complex market studied here. Namely, the market mechanism is vulnerable to private agents' anticompetitive strategies, so that, if a market mechanism were implemented to allocate water resources, competition among decentralized private owners should be safeguarded and promoted.

Notes

1. The authors gratefully acknowledge the financial support of the Fundación BBVA, Bancaixa (project P1 1B2004–28), and the Spanish Ministry of Education and Science (project SEJ 2005–07544/ECON). Kujal acknowledges financial support from the Spanish Ministry of Education and Science (Secretaría del Estado de Universidades e Investigación and grant SEJ 2005–08633/ECON). The chapter was written while Kujal was visiting ICES, George Mason University. J.C. Pernías-Cerrillo provided excellent programing assistance. Comments by the participants at the Sixth Meeting on Game Theory and Practice (Zaragoza, Spain, July 2006) and detailed discussion between Victor Galaz and the editors of this chapter are gratefully acknowledged.
2. In a recent study, Holland (2006) focuses on the costs and benefits from privatizing the construction of the Central Arizona project, reporting several shortcomings in the resulting incentive structure.
3. For example, Saal and Parker (2001) study the effects on costs and prices in privatized water and sewerage companies in the United Kingdom. Their results yield some pessimism regarding the efficiency gains from privatization, which may not offset the resulting increases in prices. Bhattacharyya et al. (1994), Noll et al. (2000), Bauer (1997), Cowan (1998), and Ogden (1997) also reach conclusions that imply that privatization measures should be taken with caution.
4. Detailed information on the European Union Water Framework Directive can be found at the corresponding European Union website: http://ec.europa.eu/environment/water/water-framework/index_en.html.
5. It is important to bear in mind the mission statement appearing in the California Water Plan update, 2005: "To develop a strategic plan that guides State, local, and regional entities in planning, developing, and managing adequate, reliable, secure, affordable, and sustainable water of suitable quality for all beneficial uses" (Department of Water Resources 2005).
6. This and other documents on the management of California's water resources can be found on the following website: www.water.ca.gov.
7. A similar suggestion on the need for adaptive management is made by Pahl-Wostl (2007), although the adaptive process there includes social learning and takes into account more complex feedback concerning both economic and noneconomic magnitudes. An even broader point of view can be adopted looking at various cross-effects between water and land use activities, as reported by Ramankutty et al. (2006).
8. LINEEX, University of Valencia (www.uv.es/lineex).
9. LEE, Universitat Jaume I, Castellón (www.lee.uji.es).
10. Extraction costs are supposed to be twice differentiable functions of quantity and stock size. First derivatives are assumed to be, respectively, positive and negative, whereas second derivatives are positive.
11. Given that their water consumption in each period is used to serve their current needs.
12. All these elements lead to a certainly complex but well-defined dynamic economic

problem that, as shown by García-Gallego et al. (2005, 2006), has an interior steady state solution for all the scenarios considered here. In any case, most of the interesting features analyzed by Dasgupta and Mäler (2004) are not present in this model.

13. The design is inspired by the hydrological system proposed in Georgantzís et al. 2004.
14. A vast literature has been dedicated to various factors that may be responsible for observed shortcomings of human behavior in complex environments, such as misperception of feedback (Paich and Sterman 1993; Sterman 1994), limitations in subjects' learning when exposed to strategic complexity (Richards and Hays 1998), or multitask decisionmaking (Kelly 1995). A number of factors that favor subjects' improvement of performance have, also, been identified. For example, trial-and-error algorithms have been shown to facilitate convergence of the strategies played by uninformed subjects toward symmetric, full-information equilibrium predictions, as shown in García-Gallego 1998 for the case of a price-setting oligopoly.
15. While full convergence near the theoretical single-product symmetric benchmark is obtained in settings such as that outlined in García-Gallego 1998, the introduction of a slightly more complex task in the multiproduct oligopolies in García-Gallego and Georgantzís 2001 or the asymmetry in García-Gallego et al. 2004 provide a sufficiently unfavorable environment for the hypothesis based on the corresponding theoretical prediction to be rejected.
16. Issues related to vertical integration are not considered in this analysis, although it could be interesting to study a scenario in which one of the two resources is run by, say, a consumer association.
17. The specific utility, extraction, and purification cost functions used are provided in Georgantzís et al. 2004.
18. Given the complex nature of the experiment 50-period sessions were run in order to give subjects a sufficiently long time for learning.
19. See the details on the organization of sessions and the instructions to experimental subjects in Appendix 10.2.
20. Given these offers, the maximal consumer rent is determined in the simulated centralized downstream market: $V(K_H,K_L) - w' k$, where w denotes the vector of sealed offers and k denotes the vector of quantities K_i, $i = H,L$.
21. This treatment is based on a version of the software requiring a number of PCs and a server forming a local network. The software can be obtained by the authors upon request and assistance is subject to usual protocols of the Laboratori d'Economia Experimental (UJI, Castellón, Spain).
22. Pilot sessions not reported here indicate that sessions lasting longer than 70 periods are required.
23. Jointly developed by the LINEEX at the University of Valencia and the Laboratori d'Economia Experimental (LEE) of the Universitat Jaume I of Castellón (Spain).
24. Detailed statistical tests supporting the findings reported here are provided in García-Gallego et al. 2005, 2006.
25. This is especially true in the case of the high-quality water (see Figure 10.3).
26. See for example the discussion by Galaz (2002) on the Chilean water market, in which the peasant farmers are awarded "theoretical rights" which are then likely to be violated by the actions of a strategically stronger urban water company.
27. The instructions to subjects were originally written in Spanish. A unified translated version for the three treatments is presented here, emphasizing the details that are specific to each one of them.

References

Bauer, C. (1997) "Bringing water markets down to earth: the political economy of water rights in Chile, 1976–95", *World Development*, 25: 639–56.

Bhattacharyya, A., Parker, E., and Raffiee, K. (1994) "An examination of the effect of ownership on the relative efficiency of public and private water utilities", *Land Economics*, 70: 197–209.

Cowan, S. (1998) "The water industry", in H. Dieter and T. Jenkinson (eds) *Competition in Regulated Industries*, Oxford and New York: Oxford University Press.

Dasgupta, P. and Mäler, K.G. (2004) "The economics of non-convex ecosystems", in P. Dasgupta and K.G. Mäler (eds) *The Economics of Non-Convex Ecosystems*, Dordrecht: Kluwer Academic Publishers.

Department of Water Resources (2005) *California Water Plan Update 2005*, Government of California. Available online: www.waterplan.water.ca.gov/cwpu2005/ (accessed April 21, 2007).

European Commission (2004) *Moving to the Next Stage in the Common Implementation Strategy for the Water Framework Directive (2000/60/EC)*, Progress and Work Programme for 2005 and 2006, European Commission.

European Commission (2005) "Common implementation strategy for the water framework directive (2000/60/EC): groundwater summary report", technical report on groundwater body characterization, monitoring and risk assessment issues as discussed at the WGC workshops in 2003–04.

Galaz, V. (2002) *Water and Equity: A Game-Theoretic Exploration of the Chilean Water Market's Social Impacts*, FLASCO draft.

García-Gallego, A. (1998) "Oligopoly experimentation of learning with simulated markets", *Journal of Economic Behavior and Organization*, 35: 333–55.

García-Gallego, A. and Georgantzís, N. (2001) "Multiproduct activity in an experimental differentiated oligopoly", *International Journal of Industrial Organization*, 19: 493–518.

García-Gallego, A., Georgantzís, N., and Kujal, P. (2005) "How do markets manage water resources? An experiment on water management with endogenous quality", mimeo LEE/LINEEX, Universitat Jaume I, Spain.

García-Gallego, A., Georgantzís, N., and Kujal, P. (2006) "Water management in an experimental dynamic duopoly with endogenous quality", mimeo LEE/LINEEX, Universitat Jaume I, Spain.

García-Gallego, A., Georgantzís, N., and Sabater-Grande, G. (2004) "Identified consumers: on the informativeness of cross-demand price effects", *Cuadernos de Economía*, 27: 185–216.

Gardner, R., Moore, M.R., and Walker, J.M. (1997) "Governing a groundwater commons: a strategic and laboratory analysis of Western law", *Economic Inquiry*, 35: 218–34.

Georgantzís, N., García-Gallego, A., Fatás, E., Neugebauer, T., and Kujal, P. (2004) "Mixture and distribution of different water qualities: an experiment on vertical structure in a complex market", *Cuadernos de Economía*, 27: 95–138.

Gordon, H.S. (1954) "The economic theory of a common property resource: the fishery", *Journal of Political Economy*, 62: 124–42.

Holland, S.P. (2006) "Privatization of water-resource development", *Environmental and Resource Economics*, 34: 291–315.

Kelly, F.S. (1995) "Laboratory subjects as multiproduct monopoly firms: an experimental investigation", *Journal of Economic Behavior and Organization*, 27: 401–20.

Mason, C.F. and Phillips, O.R. (1997) "Mitigating the tragedy of the commons through cooperation: an experimental evaluation", *Journal of Environmental Economics and Management*, 34: 148–72.

McCabe, K.A, Rassenti, S.J., and Smith, V.L. (1989) "Designing 'smart' computer assisted markets in an experimental auction for gas networks", *European Journal of Political Economy*, 5: 259–83.

McCabe, K.A., Rassenti, S.J., and Smith, V.L. (1991) "Smart computer assisted markets", *Science*, 254: 534–38.

Moench, M. (1992) "Chasing the water table: equity and sustainability in groundwater management", *Economic and Political Weekly*, 27(51–52): A171–7.

Murphy, J.J., Dinar, A., Howitt, R.E., Rassenti, S.J., and Smith, V.L. (2000) "The design of smart water market institutions using laboratory experiments", *Environmental and Resource Economics*, 17: 375–94.

Noll, R., Shirley, M., and Cowan, S. (2000) "Reforming urban water systems in developing countries", in A. Krueger (ed.) *Economic Policy Reform: The Second Stage*, Chicago and London: University of Chicago Press.

Ogden, S. (1997) "Transforming frameworks of accountability: the case of water privatization", in K. Keasey, S. Thompson, and M. Wright (eds) *Corporate Governance: Economic and Financial Issues*, New York: Oxford University Press.

Pahl-Wostl, C. (2007) "The implications of complexity for integrated resource management", *Environmental Modelling and Software*, 22(5): 561–9.

Paich, M. and Sterman, J.D. (1993) "Boom, bust and failures to learn in experimental markets", *Management Science*, 39: 1439–58.

Ramankutty, N., Delire, C., and Snyder, P. (2006) "Feedbacks between agriculture and climate: an illustration of the potential unintended consequences of human land use activities", *Global and Planetary Change*, 54: 79–93.

Richards, D. and Hays, J.C. (1998) "Navigating a nonlinear environment: an experimental study of decision making in a chaotic setting", *Journal of Economic Behavior and Organization*, 35: 281–308.

Saal, D. and Parker, D. (2001) "Productivity and price performance in the privatized water and sewerage companies of England and Wales", *Journal of Regulatory Economics*, 20: 61–90.

Sterman, J.D. (1994) "Learning in and about complex systems", *System Dynamics Review*, 10: 291–330.

Walker, J.M. and Gardner, R. (1992) "Probabilistic destruction of common-pool resources: experimental evidence", *The Economic Journal*, 102: 1149–61.

Walker, J.M., Gardner, R., and Ostrom, E. (1990) "Rent dissipation in limited access common pool resource environments: experimental evidence", *Journal of Environmental Economics and Management*, 19: 203–11.

Williams, A., Smith, V.L., and Ledyard, J.O. (1986) "Simultaneous trading in two competitive markets: an experimental examination", Working Paper, Indiana University.

Appendix: organization of experimental sessions and instructions to the subjects

The experiments whose results are reported here were organized at the Laboratori d'Economia Experimental (www.lee.uji.es) at the Universitat Jaume I (Castellón, Spain). Subjects were recruited among students of advanced Engineering and Economics courses using standard protocols aiming at a balanced gender representation and good history of compliance with the laboratory's rules in the past (punctuality, discretion, motivation). Average per subject earnings were slightly below €25. Thus, the total cost of the sessions reported here was about €6,500. Another €4,200 were spent on programming and subject rewards in the initial pilot sessions that were run to improve several feedback and interface features. Duopoly sessions were systematically more expensive. A higher exchange rate was used in the duopoly sessions to avoid significant differences in individual subject rewards across treatments (apart from the horizontal externality arising from competition, each duopolist manages only a part of the market supplied by each subject in the monopoly treatments). Before each session, the following detailed written instructions were distributed to the subjects:

Instructions to experimental subjects[27]

Treatments 1, 2, and 3

The aim of this experiment is to study how people make their decisions in certain contexts. Your decisions in the scenario explained below in detail will be directly related to a monetary reward you will receive in cash at the end of the experiment. Any doubt you may have will be clarified personally to you by one of the organizers after you raise your hand. Beyond these questions, any other communication is strictly forbidden and is subject to immediate exclusion from the experiment.

You participate in a market which is characterized by the following features:

- You are the only producer of two commodities: product *H* and product *L*. [*Treatment 2*: There are two producers (*1* and *2*) and two commodities (product *H* and product *L*).] [*Treatment 3*: You represent a social planner who produces two commodities: product *H* and product *L*.] Specifically, product *H* is *water of High quality*, while product *L* is *water of Low quality*. Products *H* and *L* are substitutes, namely, consumers may, to a certain extent, substitute one type of water with the other.
- *Treatment 2*: You are one of the two producers in this market. At the beginning of the session, the computer will indicate if you are producer *1* or *2*. Your competitor will be one (always the same) of the subjects in this room, randomly selected by the computer when the session starts.
- There are two types of consumers: *households* and *farmers*. Although

they have different preferences with respect to the two types of water, they all prefer water of high quality (product *H*) to water of low quality (product *L*). That is, they are willing to pay more for *H* than for *L*.

- The market will last for 50 rounds.

Decisionmaking

Your only decision as a producer is announcing the minimum price at which you are willing to sell each one from a maximum of 5 units you may sell for each product. Such announcements of minimum prices are called *price bids*. In order to make your decisions, you have to take into account that:

- The extraction cost per additional unit extracted and by product is included in the "table of costs" (Table 10.3). These costs are the same for the two products [*Treatment 2*: (therefore, costs conditions for you and your competitor are identical)], and they are expressed in ExCUs, a fictitious Experimental Currency Unit.
- Taking into account the costs of the table, you have to announce *five minimum prices* at which you are willing to sell each unit of the 5 units of each [*Treatment 2*: your] product you may sell in each round. Therefore, your decisionmaking consists of fixing *five price bids* for each [*Treatment 2*: your] product.
- You should have in mind that, in order not to make any losses, price bids cannot be lower than the corresponding unit costs included in the table of costs.
- Price bids *cannot be decreasing*. That is, your bid for the 1st unit cannot be higher than your bid for the 2nd unit; the bid for the second cannot be higher than the bid for the 3rd unit, and so on and so forth.
- Observe in Table 10.3 that the unit costs decrease with the stock size. At the beginning of the session, you have an *initial stock size* of *20 units* [*Treatments 1 and 3*: for each product]. At the beginning of each round, you get *3* more *units* [*Treatments 1 and 3*: for each type of water].
- Your stock size [*Treatments 1 and 3*: for each type of water] can never exceed *20 units* and, therefore, once 20 units are reached, any additional units you may receive are lost.

Table 10.3 Table of costs (expressed in ExCUs)

Stock size	20	19	18	17	16	15	14	13	12	11
Unit cost	0	0	0	0	0	1	1	2	2	4

Stock size	10	9	8	7	6	5	4	3	2	1
Unit cost	7	11	18	30	50	82	135	223	368	607

Example. Suppose that at the end of a round your stock size [*Treatments 1 and 3*: of one of your products] is 9 units. At the beginning of the new round, you get your additional 3 units (so that your stock now is 12 units). Observe in the table that, for a stock size of 12 units, the unit cost for the first 5 units extracted is the following:

- The cost of the 1st unit: 2 ExCUs.
- The cost of the 2nd unit: 4 ExCUs.
- The cost of the 3rd unit: 7 ExCUs.
- The cost of the 4th unit: 11 ExCUs.
- The cost of the 5th unit: 18 ExCUs.

In order not to make losses, each one of your bids should not be lower than the corresponding unit cost. Therefore, in this example, your bid for the 1st unit should not be lower than 2 ExCUs (cost of the 1st unit); your bid for the 2nd unit should not be lower than either 4 ExCUs (cost of this unit) or your bid for the 1st unit; your bid for the 3rd unit should not be lower than either 7 ExCUs or your bid for the 2nd unit, and so on for the rest of the units.

In case you sell 5 units [*Treatments 1 and 3*: of this product], the stock size at the beginning of the next round would be 10 units (seven you kept plus three you get in the new round). If, given your bids for the 5 units, your sales are zero, your stock would be 15 units (12 you already had plus 3 you get at the beginning of the round).

Decisions

- You make decisions on the minimum price at which you are willing to sell each unit of each one of the two products [*Treatment 2*: of your product]. You will fill in all the boxes that appear at your computer screen with your price bids [*Treatments 1 and 3*: (five bids for product *H* and five for product *L*)]. In each box, you will also get information related to the corresponding unit cost. The bids you submit have to be integer numbers between zero and 2,000.
- Although you may propose five different price bids, all units of the same product will be sold to consumers at a *single price*. This price will be your bid for the "last" unit sold of each product. The number of units sold each period is calculated by a program which simulates the optimal behavior of consumers.

Example. In the example above, assume that your bids for [*Treatment 2*: your product] one of the products are: 10 (for the 1st unit), 12 (for the 2nd), 14 (for the 3rd), 16 (for the 4th) and 20 (for the 5th). Given your bids, the program which simulates the optimal behavior of consumers determines that

3 units of this product will be sold. The price at which you will sell the 3 units will be your bid for the 3rd unit, that is, 14 ExCUs.

Only Treatments 1 and 2: the profits

- Your net profit of selling each unit of a product will be the difference between the market price at which you sold all units [*Treatment 1*: of that specific product] (your unit income) and the corresponding unit extraction cost. Total profits will be the sum of the unit profits for all periods.

Example. Taking again the previous example, if, at the beginning of a round, your stock size is 12 units, your total profits in that round will be 29 ExCUs, which are broken down as follows:

- 12 ExCUs for the 1st unit sold (14 ExCUs you receive for that unit minus 2 ExCUs it costs you extracting it).
- 10 ExCUs for the 2nd unit sold (14 ExCUs you receive for that unit minus 4 ExCUs it costs you extracting it).
- 7 ExCUs for the 3rd unit sold (14 ExCUs you receive for that unit minus 7 ExCUs it costs you extracting it).

Only Treatment 3: the aim

- As a social planner, your aim in each round is to maximize the social benefit per unit sold in this market, which is defined as the difference between the utility level generated by each unit consumed and the corresponding unit extraction cost.

The information

- During decisionmaking, the computer will provide you with a table simulating results [*Treatment 1*: (for each product)], conditional to your bid and cost for the corresponding unit in five possible scenarios: *(a)* In case you only sell the 1st unit; *(b)* If you just sell the first 2 units; . . . *(e)* In case you sell 5 units.
- *Only Treatment 3*: At the beginning of each round, the computer will provide you with a table containing, conditional to the stock size for each type of water and all possible combinations of consumption of the two products, the corresponding social benefits (measured as the difference between the utility level and corresponding extraction costs) of that round.
- At the end of each round, the computer screen will show you the total profits [*Treatment 3*: social benefits] obtained in that round, including

information about unit cost, market price and number of units sold of each product [*Treatment 2*: as well as your rival's price].

- During the experiment, you will be provided with a screen containing the history of *past rounds* (market price for each product, number of units sold [*Treatments 1 and 3*: of each product], [*Treatments 1 and 2*: total revenue and total profits [*Treatment 1*: per product]], [*Treatment 3*: and social benefit]).

Monetary reward

- Your *monetary reward* at the end of the session will be the sum of your profits accumulated in 15 rounds (randomly selected by the computer) of the total of 50 rounds, at an equivalence rate of 800 ExCUs = 1 Euro [*Treatment 2*: 500 ExCUs = 1 Euro]. You will be paid *in cash* at the end of the session.

In order to make sure you understood correctly the market described above, we will proceed next to run a *pilot session of 5 rounds*. Please, feel free to make any questions you may have during this pilot session.

Thank you for your collaboration. Good luck!

11 A fair tariff system for water management

Rita De Agostini and Vito Fragnelli[1]

Access to water will be one of the most challenging problems in coming years. It involves many aspects, one of which is considered by this chapter: the possibility of designing a tariff system that respects fairness and allows an accurate evaluation of the amount of water that a community will need in a given period. The Italian situation is analyzed, taking into account the law and the consequent organization of water management. Particular emphasis is devoted to the truthfulness and implementation aspects.

11.1 Introduction

According to many environmentalists, water management will assume an increasing relevance in the near future. In the recent past water shortage was often thought of as a problem related to developing or tropical countries, but now governments are realizing that water shortage, in the face of increasing demand, is becoming a usual situation in countries throughout the globe. Increasing competition and conflict over access to water between countries sharing the same river basin was among the factors prompting the United Nations to declare 2003 the International Year of Freshwater, which helped promote the analysis of water-related issues.

In Europe a series of hot, dry summers, for example those of 2002 and 2003, have exacerbated water shortage problems. In Italy, for example, many regions suffered from scarcity of water. In southern Italy the situation was particularly grave, with such factors as low rainfall levels, permeability of the ground, and limited number of natural or artificial basins aggravated by longstanding management problems and old and badly maintained water networks. For example, some aqueducts have a water loss larger than 50 percent.

Some of these factors are natural (for example the characteristics of the ground), and are beyond realistic remedy; other factors are historical and cultural (for example, water problems were never suitably tackled where large landowners had little interest in less productive areas); and others are related to political choices about the development and maintenance of infrastructure, better care of which could significantly reduce water supply problems.

On the other hand, the Italian northern regions, thanks to their geographical position at the foot of the Alps, and the central regions, located close to the Apennines, have suffered few water problems in the past. Both of these mountain chains represent a very large reservoir of water; the River Po, for example, which draws tributaries from both ranges, drains an area of over 70,000 square kilometers.

Analogously, the northern and southern regions of Italy differ in the quality of water, which is higher in the north than in the south. In the south pollution and infiltration of salt in coastal areas reduce the quality of water; in the northern regions, despite the large degree of industrialization, the water quality is generally good. One simple reason is the location of natural reservoirs and artificial basins close to the cities, so that the water reaches the urban network after traveling a short distance, maintaining its quality. Moreover, the most recent depuration tools (for example active carbon) allow a limited addition of chlorine derivates, improving the quality of the water in recent years. As a result, the water quality in some municipalities (for example Genoa and Florence) is higher than for some commercially produced mineral waters.

An element that reduces the availability of high-quality water is the frequent use of drinking water (about 90 percent) for purposes for which it is not needed, particularly irrigation and industrial use. A proposal suggests having two separate networks: one for high-quality water and one for low-quality water, especially for farm usage, which takes over half of the water supply, mostly for irrigation, for which water from sewage disposal could be exploited. The objections to this solution arise from the potential hazard of confusing the two types of water and from the necessity for construction of large-scale and expensive infrastructure. It has been suggested that building the required infrastructure be devolved to consortia of users, with suitable reductions in water tariffs for a period.

The climatic conditions in 2002 caused a number of problems. From the beginning of the year the reduced rainfall failed to fill the lakes and other reservoirs to expected levels; the summer was extremely hot, resulting not only in a high level of evaporation of water from the reservoirs, but also in an increased usage of water, mainly for irrigation but also for home usage. Municipalities forbade use of water for unnecessary purposes, for example garden irrigation and car washing, but such measures were insufficient to compensate for the scarcity of water.

The so-called *Legge Galli* (5/1/94 n. 36),[2] a set of technical and economic rules on water resource management, has two main targets: a new approach to management operations in order to preserve the resources and a new tariff policy. In particular, the Legge Galli states that the management of water has to be centralized, as opposed to the current decentralized system of many small managers, public and private, currently in charge of parts of the water system; the size of the area managed should be optimized, requiring the determination of suitable areas, each constituting an *ambito territoriale ottimale* (optimal territorial area) or ATO; the programing and control of the water services, devolved to the local authority or *autorità d'ambito*, are to be separated from the management of the service; and the tariff policy is a tool for increasing the efficiency and quality of the service and reducing management costs.

As regards the tariff system, the Legge Galli was improved by the technical annex of Ministerial Act 1/8/1996,[3] according to which the average (normalized) tariff for the current year has to take into account the operating costs, mortgage costs, and investments of the preceding year and the inflation rate of the current year. In other words, the tariff can be increased in order to fully cover the costs of the water system, but this increase cannot be applied in a single year; to justify such an increase, the quality of the water is a relevant parameter; nevertheless, the tariff should be reduced in special situations, such as low-income users, critical areas, and small amount used.

A note of the Interministerial Committee for Economic Programing (CIPE) specifies other targets. First, low-income domestic users are to receive particular priority; second, it is possible to identify other priority categories where relevant differences from the standard situation exist; and third, incentives for saving water should be provided.

The geographical area considered, ATO/6 (ATO n.6 of Piedmont), includes 133 municipalities of the province of Alessandria and 14 municipalities of the province of Asti, with about 350,000 inhabitants, about half of them concentrated in five main towns: Alessandria (85,000), Tortona (25,000), Acqui Terme (19,000), Novi Ligure (27,000), and Ovada (12,000).

The preceding summary outlines the political, legal, and geographical scenario within which the work described here was carried out.

Cooperation with the autorità d'ambito led to the idea of designing a tariff system that may allow a reduction in water usage and, particularly, in water waste. Clearly, such a relevant problem should be tackled through a joint effort, involving experts from all related fields: economists, engineers, mathematicians, geologists, and politicians.

This chapter proposes a tariff system that respects the current directives and is based on the declarations of the users about the amounts of water they forecast they will use each year, so it is called a "declarative tariff". The aim of this proposal is threefold: a fair tariff system, that is a system that respects social welfare features; a reduction in water waste; and the possibility of the manager obtaining in advance good data about the total water requirement for each year.

The chapter is organized as follows: in section 11.2 the actual tariff system in ATO/6 is presented; section 11.3 is devoted to the proposed tariff system, analyzing its potential features; and section 11.4 provides some final comments.

11.2 Current tariff system

As described in the previous section, Italian law gives some suggestions and rules for a fair water tariff system; in particular, the tariff should balance the quality standards offered to the users and the costs for the service, including financial costs and capital risk. Other relevant points are the incentives for optimal use of water, the reduction of water waste and the environmental impact of such measures, and the setting of a socially sustainable tariff.

The tariff system adopted in ATO/6 (see ATO/6 2002a, 2002b) already respects some of the requirements of the Legge Galli; in particular, it considers different categories of users, with different tariffs on the basis of usage ranges, taking into account geographical and local characteristics.

More precisely, the current tariff system is strongly rooted in the so-called *tariffa di riferimento* (reference tariff), which is defined in the Ministerial Act 1/8/1996 as "a tool that allows suitable levels of service to be obtained and provides incentives for developing programs, reducing the costs for users and increasing the efficiency of management".

The reference tariff is defined as $T_n = (C + A + R)_{n-1} (1 + \pi + K)$, where:

- T_n is the tariff for year n
- C is the operating costs
- A is the mortgage costs
- R is the capital costs
- π is the programmed inflation rate for the year n
- K is the *limite di prezzo* (fare bound).

The reference tariff for the starting year, T_0, can be obtained as the weighted average of the costs of the previous managers that were in charge of the water system in the area corresponding to the current ATO, including the rent fees for public water, cost of the water purchased from third parties, rent fees for the water system, costs arising from current laws, and costs for current loans.

K represents a percentage that preserves the users from an increase in tariff sufficient to completely cover management costs. This percentage is necessary as the subsidization of water costs in some areas of Italy was very high, larger than 90 percent, too great to be covered by a single increase of the tariff. For this reason the law fixes the maximum percentage increase for each year according to the costs in the previous year.

For the first year the value of K is defined with reference to the weighted average tariff per cubic meter in 1995 (WAT) expressed in Italian lire. For the following years the value of K is defined with reference to the actual average tariff per cubic meter in the previous year (AAT) expressed in Italian lire (Tables 11.1 and 11.2).

Table 11.1 Definition of K using weighted average tariff

WAT $\leq$ 1000($\approx$0.52euros)	WAT $\geq$ 1600 ($\approx$ 0.83 euros)	1001 $\leq$ WAT $\leq$ 1599
K = 25.0%	K = 7.5%	K is computed by linear interpolation

Table 11.2 Definition of *K* using actual average tariff

AAT ≤ 1100 (≈ 0.57 euros)	AAT ≥ 1750 (≈ 0.90 euros)	1101 ≤ AAT ≤ 1749
K = 10.0%	*K* = 5.0%	*K* is computed by linear interpolation

Table 11.3 Tariff system of ATO/6, 2002

Usage	*Rent fee*	*Reduced tariff*	*Basic tariff*	*First exceeding*	*Second exceeding*
		up to 60 cubic meters/year	*60–150 cubic meters/year*	*150–240 cubic meters/year*	*over 240 cubic meters/year*
House user	24.00	0.50	0.80	0.95	1.20
Nonhouse user	48.00	0.80		1.20	1.30
Farm user	48.00		0.50		
Public user	24.00		0.80		
Large user		decided by agreements			

Note: The rent fees are in euros per year; the tariffs are in euros per cubic meter.

Referring to the average use of 150 liters per inhabitant per day reported in the Decree of the Council of Ministries n.47 of 4 March 1996, a note of the Interministerial Committee for Pricing (CIP) proposed the following limits for usage ranges: 50 cubic meters per year for applying reduced tariffs and 150 cubic meters per year for applying penalties for large use.

The tariff is the sum of two parts: one fixed, namely the rent fee, and one variable, depending on the amount of water used, with increasing prices per cubic meter of usage. For ATO/6, the tariff system for 2002 was as in Table 11.3.

11.3 Proposed tariff system

The basic aim was to develop a new tariff system that fully considered and accommodated the future needs of water users. The idea was inspired by informal communication with the general manager of ATO/6, who spoke about the difficulties and the advantages of forecasting the future needs of users. As water can be easily stored for long periods, this information could be used to reduce the problems arising during periods of scarcity. Of course it is possible that the total demand forecast by users exceeds the available amount of water, but even in this case the information can be useful to the manager, who can then anticipate some restrictions on usage, making them less stringent. The system could also act as an incentive for the manager to undertake a deeper analysis of the problem, taking into account the possibility of building more efficient infrastructure in the future, or of purchasing water from other ATOs, interconnecting the different networks or reservoirs. The new tariff system refers mainly to domestic users but it can be tailored to other kinds of users.

The system works as follows. Each user commits to a forecast amount of water at the beginning of each year, on the hypothesis that a larger declared amount corresponds to a higher basic tariff per cubic meter of water. At the end of the year, penalties are assigned to those users who required a larger amount of water. The penalty system assigns larger penalties to users that made larger misevaluations. The penalty has a twofold motivation: first, it should avoid the usage of a large quantity of water; second, it should encourage truthful declarations in order to allow the water manager to calculate a required amount of water as close as possible to actual needs. The users may take into account their necessities in previous years but they may be influenced by other factors; for example, number of members in the family group, announcement of a water shortage, or imposition of a higher price for greater usage.

The following notations will be used:

d declared cubic meters
x used cubic meters
p standard price per cubic meter
p' penalty price per cubic meter ($p' > p$).

11.3.1 Alternative tariff systems

The first step in setting a declarative tariff is to ask users for the forecast amount of water needed for the following year, d, and then to apply the standard price, p, to the used quantity of water, x, up to the declared cubic meters, and the penalty price, p', to the quantity exceeding that amount, as follows:

Tariff 1

$$\begin{cases} px & \text{if } x \le d \\ pd + p'(x-d) & \text{if } x > d \end{cases}$$

This tariff system provides incentives for the users that reduce water waste, but fails in the truthfulness of the declaration; in fact the users may declare very large amounts of water required in order to be sure of paying the standard price, whatever the amount of water actually used.

The simplest way to offer an incentive for truthfulness is to ask the users to pay the whole declared amount at the standard price, as follows:

Tariff 2

$$\begin{cases} pd & \text{if } x \le d \\ pd + p'(x-d) & \text{if } x > d \end{cases}$$

It is straightforward to check the truthfulness of the declaration: if the real used amount of water is larger than the declared amount, the penalty price is applied; if it is smaller, the whole declared amount is paid anyhow. On the other hand, this tariff fails to reduce water waste, as a user may think that the water up to the declared amount can be used for free.

A possible way to achieve both the aims is the introduction of variable prices. More precisely, the standard price can be defined using an increasing function π that fixes the price per cubic meter up to the quantity d, and the penalty price using an increasing function π', with $\pi'(x) > \pi(x)$ for each x, which is applied to the exceeding quantity of water, as follows:

Tariff 3

$$\begin{array}{ll} \pi(d)x & \text{if } x \leq d \\ \pi(d)d + \pi'(d)(x-d) & \text{if } x > d \end{array}$$

This third proposal seems effective but the following example shows that it may fail.

Example 1

Let the standard price be expressed by $\pi(d) = 0.50 + 0.01d$ and the penalty price by $\pi'(d) = 0.70 + 0.015d$. If a user who needs 150 cubic meters per year declares 150, then the standard price is 2.00 euros and the penalty price is 2.95 euros, so the final tariff is $2.00 \times 150 = 300.00$ euros; but if the user declares 148 the standard price is 1.98 euros and the penalty price is 2.92 euros, so the final tariff is $1.98 \times 148 + 2.92 \times 2 = 298.88$ euros.

So, a user may gain an advantage from a nontruthful declaration. To avoid these situations, and taking into account the first two tariffs, the prices when the actual usage is larger than the declaration can be computed referring to the actual usage instead of to the declared one, as follows:

Tariff 4

$$\begin{array}{ll} \pi(d)x & \text{if } x \leq d \\ \pi(x)d + \pi'(x)(x-d) & \text{if } x > d \end{array}$$

Referring to Example 1, when the declaration is 148 but the actual usage is 150 the standard price is 2.00 euros and the penalty price is 2.95 euros, so the final tariff is $2.00 \times 148 + 2.95 \times 2 = 301.90$ euros.

It is possible to enforce the penalty for nontruthful declaration by applying the standard price $\pi(x)$ to the whole amount x and adding the penalty price for the exceeding part $(x - d)$:

$$\pi(x)x + \pi'(x)(x-d) \quad \text{if } x > d$$

Referring again to Example 1, the final tariff becomes 2.00 × 150 + 2.95 × 2 = 305.90 euros.

Tariff 4 and the modified version penalize mistakes in the declarations. If $d > x$, then for both tariffs $\pi(d) > \pi(x)$ as π is increasing; on the other hand, if $d < x$, then for tariff 4, $\pi(x)d + \pi'(x)(x - d) > \pi(x)x \Leftrightarrow \pi'(x)(x - d) > \pi(x)(x - d) \Leftrightarrow \pi'(x) > \pi(x)$ holds for the hypothesis on the penalty function π' and for the modified tariff $\pi(x)x + \pi'(x)(x - d) > \pi(x)x$ trivially holds.

So, both tariffs achieve the stated objectives, limiting water waste and providing incentives for an accurate evaluation of the total water needed through a system of increasing standard and penalty prices.

It is possible, of course, to modify the penalty function π' in order to reduce the charge for "small" mistakes, taking into account the characteristics of the current year.

11.3.2 Variable prices and the Italian law

The idea of variable prices is consistent with the Italian directives on reduced usages. In this case it is simpler to define the standard costs and the penalty costs instead of the prices per cubic meter. For example, the standard cost function C can be defined taking into account the ranges and the tariff for domestic users stated by ATO/6 (see section 11.2) as:

$$C(x) = \begin{cases} 0.50x & \text{if } x \leq 60 \\ 30.00 + 0.80\,(x - 60) & \text{if } 60 < x \leq 150 \\ 102.00 + 0.95\,(x - 150) & \text{if } 150 < x \leq 240 \\ 187.50 + 1.20\,(x - 240) & \text{if } x > 240 \end{cases}$$

In this case the function is piecewise linear, so each user has an advantage in declaring the upper bound of its range (the cost per cubic meter is constant in each range), paying the penalties only if the upper bound is exceeded.

Also, the amount of the rent fee plays an important role, as it increases the final price per cubic meter, with a higher influence on small amounts of water, but this is beyond the focus of this chapter. Anyhow, it seems reasonable that the rent fee, the reduced tariff for small usage, and the limit for reduced tariff should be correlated so that the cost per cubic meter increases if the usage is larger than the quantity fixed for the reduced tariff.

In the case of ATO/6 the average cost per cubic meter, including the rent fee, for reduced usage can be expressed as:

$$\pi(x) = \frac{24.00}{x} + 0.50 \quad \text{if } x \leq 60,$$

which leads to a reduction of the cost for increasing usage until 0.90 euros per cubic meter when the usage x is 60 cubic meters. The cost of further water, up to 150 cubic meters, is 0.80 euros per cubic meter. In order to improve fairness, the function $\pi(x)$ should assume its minimum for the upper limit of the range of reduced usage, that is for $x = 60$ in the case of ATO/6. This can be done by modifying the rent fee or the range of reduced usage or the tariff system.

For example, in the case of ATO/6 fixing the rent fee at 18.00 euros, the average cost per cubic meter for reduced usage is $\pi(x) = \frac{18.00}{x} + 0.50$ and the cost per cubic meter for $x = 60$ is 0.80 euros, that is, it is equal to the cost of the following range. Extending the range of reduced usage fee up to 80 cubic meters and using the actual tariff the cost per cubic meter for $x = 80$ is again 0.80 euros, as in the following range. Other possibilities depend on changes in the tariffs; for example, decreasing the reduced tariff to 0.40 euros per cubic meter or increasing the basic tariff to 0.90 euros per cubic meter.

11.4 Conclusion and policy implications

The results of the declarative tariff system were satisfactory, not only from a theoretic point of view, but also in the evaluation of the technical experts of ATO/6, compared with the current tariff system. In order to increase the water savings, it is possible to introduce not only a penalty in case of larger use but also a prize in case of less use.

It is necessary to take into account that a strong pricing policy, aimed at reducing water wasting, cannot be carried out. Water is essential for health, so its supply has to be guaranteed to all, both by providing a connection to the network and by maintaining a low cost per cubic meter. This means that it is not possible to raise the price of water by any great amount (10 euros or more per cubic meter), as users may, as a consequence, reduce usage for essential purposes as well as for inessential purposes such as car washing.

A comparison was also made with another game-theoretic approach, the Rabbi rule or contested garment (Young 1995); in this case the total monetary amount that has to be covered by the users is assigned as a claim to the highest user and the claims of the other users are obtained via a linear approximation of their real usage or of the range of usage they belong to. In this way the total amount is divided into several intervals; the amount corresponding to each interval is equally shared among the users that appear in the interval. The following example may make the procedure clearer.

Example 2

Consider four users that require 40, 100, 130, and 210 cubic meters of water, respectively (see Figure 11.1); let the total amount of money required for the current period be 630 euros, taking into account also the total amount of 480 cubic meters used.

The amount of 630 euros is associated to the largest usage of 210 cubic meters and, correspondingly, using a linear approximation, the amounts

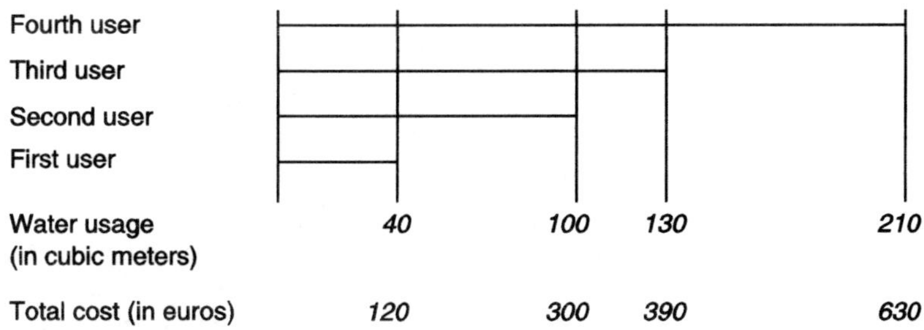

Figure 11.1 Contested garment approach.

of 120, 300, and 390 euros are associated to usage of 40, 100, and 130 cubic meters, respectively.

In this case there are four intervals that correspond to the monetary amount associated to each level of usage. All the four users are in the first interval, so the monetary amount of 120 is divided among all the users; in the second interval only the last three users appear, so the corresponding amount of 180 is divided among three users; the third interval involves only the third and the fourth users, so the amount of 90 is divided among the two users; finally, only the fourth user is concerned with the last interval and pays the amount of 240. Summing up, the four users pay 30, 90, 135, and 375 euros, respectively, and the costs per cubic meter are 0.75, 0.90, 1.04, and 1.78 euros, respectively.

This approach could make the cost per cubic meter too great, depending on the amount needed. It should be clear that, considering a reasonable number of users, those using smaller amounts of water, or those in the first ranges, have a low cost per cubic meter, while those using larger amounts of water may be penalized by this tariff system, if their numbers are small. A simple possibility is to use a nonlinear approximation of the usages, but this system is difficult to implement in a fair way.

Returning to the proposed tariff system it is possible to remark that the high level of freedom in determining the cost function π and the penalty function π' allows matching many possible requirements and constraints, both technical and normative.

Another relevant question is the fixed tariff, namely the rent fee; the underlying idea is that the fixed costs of the water system should be equally paid by all the users. Nevertheless this approach does not take into account other features of the problem, such as the amount of the rent fee could be significant for low-income users, but could be a very small fraction of income for other users; or it is possible that for very small water usage the rent fee is larger than the amount for water and in this case it is necessary, at least from a social welfare point of view, to distinguish between people living alone that try to save water (and, consequently, money) and people that use their house only for short periods of the year. These particular situations may be at odds with the analysis in the previous section, and further analysis may be required.

Again game theory may be a valid instrument for determining a fair tariff; it is possible to take into account the different requirements of the agents involved using bankruptcy approaches (Young 1987). More precisely, it is possible to define a bankruptcy (or taxation) problem where the users have the role of the agents, the estate to be divided is represented by the total amount of money required for the current period, and the claims are represented by the needs of the users. The related literature provides several rules for computing the tariff for each user. Another game-theoretic approach may involve the class of infrastructure cost games that were successfully applied to railway infrastructure (Fragnelli et al. 1999) and to urban solid waste consortia (Fragnelli and Iandolino 2004). The tariff is computed by defining a suitable game that considers the users divided into different groups with different needs and applying a game-theoretic solution of the game; for example, the Shapley value (Shapley 1953) or the Owen value (Owen 1977). In the case under study this approach could easily be applied as the users are naturally divided into groups according to usage range, income, or geographical location. This approach may be applied both to determine fair rent fees and to compute usage costs.

Notes

1. The authors are grateful to the anonymous internal reviewers and external referee for useful suggestions. The cooperation of Ing. Renzo Tamburelli, general manager of ATO/6, is also gratefully acknowledged.
2. See *Legge 5/1/1994 n. 36, Disposizioni in materia di risorse idriche* (in Italian), at website www.cipecomitato.it/Documentazione/Normativa/Tariffe_idriche/L.36-94.htm.
3. See *Decreto Ministeriale 1/8/1996* (in Italian), at website www.cipecomitato.it/Documentazione/Normativa/Tariffe_idriche/D.M.1.8.1996.htm.

References

ATO/6 (2002a) "Piano d'ambito: programma di infrastrutture, piano economico-finanziario e dinamica tariffaria per l'erogazione del servizio idrico nell'ATO/6", Rapporto Tecnico (in Italian).

ATO/6 (2002b) "Regolamento di utenza per l'erogazione del servizio idrico integrato nell'ATO/6", Rapporto Tecnico (in Italian).

Fragnelli, V. and Iandolino, A. (2004) "A cost allocation problem in urban solid wastes collection and disposal", *Mathematical Methods of Operations Research*, 59: 447–64.

Fragnelli, V., García-Jurado, I., Norde, H., Patrone, F., and Tijs, S. (1999) "How to share railway infrastructure costs?" in I. García-Jurado, F. Patrone, and S. Tijs (eds) *Game Practice: Contributions from Applied Game Theory*, Amsterdam: Kluwer.

Owen, G. (1977) "Values of games with a priori unions", *Lecture Notes in Economic and Mathematical Systems*, 141: 76–88.

Shapley, L.S. (1953) "A value for n-person games", *Annals of Mathematical Studies*, 2(28): 307–318.

Young, H.P. (1987) "On dividing an amount according to individual claims or liabilities", *Mathematics of Operations Research*, 12: 398–414.

Young, H.P. (1995) *Equity: In Theory and Practice*, Princeton: Princeton University Press.

12 Game-theoretic modeling of water allocation regimes applied to the Yellow River basin in China

Xiaokai Li, Haifeng Shi, and Xueyu Lin

This chapter presents a review of water allocation modeling studies and discusses sustainable water allocation in a river basin context from the perspective of game theory. The Nash–Harsanyi negotiation model is applied to the examination of several water allocation and management options, including unregulated withdrawal, allocation through agreed quota, and water rights and water pricing mechanisms. Some conclusions and policy implications are derived from the case study of the Yellow River basin and are applicable to other river basins for improving water management.

12.1 Introduction

The sustainable socioeconomic development and well-being of China are increasingly under threat from such problems as floods, droughts, water pollution, and deterioration of the eco-environment in major river basins (Ministry of Water Resources 2001). As the country continues to grow economically and in population, accompanied by higher levels of urbanization and industrialization, water scarcity and pollution, predominantly in the north and northwest, are becoming worse. Therefore, how to allocate very limited water resources effectively in those river basins has become one of the most critical challenges facing the water sector (Zhang 1999).

The current practice of water allocation in China is dominated by administrative mechanisms for most rivers. While such a regime may serve to maintain a reasonable equity of water sharing among different user sectors and regions, many question the true fairness of this approach because it unintentionally encourages waste and misuse of water by making water a resource of free access. It is against the basic principles of water allocation for maximizing returns on water use. Consequences in the Yellow River basin include frequent drying up of the river and water shortage crises, with enormous social and economic costs; raising of the riverbed in the lower reaches of the river; and irrigation efficiency lower than 30 percent in many parts of the river's upper reaches. An unsustainable situation thus prevails, which necessitates further investment to augment water supply capacity through giant operations such as the South–North Water Transfer project. Such situations can be attributed to lack of an effective allocation and management mechanism.

Aware of escalating conflicts over transboundary water use worldwide, particularly in many parts of Asia and in Africa, professionals in China's water sector are devoting more effort toward developing a fairer and more efficient water allocation regime in China (Liu and He 1999; Guan 2002). Naturally, the troubled Yellow River becomes a focus of research by many.

The Yellow River is the major water source for north and northwest China. It irrigates 15 percent of China's total cultivated land and supplies water to 12 percent of China's total population with water resources accounting for 2 percent of China's total average annual runoff. The water resource development in this basin has reached an alarming 70 percent, and consequently water security and the eco-environment in the river system, particularly in the lower reaches, are at high risk. For example, in 21 (eight of which occurred in the 1990s) of the 27 years from 1972 to 1998, the river ran dry for a total of 1,050 days at Lijin station in the lower reach. The negative impact was enormous, on flood management, ecological environment, drinking water supply, and economic and social services.

The situation described above can be well explained using the arguments of the "tragedy of the commons" (Hardin 1968). According to Hardin's theory, if there is no right defined to a property (water in this case), even when the marginal benefit of water use is close to zero and tomorrow's return on water is lower than today, it is still a "rational" decision for water users to use more water today because the water is of no or negligible value to them and nobody could guarantee them that water available today would still be available to them tomorrow. Such rational decisions by individual water users, taken together, ultimately lead to such disasters as the long-lasting drying up of the Yellow River lower reaches. The question now is how such tragedies can be prevented from happening by improving the way water resources are allocated and managed.

Section 12.2 provides a brief review of the various approaches of water allocation modeling studies undertaken by different scholars. Section 12.3 introduces game theory and its application in mathematical modeling of water allocation through a case study of the Yellow River basin in China, looking at several management options. Section 12.4 concludes the chapter with discussion of the policy implications of the modeling analysis.

12.2 Review of water allocation studies

A literature review of previous modeling studies of water allocation issues and water conflict management indicates that the approaches for water allocation modeling fall into four broad categories: (a) system simulation; (b) optimum allocation based on optimization theory or economics theory; (c) optimum allocation based on eco-economics theory; and (d) optimum allocation based on institutional economics theory or game theory.

The system simulation approach, category (a), is the earliest in application. It simulates the pattern, character, and essence of a prototype water resource system using computer, network, and 3S technologies. Examples include the five-reservoir basin model used in the Harvard water resource research program (Maass et al. 1962), and the water resource model developed by the US Corps of Army Engineers for operational study of cascade reservoirs on the Missouri River (Hall and Dracup 1970). Such an approach can vividly present the movement and transformation of water systems, and is usually used with relatively certain system management rules and regimes. It is, however, difficult to adapt to effective evaluation of a water allocation regime.

Category (b) includes optimization of allocation based on the theory of optimization and on economics theory, particularly Pareto optimization theory. The most common approach of the first variety is the so-called linear and dynamic programming model, which many scholars worldwide use in analysis of water resource allocation issues. The Institute of Water and Hydropower Research and Tsinghua University in China used this method in a North China Water Resources Study (Xu et al. 1997). The second variety is also referred to as the marginal analysis method. An example of this application is the Yellow River Basin Investment Planning Study jointly completed by the Ministry of Water Resources and the World Bank using a GAMS-programmed model (World Bank and Ministry of Water Resources 1993). The limitation of the first optimization approach is that it is hard to achieve optimization of an entire system, while that of the second is that it cannot yet take into account the social factors and assess the impact of such important factors as allocation regime, planning, and tradition in water allocation, although it can to a certain extent assess the efficiency of allocating water as a marketable scarce resource.

The next approach, category (c), is optimum allocation based on eco-economics theory. It treats water resources as an element of an integrated nature–society–economy system, and aims at maximizing ecological service value or total system ecological benefits with minimum material inputs. The core concept of this method is to minimize water consumption intensity per unit of service. Based on the comprehensive service capacity of water resources available, the level and limit of nature–society–economy system development can be scientifically assessed. The main shortcoming of this approach is that it lacks an established theory and detailed rules in its evaluation criteria and research application, and many aspects such as the water–environment interaction mechanism and the roles of water in the integrated system are still being explored. However, there are cases where researchers have tried to apply this method to analyze water allocation and use efficiency issues; for example, Liu explored ways of assessing the value of sustainable water resource development using the so-called marginal ecological utility theory (Fu and Hu 2004).

The fourth approach, category (d), is optimum allocation based on institutional economics theory or game theory. Such an approach has a broad range of interactive analysis and involves more variables (monetary and nonmonetary) than other approaches, and focuses more on integrity and evolution of the issues, and the influence of regime and rights on the market. In studying water allocation, the approach can take into consideration both economic and social system factors. Nevertheless, its application in this field, especially quantitative study, is still at an early stage. Carraro et al. (2005), in their discussion of noncooperative game model applications, pointed out the limitation of negotiation support systems developed by different researchers in solving practical problems, and concluded that development of negotiation models appropriate for studying issues of multiple objectives and n-players with incomplete information is feasible. Similarly, Wang (2003a) provided some preliminary recommendations for improvement of the water allocation regime of the Yellow River, on the basis of a qualitative analysis of water allocation evolution history.

This method tackles sustainability of water allocation through analyzing the system stability using game theory. It considers the roles of market, regime, and tradition, and accommodates the principles of efficiency, fairness, and stability by satisfying market, social value, and sustainable development requirements in relation to water allocation. Therefore it is, at the present stage, a comparatively more comprehensive and effective way of presenting water allocation issues. This approach is thus adopted in the discussions and case study presented in this chapter.

12.3 Game-theoretic modeling of water allocation and case study of the Yellow River

12.3.1 Game theory and its application to water allocation analysis

Game theory is a good mathematical tool for analyzing resource use conflict issues. Unfortunately its application in the water sector to date is very much limited to qualitative study of water quality and environment-related conflicts, and cooperation in sharing of transboundary water. Early applications of the game theory approach include studies of international river issues by Rogers (1969) and Dufournaud (1982). Starting from the late 1980s, game theory found wider applications in water resource planning, such as benefit-sharing analysis (Dinar et al. 1986) and a cooperative game approach to costs and benefit distribution (Dufournaud and Harrington 1990). Scholars in China have begun to tap game theory for qualitative assessment of water allocation conflicts and water market issues in recent years (Liu et al. 2002; Kong et al. 2005).

12.3.2 Mathematical representation of water allocation

Before discussing the mathematical model for water allocation analysis, it is important to clearly define rational water allocation. While there are different descriptions in the literature, Chen and Wang (1996) provided a very clear definition, that is, rational water allocation refers to distribution and dispatching of water resources among different water users, in a river basin or hydraulic unit, following the principles of efficiency, fairness, and sustainability, through various structural and nonstructural measures and the market mechanism, to rationally curtail water demand and ensure effective water supply and preservation of river ecology. With this definition, the following describes the mathematical presentation of water allocation. For the purpose of illustration, a water supply system is used as an example.

Water supply system. If the total amount of exploitable water is Z_0, of which x_0 ($x_0 < Z_0$) is diverted by a water supply system without delivery loss, and x_i represents the amount of water supplied to region i, then the water balance and water consumption of each region can be calculated.

Assuming that the total return water, represented by k as part of total water use ($0 < k < 1$), goes into the aquifer, $(1 - k)$ will be the evapotranspiration (ET). If k remains constant across regions, then G_i, the changes in groundwater arising from groundwater withdrawal and recharge from return surface water in region i, can be derived accordingly.

Economic benefit of a water supply system. The costs of water supply include fixed (C_0) and variable costs. The variable costs for delivering water x_i to region i, which is d_i away from the water source, is $\zeta_0 d_i x_i$, where ζ_0 is constant. Assuming the water supplier is nonprofitable and it provides water to region i at price p_i, then its total costs equal to water supply revenue plus government subsidy S, namely:

$$C_0 + \zeta_0 \Sigma\, d_i x_i = \Sigma\, p_i x_i + S, \tag{12.1}$$

where S ($S < 0$) is net tax. The costs of groundwater withdrawal increase with the water amount pumped and the fall in the water table Q. The groundwater withdrawal cost function of region i will then be $C_i(F_i, Q)$, subject to

$$\frac{\partial C_i}{\partial F_i} > 0,\ \frac{\partial C_i}{\partial Q} < 0,\ \frac{\partial^2 C_i}{\partial F_i^2} > 0,\ \frac{\partial^2 C_i}{\partial Q \partial F_i} < 0,\ \frac{\partial^2 C_i}{\partial Q^2} < 0.$$

Obviously the marginal cost of groundwater withdrawal declines as the water table drops. Within a particular region, it does not incur additional costs to transfer water or pump water. Because a region is considered a single and unified decisionmaking unit, and it only serves water users within the region, the production function of region i is

$$f^i = f^i(A_i, I_i, y_i),$$

where y_i represents the level of other inputs. Assuming that there exist secondary differential and concave production functions, and x_i is not allocated based on existing water rights and not subject to water rights, and the total amount of water inflow is decided by region i itself, then the aggregate net production function of each region will become the following:

$$\Pi_i = f^i(A_i, I_i, y_i) - C_i(F_i, Q) - p_i x_i - y_i.$$

Social system description. There are $(n + 2)$ players in this game: n regions, the water pricing agency, and the government. It is well known that water pricing is a major political matter because of its effect on the welfare of all parties, and is an important means for government to manage water resources. Thus any water price the government is not happy with cannot be passed.

A water resource allocation management organization or water supply organization is a nonprofitable legal entity. Its performance is normally judged by the cost-effectiveness of its operation. Concern for a water supply organization can also develop into an interest in economic efficiency in general. Accordingly, the organization's social objective function can be defined as

$$u_{0i} = V_i(p, Q) = \{f^i[A_i(Q_i), I_i, y_i] - C_i(F_i, Q) - y_i\} - \zeta_0 \sum_{i=1}^{n} d_i x_i - C_0. \quad (12.2)$$

However, the decisionmakers in a water supply organization are also interested in such personal benefits as political responsibility, personal promotion, better material welfare, and succeeding in competition. To obtain such personal benefits, they need to win support and avoid the criticism of other parties or players. However, this is not a one-sided relationship, as the decisionmakers in the organization are able to reward and penalize the other parties or players. These relationships are introduced into the model by the devices of strength of power and cost of power function, so the extended objective function of a water supply organization can be written as

$$U_{oi} = u_{oi} + s_i(c_i^0, \delta_i) + S_{n+1,o}(c_{n+1}^0, \delta_{n+1}^0) - c_0^{n+1}, \quad (12.3)$$

in which s_i is the power of player i toward the water supply organization; $S_{n+1,0}(c_{n+1}^0, \delta_{n+1}^0)$ is that of government toward the organization; c_i^0 is the cost of player i in influencing the organization's decision; and δ_i is a variable reflecting the tactics of player i toward the organization, that is,

$$\delta_i = \begin{cases} a_i \ \textit{when} \text{ player } i \text{ rewards WSO} \\ \beta_i \ \textit{when} \text{ player } i \text{ penalizes WSO} \end{cases},$$

where c_{n+1}^0 is the cost of government in influencing the water supply organization; δ_{n+1}^0 is a variable reflecting the tactics of government toward the organization; and c_0^{n+1} is the cost of the organization in influencing government's decision through nonprice factors. Note that all regions ($i = 1,2,3,...,n$) and the government ($i = n + 1$) have influence over the water supply organization's decisionmaking.

The objective function of region i can be defined as its net income, namely,

$$u_i = \Pi_i(p_i, Q). \tag{12.4}$$

Accordingly its expanded objective function will be

$$U_i = u_i - c_i^0 - c_i^{n+1}, \tag{12.5}$$

where c_i^0 and c_i^{n+1} are costs of region i in influencing the water supply organization and government respectively.

The values of S_i, $S_{n+1,\,0}$ and C_i can be determined as weighted averages based on the initial values provided by experts.

Water resource allocation. The relevant policy measures are confined to water price p and net subsidy S. p is positive and subject to the amount of water disposable by the water supply organization, that is, $\Sigma\, x_i(p_i, Q) \leq Z_0$.

It is obvious from equation (12.5) that p and S are interdependent. The objective function of water allocation for region i can be defined as follows:

$$W_i = u_{0i} + b_i u_i = V_i[p, Q^s(p)] + b_i\, \Pi_i(p_i, Q^s). \tag{12.6}$$

The long-term goal of water resource allocation in a stable system is to maximize the value of the above management function W and the net social gains or benefits of the entire water resource system, for example a river basin.

12.3.3 Nash–Harsanyi bargaining model

A standard game theory approach assumes that every player possesses the same information at the beginning of the game. In reality, information asymmetry among players is a common phenomenon. To accommodate the need for modeling such circumstances, Harsanyi developed in 1967–68 the Bayesian model for incomplete information, and defined the consistent Bayesian game as all players sharing the same prior belief, their different beliefs at the start of the game originating from the observation of different stochastic variables, namely the different experiences of the players. As a result, the general consistent Bayesian model developed by Harsanyi became the standard analysis framework of information economics (Zhang 1996).

On the issue of fairness in a society, Harsanyi expressed the view that if sufficiently informed any group of individuals born with the same preference are able to judge the fairness of a society for themselves individually and for others. If one assumes that all players in a game are born with the same preference, each player will have its individual utility function U_i. Then one can compare the utilities of different players and sum them up to arrive at the so-called social welfare function:

$$W = \sum_{i \in I} U_i.$$

Later Harsanyi, through his second theorem, changed the equal factor ($1/n$) into a variable, assuming that the probability of certain events for some players is higher than for others, that is, the weight of player i (a_i) varies. So the social welfare function can be rewritten as

$$W = \sum_{i \in I} a_i U_i.$$

Furthermore, according to the Nash–Harsanyi theorem on the independence of independent solutions, for a known negotiation problem, if one or more original or inappropriate solutions are dropped from the solution pool, the final solutions to the original problem and problems so modified are consistent (Zhang 1996). Therefore, the Nash–Harsanyi negotiation model can be simplified to a matter of maximizing the value of an objective or management function. Similarly, a public choice problem could be converted into that of finding a solution to maximizing an objective function. Considering the multiplayers involved and the highly interactive nature of water allocation, the study in this chapter adopts the Nash–Harsanyi negotiation (or bargaining) model for evaluation of water allocation regimes.

12.3.4 Yellow River water allocation regime analysis

Water allocation among the 11 different provinces of the Yellow River basin, including Tianjin municipality, dates back to 1954, when 47 billion cubic meters of the total 54.5 billion cubic meters natural runoff were allocated for irrigation. Adjustments were made in 1959, 1961, 1970, and 1987 according to various criteria. The amount of water allocated for consumptive use approved by China's State Council in 1987 was 37 billion cubic meters, against the estimated annual average runoff of 58 billion cubic meters.

However, in the Yellow River basin, the lower reaches, particularly the delta area, began to run dry for many miles over a long period of each year in the 1990s. There is no doubt that natural factors and human activities all contributed to this phenomenon, but the root cause was lack of clear water entitlements and an appropriate water allocation and management mechanism.

Prior to 1999, the major water control works and canal intakes of large irrigation schemes along the Yellow River belonged to different jurisdictions and sectors. Despite that a river basin commission, the Yellow River Conservancy Commission, was established with a somewhat vague legal status, hardly able to fulfill its mandate of effective water allocation and dispatchment in the basin under the hybrid basin and jurisdiction management system (Wang 2003b). Consequently, the lower reaches of the river dried up each and every year, leading to social crises, heavy economic losses, siltation of the main river course, and a deteriorating eco-environment in the river delta.

To address these issues, the national government authorized the Yellow River Conservancy Commission to undertake unified water resource management and water quantity dispatchment with an upper ceiling of water use for each jurisdiction (province) in the basin. The impact was prominent and far reaching. From 2000, no drying up of the river has occurred up to the time of writing, even under consecutive years of drought. Some argue that this has resulted from the operationalization of a new giant multipurpose reservoir on the main stream in 2000. However, studies indicate that a key factor has been the enforcement of unified water management and water use quota measures (Qian et al. 2001).

12.3.5 Allocation regime options

Based on historical surface water allocation practices and water conflicts in the basin, this chapter endeavors to examine three options of water allocation regimes commonly used.

Option 1: Water allocation through unregulated withdrawals. Under such a regime, each region withdraws water freely according to its needs. The actual water withdrawal depends mainly on its diversion capacity and the costs of increasing diversion capacity.

Option 2: Water allocation through prior agreed water quotas. Each region withdraws water, under the oversight of a river basin organization, within the cap of a predetermined quota that specifies minimum ecological flows and limits on pollution loads discharged to the basin (both point and nonpoint). The upper limit (quota) of water use for each region is determined through an intensive consultation process. Water supply will be stopped once withdrawal reaches the allocation quota for that region.

Option 3: Water allocation through a water market with clearly defined (tradable) water rights and limited regulation of public good usage. Within such a regime, regions or use sectors have well-defined water rights, including minimum ecological flows, for their respective river segments. They can withdraw water freely within their entitlement limit. When they need more water than they are entitled to, they need to first purchase the additional use right. The basin organization takes the responsibility of ensuring river ecological environment preservation and sustainable development. Water users can trade their use rights through a market mechanism under the limited regulation of basin organization. Water pricing is determined by the market.

12.3.6 Model development and analysis

The following Nash–Harsanyi bargaining model was used in developing the model for the Yellow River case analysis:

$$Max \prod_{i=1}^{n} (W_i - c_i)$$

$$\text{s.t. } W_i \geqslant c_i \quad i = 1,2,\ldots,n$$

$$\sum_{i=1}^{n} x_i \leq Q_c$$

$$\sum_{i=1}^{n} x_i^0 \leq Q_c$$

$$x \in R \tag{12.7}$$

where c_i is current net benefit of user i; W_i is negotiated net benefit of user i; x_i^0 and x_i are current and negotiated water uses respectively; and Q_c is total amount of exploitable water.

Therefore, the utility function of water users can be described similarly to equation (12.6), namely:

$$W_1 = W(x_1)$$

$$W_2 = W(Q_c - x_1, x_2)$$

$$\ldots$$

$$W_n = W(Q_c - x_1 - x_2 - \ldots - x_n, Q_c - x_2 - \ldots - x_n, \ldots, x_n)$$

By computing the values of s_i, $S_{n+1,0}(C_{n+1}^0, \delta_{n+1}^0)$, and c_i^0, and making assumptions about δ_i and c_i corresponding to different regime options, one can convert the public choice problem of multiple water users (players) to a problem of maximizing the water allocation management function:

$$\prod_{i=1}^{n} (W_i - c_i).$$

To solve this problem, one needs to compute the available water amount Q_{ai} for each water use region under every option of water allocation and ensure that actual water use of each region Q_i is no more than Q_{ai} (that is, $Q_i \leq Q_{ai}$) to allow for ecological flows.

For option 1:

$$C_i^0 = \begin{cases} = \text{water supply cost} \times \text{water supply amount, when water use is smaller than existing withdrawal capacity} \\ = \text{water supply cost} \times \text{water supply amount} + \text{incremental supply cost, when water use is bigger than existing withdrawal capacity} \end{cases}$$

$$c_i^{n+1} = 0.$$

Since costs only increase with water use and withdrawal capacity, users will continue to increase their water use until benefits from increased water use fall below the corresponding costs. As equilibrium can be reached under the bargaining model when no user is worse off than at present, an allocation agreement acceptable to all users can only be found under this option when there is abundant water in the basin.

For option 2:

$$C_i^0 = \begin{cases} = (\text{water supply cost} + \text{water price}) \times \text{water supply amount, when water use is smaller than existing withdrawal capacity;} \\ = (\text{water supply cost} + \text{water price}) \times \text{water supply amount} + \text{incremental supply cost, when water use is bigger than existing withdrawal capacity} \end{cases}$$

$$c_i^{n+1} = (-\delta_i + \text{management cost}) \times \text{water supply amount.}$$

Under this option, management facilities and personnel are required to ensure users stay within their water use quota. That means there will be additional costs of monitoring and enforcement. The amount of water use will change with the water allocation quota and water price. Users' response to water quota and water price under this option was analyzed and discussed in a study of north China's river basins (World Bank 2002).

For option 3:

$$C_i^0 = \begin{cases} = (\text{water supply cost} + \text{water price}) \times \text{water supply amount} - \text{water right trade revenue, when water use is smaller than water right;} \\ = (\text{water supply cost} + \text{water price}) + \text{incremental water supply cost} + \text{water right trade revenue, when water use is bigger than water right} \end{cases}$$

$$c_i^{n+1} = -\delta_i \times \text{water supply amount.}$$

Under this option, lower water use means lower costs, so that users will intentionally reduce their water use when the cost savings from using less water exceed the benefits from increasing water use. The values of c_i^0 and c_i^{n+1} for all three options can be obtained, and then U_i and W_i can be derived from equations (12.5) and (12.6).

For modeling purposes, the whole basin is divided into a number of regions, based on the natural geographic locations, socioeconomic features, and administrative boundaries (Figure 12.1). Equation (12.7) is adopted to develop the mathematical model for the water allocation. The main data and parameters used are summarized in Tables 12.1 to 12.4.

For the purposes of illustration, the modeling results are shown in Tables 12.5, 12.6, and 12.7 in terms of water use and net social benefits, and the corresponding figures, 12.2, 12.3, and 12.4, respectively show net social benefits across regions for the three allocation regime options.

It is evident from the above calculations that no rational results can be derived for option 1 from the model in extremely dry years, which helps explain the phenomenon of consecutive years of drying up of the lower

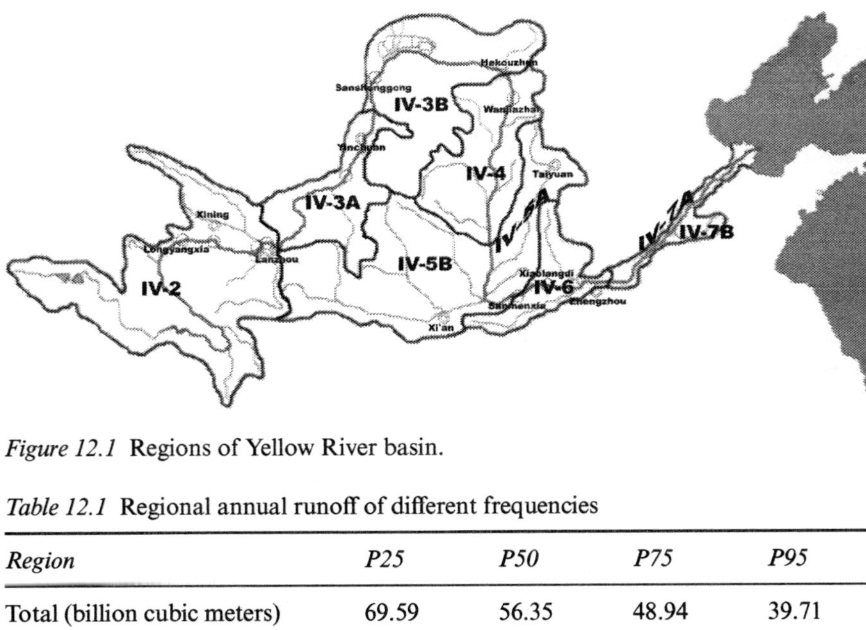

Figure 12.1 Regions of Yellow River basin.

Table 12.1 Regional annual runoff of different frequencies

Region	*P25*	*P50*	*P75*	*P95*
Total (billion cubic meters)	69.59	56.35	48.94	39.71

Table 12.2 Water demand 2000

Region	*IV-2*	*IV-3A*	*IV-3B*	*IV-4*	*IV-5A*	*IV-5B*	*IV-6*	*IV-7A*	*IV-7B*
2000 (billion cubic meters)[a]	1.363	2.049	0.000	0.375	3.746	0.000	1.381	0.901	0.000

a. Excluding irrigation

Table 12.3 Effective irrigated area of regions 2000

Region	*IV-2*	*IV-3A*	*IV-3B*	*IV-4*	*IV-5A*	*IV-5B*	*IV-6*	*IV-7A*	*IV-7B*
2000 (million mu)[a]	3.313	8.690	13.727	5.079	8.130	14.750	4.548	5.762	3.507

a. 1 mu = 0.067 hectare

Table 12.4 Regional cropping patterns (% of cropping area)

	Region					
Crop	*IV-2*	*IV-3A*	*IV-4*	*IV-5A*	*IV-6*	*IV-7A*
Winter wheat			35	50	65	67
Summer wheat	68	33	2.5	0.6		
Spring maize	32	49	38	30	6	
Summer maize		15	37	38	58	60
Rice		5.7			4.1	8.9
Cotton			5	9	13	12
Vegetables	3.5	1	2	4.6	3	4.1
Peanuts		9.2				
Soybeans	6.5	2.3			14.2	13
Millet			14.8	11		
Sesame				4		

Note: Totals for each region exceed 100 due to double cropping.

Table 12.5 P50 water use and net social benefit

	Option 1		*Option 2*		*Option 3*	
Region	*Water use*[a]	*Net social benefit*[b]	*Water use*[a]	*Net social benefit*[b]	*Water use*[a]	*Net social benefit*[b]
IV-1	0.30	0.11	0.26	0.14	0.27	0.09
IV-2	3.80	2.95	3.40	2.63	3.22	3.10
IV-3	18.10	13.80	18.17	13.18	17.86	14.70
IV-4	1.10	1.02	1.25	0.91	1.30	1.54
IV-5	10.70	8.54	10.61	9.33	11.50	12.10
IV-6	3.40	3.31	2.85	3.98	2.40	3.80
IV-7	4.23	5.13	4.62	7.23	5.80	10.00
IV-8	0.48	0.67	0.34	0.43	0.45	0.65
Total	42.11	35.54	41.50	37.84	42.80	45.98

a. Billion cubic meters
b. 10^9 yuan

reaches of the Yellow River in the 1990s. It can also be observed that the total net social benefit is the highest under option 3, followed by option 2. Nevertheless, it is not very obvious whether option 3 is better than option 2 because the transaction costs of option 3 are not easy to quantify at this stage.

Table 12.6 P75 water use and net social benefit

Region	*Option 1*		*Option 2*		*Option 3*	
	Water use[a]	*Net social benefit*[b]	*Water use*[a]	*Net social benefit*[b]	*Water use*[a]	*Net social benefit*[b]
IV-1	0.31	0.14	0.3	0.14	0.3	0.14
IV-2	4.89	3.09	4.6	3.02	4.01	3.47
IV-3	21.76	15.14	19.9	14.45	17.98	15.14
IV-4	2.98	1.95	2.4	2.09	2.42	2.34
IV-5	11.50	9.33	12.7	9.76	13.1	11.59
IV-6	3.71	3.34	3.6	3.56	4.23	4.31
IV-7	4.87	8.32	5.7	9.33	7.23	12.02
IV-8	1.34	1.21	0.46	0.60	0.45	0.60
Total	51.36	42.52	49.66	42.95	49.72	49.61

a. Billion cubic meters
b. 10^9 yuan

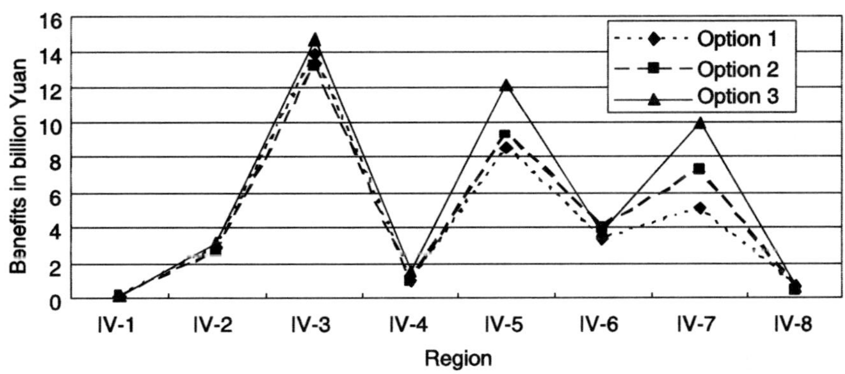

Figure 12.2 Net social benefit in the case of P50.

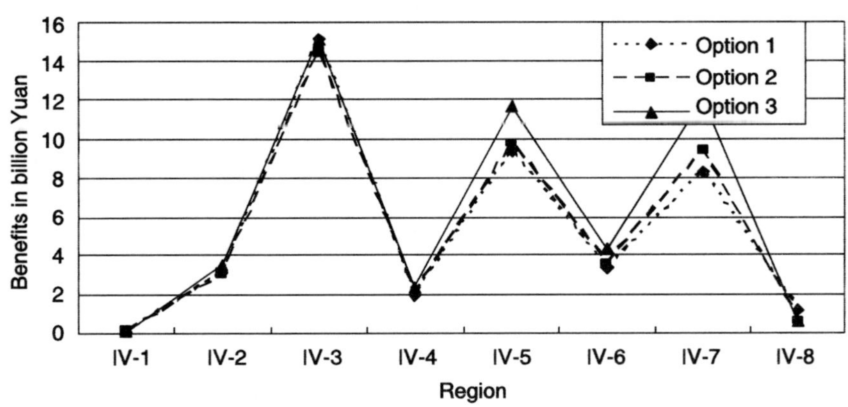

Figure 12.3 Net social benefit in the case of P75.

Table 12.7 P95 water use and net social benefit

	Option 1[a]		*Option 2*		*Option 3*	
Region	*Water use*	*Net social benefit*	*Water use*[b]	*Net social benefit*[c]	*Water use*[b]	*Net social benefit*[c]
IV-1	n.a.	n.a.	0.26	0.30	0.26	0.30
IV-2	n.a.	n.a.	3.30	3.55	3.29	3.55
IV-3	n.a.	n.a.	15.72	16.22	14.60	15.85
IV-4	n.a.	n.a.	0.95	1.20	0.95	1.20
IV-5	n.a.	n.a.	9.26	13.18	9.36	13.49
IV-6	n.a.	n.a.	2.94	3.39	2.93	3.24
IV-7	n.a.	n.a.	3.68	5.62	3.73	6.46
IV-8	n.a.	n.a.	0.32	0.83	0.69	0.83
Total	n.a.	n.a.	36.82	44.30	35.83	44.92

n.a. Not applicable.
a. There are no rational results that can be derived for option 1 from the model in extremely dry years
b. Billion cubic meters
c. 10^9 yuan

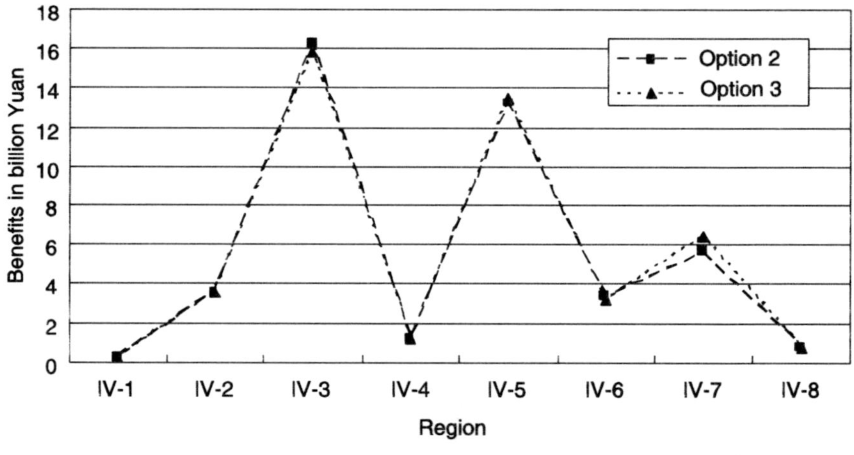

Figure 12.4 Net social benefit in the case of P95.

Moving from option 1 to option 2, the general trend is that the net social benefits decrease in upper reach regions, but not to the extent by which the benefits increase in the lower reaches. From option 1 to option 3, the net social benefits virtually increase throughout the basin, although the transaction costs of water rights are not fully accounted. Theoretically the modeling results tend to suggest that option 3 would be the most ideal option. However, for the Yellow River basin, the fact of the matter is that a water market is just emerging, and there is an extremely long way to go before such an allocation regime could find its roots there.

12.4 Conclusion and policy implications

On the basis of the analysis summarized above, some preliminary conclusions can be drawn. It should be noted that despite the fact that the analysis is basin specific, the game theory approach to water allocation regime analysis as a decision support tool can be extended to other river basins with adaptation to local water and socioeconomic conditions and political systems.

The amount of water available in the Yellow River basin at present is not sufficient to meet the increasing water demands of domestic use, production, and the eco-environment. Adoption of a much more effective system of water allocation is the best solution to sustainable water resource development in the basin, which requires changes to the present allocation regime (for example in design, enforcement, and monitoring mechanisms), the operating environment, and behaviors of water users.

Of the three options of water allocation regime discussed, the first option, unregulated water withdrawal and diversion, is the least desired and has the lowest total net social benefit, and would be harmful to river ecology and the water environment, and to management of water use conflicts among different regions of the basin. Such an allocation regime would encourage users to pursue their short-term benefits, leading to increasing water shortage and waste of water resources. One should be very cautious in applying this allocation mechanism, because even for a water-rich basin, it is likely to lead to a further decline in water quality, an increase in salinization, and deterioration of the ecological environment because of uncontrolled overuse of water.

The second option, water allocation through prior agreed water quotas for different users, including the minimum ecological flows and limits to the pollution load discharge to the river, is practiced in some parts of the world. This allocation mechanism is close to the present allocation arrangement in the Yellow River basin, and it requires a strong management authority in place. The need for management facilities and management costs would increase with the level of competition among water uses. From the perspective of total basinwide benefits, this may not be an optimal allocation regime, especially for a water-scarce basin such as the Yellow River basin, where there is very strong competition for water among various users.

This can, however, be a good intermediate allocation mechanism for some years to come, before the government and the river basin organization are able to work out and establish a more rational water entitlement system, particularly for allocation of irrigation water, given the great difficulties and long path linked with charging irrigation water in most developing nations (Dinar et al. 1997). If the water quotas are determined with adequate consultation and participation of the stakeholders and enforced effectively, this can be a very good water allocation mechanism even in the long run.

The third option, allocation through clearly defined water rights, serves to promote water conservation and efficient use. It is the most favorable regime in terms of total social benefit for the basin. For the Yellow River basin, it is still premature to adopt such an allocation mechanism. Some of the most crucial requirements to lay the foundation for transition into water market allocation are strict control of the monitoring and supervision costs of the government and river basin organizations; making best use of market forces in optimizing water allocation; and limitation of government regulation to the bare minimum for managing the perceptions and behavior of water users and involving them in allocation decisions.

Nevertheless, one must remember that it takes tremendous effort to have a functional water market. At times it could become a close to impossible task, especially in the case of irrigation, as international experiences have demonstrated (Bruns et al. 2005). Proper functioning of water markets requires a good operating environment and effective monitoring and enforcement, involving high transaction costs. For instance, in the case of water management in the Murray–Darling basin of Australia, it has proven an enormous effort for the water authorities to enforce water withdrawal control, in particular control over groundwater pumping. Furthermore, there is a risk of an increase in actual water use resulting from a water market without proper regulation and stakeholder participation. Also, minimum ecological flows would have to be enforced through quotas because of natural variations in water flows and possible failures in enforcing water rights. Despite all of these difficulties, this option can be the preferred model for most river basins in the long run, especially in countries with well-established market economies.

Finally, analysis of the modeling results indicates that the development history of water allocation regimes in the basin and the perceptions of different stakeholders have a profound effect on allocation regime evolution. Quantitative analysis of such effect calls for further modeling study using cognition theory.

References

Bruns, B.R., Ringler, C., and Meinzen-Dick, R. (eds) (2005) *Water Rights Reform: Lessons for Institutional Design*, Washington, DC: International Food Policy Research Institute.

Carraro, C., Marchiori C., and Sgobbi, A. (2005) "Application of negotiation theory to water issues", Policy Research Working Paper 3641, World Bank, Washington, DC.

Chen, J. and Wang, H. (1996) *Fundamental Principles and Theory of Water Resources*, Beijing: Water and Hydropower Press.

Dinar, A., Rosegrant, M.W., and Meinzen-Dick, R. (1997) "Water allocation mechanisms: principles and examples", Policy Research Working Paper 1779, World Bank, Washington, DC.

Dinar, A., Yaron, D., and Kannai, Y. (1986) "Sharing regional cooperative gains from reusing effluent for irrigation", *Water Resource Research*, 22(3): 339–44.

Dufournaud, C.M. (1982) "On the mutually beneficial cooperative scheme: dynamic change in the payoff matrix of international river basin schemes", *Water Resource Research*, 18: 764–72.

Dufournaud, C.M. and Harrington, J.J. (1990) "Temporal and spatial distribution of benefits and costs in river basin schemes: a cooperative game approach", *Environment and Planning* A22: 615–25.

Fu, C. and Hu, Z. (2004) "A literature review of water right studies", *Advances in Water Science and Technology*, 2004(2).

Guan, Y. (2002) "Essence of rational water resources allocation approach", *China Water Resources*, 2002(5).

Hall, W.A. and Dracup, J.A. (1970) *Water Resources Systems Engineering*, London: McGraw Hill.

Hardin, G. (1968) "The tragedy of the commons", *Science*, 162: 1243–8.

Kong, K. et al. (2005) "An analysis of water market games", *Academic Journal of Water Resources*, 2005(4).

Liu, C. and He, X. (1999) *China Water Strategies for the 21st Century*, Beijing: Press of Science and Technology.

Liu, W. et al. (2002) "Game theoretic analysis of water distribution conflicts", *System Engineering Theory and Practice*, 1: 16–25 (in Chinese).

Maass, A., Hufschmidt, M.M., Dorfman, R., Thomas, H.A., and Marglin, S.A. (1962) *Design of Water Resource Systems*, Cambridge, MA: Harvard University Press.

Ministry of Water Resources, China (2001) "Water resources", in "Country report: China sustainable development report", unpublished internal report.

Qian, Z., Zhang, G., et al. (2001) *Reports on Water Resources Strategy Study for China's Sustainable Development, Volume 1*, Beijing: Water and Hydropower Press.

Rogers, P.P. (1969) "A game theory approach to the problem of international river basins", *Water Resource Research*, 5(4): 749–60.

Wang, Y. (2003a) "On improvement of water allocation regime in the Yellow River basin", *People's Yellow River*, 25(4).

Wang, Y. (2003b) "Water dispute in the Yellow River: challenges to a centralized system", *China Environmental Series*, 3: 94–8, Woodrow Wilson Center, Washington, DC.

World Bank (2002) *China: Agenda for Water Sector Strategy for North China*, World Bank Report 22040-CHA.

World Bank and Ministry of Water Resources, China (1993) *China: Yellow River Basin Investment Planning Study*, World Bank Report 11146-CHA.

Xu, X. et al. (1997) *Theory and Methodology of Water Resources Planning for North China*, Zhengzhou: Yellow River Press.

Zhang, W. (1996) *Game Theory and Information Economics*, Shanghai: Shanghai People's Press.

Zhang, Y. (1999) *China Water Resources and Sustainable Development*, Beijing: Water and Hydropower Press.

13 Contributions of game theory to the analysis of consumer boycotts

Philippe Delacote

Consumer boycotts are commonly used by citizens protesting against unfair environmental, social, or health practices by firms. Game theory offers simple tools to analyze the impact and the potential for success of such actions. Consumer boycotts may be viewed as a war of attrition between a group of consumers and the targeted firm. Focusing on environmental boycotts, this chapter shows that a simple trade-off between the consumers' hurting capacity and their opportunity cost of boycotting makes the potential for success of a boycott quite low.

13.1 Introduction

A new consumption pattern has emerged recently whereby citizens use consumption as a political act, "a new way to save the world" (McLaughlin 2004). These political consumption practices constitute a way to signal preferences and to conciliate consumption with social, environmental, or health considerations. Consumer boycotts may be considered by dissatisfied consumers as a means by which they can actively compensate for governmental inactivity, in effect acting as a substitute for public policies. The objective is to put sufficient pressure on the target to make it adopt more acceptable practices.

Most researchers have focused on field studies (Miller and Sturdivan 1977; Pruitt and Friedman 1986; Garrett 1987; Koku et al. 1997; Teoh et al. 1999) or the history of consumer boycotts (Friedman 1985, 1991, 1995; Smith 1990). Tyran and Engelmann (2005) provide an experimental analysis of consumer boycotts. Overall, most papers find a weak impact of consumer boycotts on firms' behavior.

Only few papers provide theoretical analysis of consumer boycotts. Innes (2006) considers a duopoly choosing between a clean and a dirty technology, where environmental organizations may invest in consumer boycotts to deter choice of the dirty technology. The boycott effectiveness is determined by the environmental organization's investment. Baron (2002) considers that the action of boycotting by some consumers provides information to other citizens about the seriousness of a situation. Boycotting constitutes a way for consumers to signal their private information. Diermeier and Van Mieghem (2005) describe coordination between boycotting consumers as a stochastic process with threshold effects.

Applying a simple game theory tool, this chapter makes a simple point: boycott successes are quite unlikely, because of a simple trade-off between the opportunity cost of boycotting and the boycott potential to hurt the firm's profit. A boycott is here considered effective if it induces a change in the targeted firm's behavior consistent with the boycotting group's objective. Therefore, cases of boycotts that aim only to signal disapproval to their targets are not considered.

The focus of the chapter is on environmental boycotts, though consumer boycotts on social and health grounds follow roughly the same analysis. Consumer boycott based on environmental arguments is a strategy commonly used by environmental nongovernmental organizations. An example is the boycott of cosmetic firms (for example Procter and Gamble, Colgate-Palmolive) because of their use of animal testing. Another case is the boycott of major oil companies (for example Total, Esso, Shell) for their alleged environmental damages and lobbying efforts to deter climate change policies. Some large fast-food companies (for example McDonald's) have been targeted by boycott campaigns because of their supposed environmentally unfriendly methods of producing meat. Finally, some nongovernmental organizations support the boycott of noncertified tropical timber to protest against unsustainable harvest practices.

Consider a firm producing a good with a polluting technology, with no government intervention to internalize the negative externality. This firm could opt for another technology, less or not polluting, but more expensive. The choice of the cheap and polluting technology would result from a desire to maximize profits. The success of an environmental boycott is therefore determined by its capacity to hurt the firm's profit sufficiently to make the second technology more profitable. The main factor determining the success of the boycott is therefore consumer preference, which determines the demand structure.

The conditions for success of an environmental boycott depend on several market characteristics. First, the consumers' environmental preferences may create some scale for ecological certification and product differentiation. With free entry, a second firm may enter the market and provide the good using clean production methods. Market structure is not considered explicitly in this chapter. Only one firm is boycotted and the existence of an imperfect substitute, which produces more cleanly but provides lower utility, is considered.

Second, information is crucial to both producers and consumers. On the one hand, a firm needs, if it is to maximize its profits, complete and perfect information about demand and consumer preferences (otherwise, there is still room for consumers to signal their preferences through a boycott). Consumers, also, need information on demand characteristics, available technologies, and boycott modalities.

Finally, coordination issues and strategic considerations need to be taken into account. Even a potentially successful boycott may fail because of coordination failures. Moreover, boycotting is subject to free riding. Any individual consumer, even if dissatisfied with the use of the polluting technology and hoping for the boycott to succeed, has an incentive to free ride and to consume the good anyway. Anonymity of consumer behavior reinforces this incentive.

Of course, with perfect information, no coordination issue, and no free-rider behavior, one could only witness successful boycotts. If such was the case, the perfectly informed consumer would only participate in a boycott if its success was certain. This ideal scenario is, however, considered here in order to determine which patterns of demand provide room for successful environmental consumer boycotts. It is assumed therefore that both consumers and the firm have perfect information about demand patterns and the production process. Moreover, by assumption, the dissatisfied consumers behave as one community, which avoids coordination failures and free riding. The addition of coordination issues and strategic behaviors would only decrease the likelihood of the boycott being successful.

In the situation described, an environmental consumer boycott would be a complete information war of attrition with asymmetric preferences between the targeted firm and the boycotting consumers. Complete information war of attrition models were first introduced by Maynard Smith (1974, 1982) in studies of animal behavior. Economic applications of war of attrition models include predatory pricing (Roth 1996), exit in oligopoly (Fudenberg and Tirole 1986), and the provision of public goods (Bilodeau and Slivinsky 1996). Kornhauser et al. (1989) and Fudenberg and Tirole (1986) proposed criteria for selection among potential perfect equilibria.

Overall, the potential for success of a consumer boycott depends mainly on the trade-off between the hurting capacity of the boycott group and the opportunity cost of boycotting. The consumers most able to hurt a firm's profit have large amounts of consumption. Thus, their opportunity cost of boycotting is high. Overall, this simple trade-off makes the likelihood of boycott success quite low.

Section 13.2 describes consumer boycott as a complete information war of attrition. Section 13.3 analyzes which factors influence the outcome of the game. Finally, the analysis is applied to real-life boycotts in section 13.4. Section 13.5 concludes.

13.2 Boycott as a war of attrition with perfect information

A war of attrition is a model of aggression between two players. The game takes the form of a succession of identical periods. In each period, the two players choose simultaneously between remaining in the game or withdrawing. The model is stationary: each period represents the same type of problem for both players, with no information gain or change in costs or benefits. The winning player is the one able to remain longer in the game. In the context studied here, the two players are a group of consumers and a firm. Some consumers refuse to consume the firm's good as long as it is produced with a polluting technology. Boycotting consumers act as a single group. Potential coordination failures between consumers are not considered explicitly.

Overall, both players compare their maximum conflict duration, which is the point in time after which they would never plan to remain in the game. Indeed, with net cumulative payoffs decreasing with time, there is a point in time at which these payoffs become negative. Basically, the boycott succeeds if the consumers' maximum boycott duration is longer than the maximum conflict duration of the firm.

13.2.1 Technology choice and consumers behavior

In the first stage, the firm chooses between two available technologies. The first is cheap but polluting, while the second is clean but more expensive. In this case the firm chooses the dirty technology, which implies that it generates larger profit than the clean one. The firm does not consider potential boycotts when choosing its technology.

A group of environmentalist consumers would prefer the firm to use the clean technology. An environmental organization announces a consumer boycott, requiring any consumer dissatisfied with the use of the dirty technology to stop consuming the good. Boycotting consumers switch their consumption of the good to consumption of an imperfect substitute that provides lower utility, but for which the method of production is clean. Thus boycotting is costly in terms of welfare, because it decreases the boycotting consumers' utility. In an extreme case the targeted firm is in a monopoly position, and there is no substitute available on the market. At another extreme, if the market is very competitive and fragmented there is room for ecological certification; a firm may provide the good with clean production. In that case, boycotting is welfare improving. More generally, a better substitute generates lower utility loss. The opportunity cost of boycotting is therefore the difference between the status quo utility (that is, consuming the good produced using dirty technology and not boycotting) and the utility of boycotting (and consuming the substitute).

The aim of the boycott is to decrease the firm's profit sufficiently to make the clean technology desirable. Success depends essentially on the size of the boycotting group and its capacity to coordinate. A consumer boycott therefore represents a kind of war of attrition, with an asymmetry in the players' motivations.

The set of strategies is as follows. In each period, the environmentalists choose whether to continue or discontinue the boycott, while the firm chooses whether to keep on using the dirty technology or to switch to cleaner production. It is assumed that switching technology is costless.

13.2.2 Structure of the game

The game proceeds as follows. Both players consider how long they could stay in the game without making loss. The consumers' maximum boycott duration and the maximum conflict duration of the firm are the points in time at which their cumulative net payoffs become negative. More precisely, the firm's maximum conflict duration is the point at which it would be worse off than if it had chosen the clean technology since the beginning. Conversely, the boycott maximum duration is the point at which boycotting consumers are worse off than if they had decided not to boycott at all. Obviously, no player would consider staying longer in the game than its maximum duration. Therefore the boycott is successful if the maximum boycott duration is longer than the maximum conflict duration.

This setup ignores coordination patterns and free-riding possibilities. However, it describes the necessary conditions of the demand patterns for a successful boycott. Introducing coordination issues would only decrease the boycott likelihood for success.

13.2.3 Outcome of the game

The outcome of the game depends on the two maximum durations. If the consumers know that they cannot stay long enough in the game to induce a change in the firm's behavior, they will choose to withdraw immediately and will never boycott. Conversely, if the firm knows that it cannot stay longer in the conflict than the consumers, its best response is to switch immediately to the clean technology.

There are several extreme cases, which lead to different outcomes. First, if the residual profit, that is the profit derived when the good is boycotted, is larger than the profit derived with the clean technology, the boycott is not costly enough to induce the technology change. Indeed, if the decrease in the firm's profit is too small to induce change, the firm always chooses to keep the polluting technology whatever the behavior of the environmentalists. In this case, if the opportunity cost of boycotting is positive, the environmentalists know that their pressure is too weak to induce the technology change, and they never boycott.

Second, if boycotting derives higher utility than consuming the good when it is produced with the dirty technology, the environmentalists always boycott, whatever the firm's strategy. In this case the opportunity cost is negative, which means that consumers derive positive net utility from boycotting. This case can explain why one may often witness unsuccessful boycotts that never end. If boycotting is costless for some consumers, they will always participate. But they are likely to have small amounts of consumption, which generate too small a decrease in the firm's profit to make it change its behavior.

Most commonly, however, boycotting is costly for the consumers and the boycott hurts the firm's profit. In such a war of attrition playing is costly for both players. In this case, the player able to remain longer in the game wins the conflict. It is possible then to determine which factors influence the likelihood for a boycott success.

13.3 What make a boycott successful?

Two main characteristics determine the boycott potential for success. The first refers to the market structure and especially the existence of a substitute good, while the second refers to the structure of the demand and thus the consumers' preferences.

13.3.1 Quality of the substitute

The quality of the substitute increases the potential for success by decreasing the opportunity cost of boycotting. First, market structure is important. If the firm is in a monopoly position, there is no substitute and boycotting is therefore very costly. At another extreme, if the firm plays in a very fragmented market, there is room for eco-certification or labeling, and another firm may enter and provide the good using clean production technology. Then the environmentalists would always choose to boycott, because boycotting would induce no utility loss. Therefore, boycotts are more likely to succeed if the targeted firm plays in a very fragmented and competitive market than if the firm is a monopoly, because the opportunity cost is likely to be smaller. Moreover, boycotting a single firm should be more efficient than boycotting an entire sector, because it increases the chance of finding an acceptable substitute.

Second, the substitute, even if of good quality, may be quite difficult to find on the market, which creates potentially important transaction costs and thus reduces the utility of boycotting. On the other hand, boycotting may have utility per se. Indeed, participation in collective action to improve the quality of the environment may provide positive utility for an environmentalist. The utility of boycotting is likely to be positively correlated with the number of consumers participating in the boycott. Being part of a large community with noble objectives may increase a consumer's utility.

13.3.2 Demand structure

Overall, the chance of success of a boycott depends on the consumers' ability to hurt the firm's profit sufficiently to induce change. Thus, if the share of environmentalists' consumption in the firm's demand is large, the residual profit, that is the profit received when the good is boycotted, will be low, because the boycott deprives the firm of a large share of its profit.

First, a large number of environmentalists unambiguously raise the boycott's potential for success. A large boycotting population therefore increases the utility of boycotting. Thus, it is negatively correlated with the opportunity cost of boycotting and positively correlated with the maximum boycott duration of the consumers. Moreover, a large boycotting population decreases the residual profit of the firm. Therefore, it decreases the maximum duration of the firm. Even if coordination issues are not considered here, one may consider that a large number of potentially boycotting consumers increases coordination problems.

Second, the amount consumed by environmentalist consumers is likely to decrease the maximum conflict duration of the firm because it increases the pressure of the boycott. However, this amount is also likely to decrease the maximum boycott duration, as a consumer used to consuming large amounts of the boycotted good will have a larger opportunity cost than one consuming small amounts, simply because that consumer has a larger amount to renounce.

Overall, it appears that the consumers most able to hurt the firm's profit are also those with the highest opportunity cost. Therefore, they are less likely to participate in the boycott. In the light of this proposition, it is easier to understand the existence of infinite consumer boycotts that never succeed. Indeed, people participating in boycott campaigns are often the minority of the population who are most aware of and most sensitive to their own pollution. Moreover, they will probably consume relatively small amounts, because they take into account the pollution induced by their own consumption. Boycotting is almost costless for them, but their consumption only represents a marginal share of the targeted firm's profit, and thus its withdrawal has little effect on that profit.

Take the example of the boycott of major oil companies because of their lobbying effort to deter climate change policies. Consumers most likely to boycott these companies are those who feel the highest negative utility from pollution. Even if no boycott is announced, these consumers may be expected to prefer alternative transport (cycles, public transport) to the frequent use of their cars, and their capacity to hurt the companies' profits is small. Conversely, consumers most able to hurt the firm's profit consume large amounts of petrol, and thus have high opportunity cost, which makes their participation in a boycott unlikely. It would seem therefore to be to the advantage of nongovernmental organizations willing to implement an environmental boycott to work on informing and educating nonenvironmentalist consumers to increase their awareness of and sensitivity to their responsibility in the degradation of their environment.

Finally, the maximum durations depend on the players' discounting rates. This gives the insight that a more patient player would have more chance of winning the game, since it places greater value on the future. Consumers will tend to be more patient than the firm if they consider intergenerational well-being while the firm only considers short-term profit. Moreover, it is possible that consumers are more patient when considering durable goods than other types of goods. Thus boycotts should be more successful when targeting durable goods.

In the light of these propositions, it is possible to analyze a few case studies of real-life boycotts.

13.4 Case studies

Several cases of real-life boycotts may be better understood in the light of the propositions made. First, the Shell and the Brent Spar case reflects a good example of boycott success. Conversely, the boycott of cosmetic firms represents a case where a boycott success is quite unlikely. Finally, the boycott of noncertified tropical timber is a more complex case.

13.4.1 Shell and the Brent Spar case

In 1995, the Shell Oil Company was planning to sink Brent Spar, a decommissioned 14,500-tonne oil platform, in the North Atlantic Sea. The environmental organization Greenpeace initiated a large-scale protest movement in opposition to this proposal. Activists occupied the Brent Spar platform and damaged 50 Shell service stations in Germany, and a widespread boycott of Shell took place. After a few months, Shell cancelled its plan for deep-sea disposal and decided to recycle the entire structure.

Several insights given in this chapter can help to explain this boycott success. First, oil is a relatively homogeneous good, and petrol stations are easy to find almost anywhere. Therefore, one can consider that the nonpolluting substitute (oil companies not sinking the platform) is perfect, and the only transaction cost is going from any Shell station to the nearest non-Shell station, which is likely to be quite low. Overall, boycotting Shell was costless. Thus, the maximum boycott duration was virtually infinite.

Moreover, the cost of sinking the platform was estimated at £11.8 million, while the cost of alternative disposal was estimated at £46 million. Given that Shell is a worldwide multinational, this difference in costs may be considered small compared to the size of the boycotting population, reducing the maximum duration of the firm.

In other words, Shell was almost costless to boycott and easy to hurt, which can explain why the Brent Spar case is often considered as an example of successful boycott.

13.4.2 Cosmetic firms and animal testing

Animal testing (on invertebrates, rabbits, primates) is a commonly used practice in several industries (for example cosmetics, pharmaceuticals). The practice is considered incompatible with animal rights by many environmentalists. Several environmental organizations provide lists of companies using animal testing in order to encourage consumer boycotts.

The analysis in this chapter indicates that this type of consumer boycott has little chance of success. Indeed, boycotting firms using animal testing is almost equivalent to boycotting the entire cosmetic sector. Good substitutes (cosmetic firms not using animal testing) are therefore difficult to find and transaction costs are likely to be high. For example, Ahimsa, a French organization lobbying for animal protection, lists more than 200 firms testing their products on animals (cosmetic firms and others). Note first that it is difficult to memorize completely a 200-firm list; there is therefore a problem of clarity of the boycott, which reduces considerably the utility of boycotting. It is possible to find good substitutes on the market, but this may involve high transaction costs, as those substitutes may be difficult to find, and searching for related information may be time consuming.

Overall, boycotting firms using animal testing is not likely to be very effective, because of high transaction costs, a lack of clarity in the boycott, and the difficulty of purchasing good (non-animal testing) substitutes. It is thus likely that only strong environmentalists participate in this type of boycott and their hurting capacity is probably quite small. Moreover, alternative strategies to animal testing, although an important research topic,[1] are still far from achieving profitability.

13.4.3 Boycott of noncertified timber

Several nongovernmental organizations argue for the boycott of noncertified tropical timber, as illegal logging in developing countries plagues local development and degrades forest resources. This type of boycott at first appears to have a high probability of success. Timber is a relatively homogeneous good, and ecological timber certification offers good substitutes. Overall, the opportunity cost of boycotting noncertified tropical timber seems to be quite low.

However, further consideration modifies this first impression. First, quite a few ecological labels exist (SmartWood, Scientific Certification Systems, Certified Wood Products Council, Good Wood, Forest Stewardship Council), which may create confusion and decrease the clarity of the boycott. Consumers may be uncertain which labels are most environmentally friendly, and an indirect cost may be incurred searching for information.

Moreover, boycotting consumers are found mainly in developed countries, while the major part of tropical timber is consumed in the country of production. The World Resources Institute estimates that only 20 percent of the wood produced is exported (Rezende de Azevedo et al. 2001). The potential impact of the boycotting population is thus fairly small as tropical timber offers multiple market options, reducing the influence of the boycott.

Finally, individual consumers buy more processed goods using timber than timber directly. These goods are less homogeneous than timber, and certification might be more difficult to implement. Boycotting this type of good should thus be more difficult.

Overall, the boycott of noncertified timber, although presenting small opportunity cost, does not offer much potential for success, mainly because the concerned population is small and has little potential for impact.

13.5 Conclusion and policy implications

This chapter has explored the conditions under which a consumer boycott based on environmental considerations may be successful. A boycott is presented here as a war of attrition between a firm and a group of consumers over the choice of the producing technology.

The model presented assumes perfect information and ignores coordination issues or free-rider problems. However, even with this very simple setup, some interesting implications can be derived. The ability of the boycotting group to hurt the firm's profit sufficiently to induce change is the main element determining the chance of success of this type of action. Thus, the share represented by the boycotting group in total demand is crucial. A large proportion of environmentalist consumers would tend to increase the likelihood of success (but would probably make coordination failures more likely). Again, the hurting capacity of the boycotting consumers is greater if their expenditure on the targeted product is relatively high. However, such consumers usually have high boycotting costs as they renounce a higher utility of consumption, and are thus less likely to participate in the boycott.

Overall, it appears that this trade-off makes a consumer boycott unlikely to succeed. This might explain why so few successful boycotts are witnessed in real life: boycotting groups are usually composed of consumers with small opportunity costs, whose boycott does not hurt the targeted firm's profit sufficiently to induce it to change its behavior.

A potentially more efficient policy for nongovernmental organizations would be to work on the share of the population sensitive to the quality of the environment. Indeed, the game presented here is static, but informing and educating consumers may increase their awareness of environmental degradation, especially the degradation for which they are responsible. Such a policy would potentially have two main consequences in the long run. First, it could lead to a decrease in overall polluting consumption, which would in turn reduce environmental degradation. Second, this could result in an increase in the population likely to participate in the boycott. In the long run, the combination of education and boycott would increase the potential for environmentally friendly technology adoption.

Although this model does not explicitly consider the market structure, it seems reasonable to assume that competition increases the chances of the clean technology being present on the market. Indeed, if there is free entry, there is room for ecological certification and green labeling; a firm may choose to enter the market and to produce the good with the clean technology, if it is profitable. In that case, there is a perfect substitute on the market and boycotting is not costly in terms of welfare. In a monopoly case the boycott is less likely to succeed, because there is no good substitute to which the environmentalists could switch their consumption. Even if the demand structure allows for boycott success, consumers need to coordinate and avoid free riding.

Relaxing the perfect information assumption may modify the results; for example, where the firm has imperfect information about the demand structure and the environmentalists' behavior, boycotting consumers may be able to mimic a long-term boycott.

From the viewpoint of public choice, it is tempting for governments to be sufficiently confident in such consumer actions that they allow citizens to take their destiny into their own hands to induce firms to adopt environmentally friendly practices. However, in the light of this chapter, it seems that consumer boycotts do not constitute a good substitute for public policies. Though the emergence of political consumption practices such as consumer boycotts may be a useful tool for signaling citizen preferences, their effectiveness in changing the practices of firms is doubtful.

However, it is often difficult for effective environmental policies to be accepted politically and implemented. The implementation of lighter environmental policies may, however, encourage the emergence of political consumption practices. For example, taxes on polluting technologies, or subsides on clean technologies, would reduce the difference between profits derived using dirty or clean technology, decreasing a firm's maximum conflict duration. Another potentially effective policy for governments wishing to sanction an increase in the influence of political consumption could be to facilitate the emergence of credible and trustworthy ecological certification, with comprehensive and clear sets of rules defining labeled products. Overall, environmental policies and consumer boycotts do not seem to be good substitutes, but they may be effective complements.

Note

1. See Johns Hopkins Center for Alternatives to Animal Testing website: http://caat.jhsph.edu/.

References

Baron, D. (2002) "Private politics and private policy: a theory of boycotts", Research Paper 1766, Stanford University.

Bilodeau, M. and Slivinsky, A. (1996) "Toilet cleaning and department chairing: volunteering a public service", *Journal of Public Economics*, 59: 299–308.

Diermeier, D. and Van Mieghem, J. (2005) *A Stochastic Model of Consumer Boycotts*, Department of Managerial Economics and Decision Sciences, Kellogg School of Management, Northwestern University.

Friedman, M. (1985) "Consumer boycotts in the United States, 1970–1980", *Journal of Consumer Affairs*, 19: 98–117.

Friedman, M. (1991) "Consumer boycotts: a conceptual framework and research agenda", *Journal of Social Issues*, 47: 149–68.

Friedman, M. (1995) "American consumer boycotts in response to rising food prices: housewives' protests at the grassroots level", *Journal of Consumer Policy*, 18: 55–72.

Fudenberg, D and Tirole, J. (1986) "A theory of exit in duopoly", *Econometrica*. 54: 943–60.

Garrett, D. (1987) "The effectiveness of marketing policy boycotts: environmental opposition to marketing", *Journal of Marketing*, 51: 46–57.

Innes, R. (2006) "A theory of consumer boycotts under symmetric information and imperfect competition", *Economic Journal*, 116: 355–81.

Koku, P.S., Akighbe, A., and Springer, T.M. (1997) "The financial impact of boycotts and threats of boycott", *Journal of Business Research*, 40: 15–20.

Kornhauser, L., Rubinstein, A., and Wilson, C. (1989) "Reputation and patience in the 'war of attrition' ", *Economica*, 56: 15–24.

Maynard Smith, J. (1974) "The theory of games and the evolution of animal conflicts", *Journal of Theoretical Biology*, 47: 209–21.

Maynard Smith, J. (1982) *Evolution and the Theory of Games*, Cambridge: Cambridge University Press.

McLaughlin, K. (2004) "Is your grocery list politically correct?" *Wall Street Journal*, February 17, 2004.

Miller, K. and Sturdivan, F.D. (1977) "Consumer responses to socially questionable corporate behaviour: an empirical test", *Journal of Consumer Research*, 4: 1–7.

Pruitt, S.W. and Friedman, M. (1986) "Determining the effectiveness of consumer boycotts: a stock price analysis and their impact on corporate targets", *Journal of Consumer Policy*, 9: 375–87.

Rezende de Azevedo, T., Giacini de Freitas, A., and Donovan, R. (2001) *Forest Certification: A Catalyst of Forest Conservation?* Available online: www.rainforest-alliance.org/forestry/documents/catalyst.pdf (accessed March 3, 2007).

Roth, D. (1996) "Rationalizable predatory pricing", *Journal of Economic Theory*, 68: 380–96.

Smith, N.C. (1990) *Morality and the Market: Consumer Pressure for Corporate Accountability*, London: Routledge.

Teoh, S.H., Welch, I., and Wazzan, C.P. (1999) "The effect of socially activist investment policies on the financial markets: evidence from the South African boycott", *Journal of Business*, 72: 35–89.

Tyran, J. and Engelmann, D. (2005) "To buy or not to buy? An experimental study of consumer boycotts in retail markets", *Economica*, 72: 1–16.

14 How does environment awareness arise?

An evolutionary approach

Paloma Zapata-Lillo

The action of large human conglomerates is behind many of the events that have caused significant environmental damage in recent decades. However, these same conglomerates are the main actors in some important environment protection battles. The selfish interests of many individuals lead too often to a tragic collective result desired by nobody. Sometimes, however, they lead to the construction of social organizations whose members are willing to compel them to protect nature. Evolutionary game theory studies this sort of contradiction between many individual actions and the emerging social consequences. In this chapter, that theory is proposed as a means of analyzing the possibility of the emergence of social awareness and environment protection activities within communities.

14.1 Introduction

Throughout the globe, deforestation caused by irrational human overexploitation of forest resources is transforming natural paradises into deserts. It is witnessed in the appearance of ghost towns, abandoned as the inhabitants have left in search of work instead of staying and keeping their traditional activities and culture; and in the increase in pollution and natural degradation, almost unknown in the past but now provoked by the decisions of people, industries, and governments.

While degradation of the environment has been occurring for centuries and has not been created by globalization, it has been significantly worsened by the process. Perhaps the worst consequence of globalization has been the breaking of community links that used to help maintain a healthy relationship with the environment and established a virtuous circle between nature and community. Free market competition has driven the members of many peasant communities of poor countries to ruin and to the loss of perspectives inside their communities, leading to unprecedented migration to large cities. For example, in Mexico, according to official data from SAGARPA (the Ministry of Agriculture, Cattle Raising, Rural Development, Fishing and Nutrition), in the few years after the North American Free Trade Agreement came into effect, emigration from Mexican rural communities toward the

United States doubled. Besides the negative economic and social consequences, there are two further effects of this massive emigration on ecology and the environment. First, the old community lands and natural resources are left unprotected against adverse natural phenomena and especially against the incursion of huge corporations that exploit the resources, causing great environmental damage. Second, large groups of migrants are established in places where they do not have any roots and where they do not intend to settle, so they have little incentive to preserve the local environment.

For game theorists, such environmental conflicts are typical examples of a game-theoretic model that, though very simple, captures the heart of the problem. That model is known as the *n*-person prisoner's dilemma (Hardin (1968) generalized the model to *n* players and called it the "tragedy of the commons"). Authors who have been modeling environmental and other similar conflicts using the prisoner's dilemma include Weissing and Ostrom (1991), Ostrom (1990, 1998), Okada and Sakakibara (1991), Maruta and Okada (2001), and Glance and Huberman (1993, 1997).

However, in the real world, it is not true that all people are apathetic about environmental and ecological problems. Communities involved in environment-related conflicts face their undesired effects and thus try to achieve cooperation. In order to do that, they create organizations (such as community assemblies, trade unions, and ecological organizations) that compel their members to cooperate. During the 1960s the ecological movement, previously almost nonexistent, started to burgeon. A multitude of activists and organizations emerged around the planet, trying to create consciousness of the environmental problems facing humankind, and taking action to solve these problems. Perhaps more importantly, numerous communities that had succeeded in surviving in the face of globalization organized themselves to resist the environmentally damaging tendencies of the modern world. These communities can be found in Africa, Asia, and Latin America, as well as in developed countries. Examples follow.

In Kenya, poor rural women organized the Green Belt Movement to raise environmental awareness in the country. In 1989 they led protests against a 60-story business complex planned for construction in the heart of Uhuru Park in Nairobi, succeeding in their battle. The movement has planted thousand of trees, combating deforestation and soil erosion.

In north India, people from Mehdiganj communities have mobilized to demand the closure of a large soft drink bottling plant that has created severe water shortages affecting over 20 villages, and has caused pollution of agricultural land and groundwater. Their action has so far been unsuccessful.

Indigenous communities in Latin America have been involved in many struggles related to protection of the environment. In Mexico, for example, during the period 1994–97, in the main case study used in this chapter, the community of Tepoztlán, in the state of Morelos, organized the Comité de Unidad de Tepoztlán (Unified Tepoztlán Committee) and carried out a successful campaign against powerful interests to stop the construction of a golf club. The construction would have damaged local flora and wildlife and polluted groundwater reserves, and would have concentrated the water resources of the area in the hands of the tourist complex and golf course.

Other examples from Mexico illustrate the range of environment-related conflicts in which rural communities are involved. At the time of writing, the community of La Parota, in the state of Guerrero, is continuing its opposition to the construction of a dam. The Zapatista indigenous communities of Chiapas are actively engaged in the defense of water, biodiversity, and other natural resources, such as forests. The resistance of the ecologist groups of the Sierra de Guerrero to powerful interests that have threatened the existence of community forests through overexploitation has attracted international attention, with some of the leaders being imprisoned for over a year. Throughout Mexico numerous communities have attempted to carry out various forms of ecological protection, some of which have solidified into communal projects, for example involving ecotourism.

But why do these organizations appear? Game theory provides useful tools to answer that question. Nash equilibria can represent the behavior that rational theory anticipates for organizational conflicts, though it cannot be convincingly argued that people necessarily act according to those equilibria.

Organizations are not the result of inspirational leadership, or forces operating from outside the communities. On the contrary—they emerge, in large populations, as the free choices of uncoordinated and not very rational individuals who interact with one another over long periods of time. It is necessary to study real people involved in conflicts, not theoretically rational individuals, in order to answer the question of why organizations arise. It is also necessary to identify the social patterns that emerge as a result of the behavior of a large number of these normal individuals. In this chapter it is considered that evolutionary theory is an adequate approach to study the emergence of environment-related organizations. Why should an evolutionary approach be meaningful in this setting? Because social organizations are shaped by historical processes. That is, they are shaped by the cumulative effects of interactions among the members of large populations. Real individuals are not concerned with achieving the optimum in the long run. Furthermore, they are not concerned with the social effects of their decisions. People change their behavior according to experience, but they have limited rationality, and can only improve their payoff according to that experience. In addition, they experiment and make mistakes. They also have poor information. The emergent patterns of behavior (organizations, for example) are equilibria of such dynamic processes. But the dynamic process never stops, because, in each period, experiments and mistakes move the process from each state, and evolutionary dynamics never settle down. These processes are stochastic, since chance plays an important role in them. In each encounter, people who interact with other people are drawn at random from a large population. They also gather information, experiment, and make mistakes at random. The followed course cannot be predicted, but the probability of each pattern happening in the long run can be estimated. Some states occur much more frequently than others in the long run. They are the stochastically stable equilibria (Young 1993, 1998). Sometimes those states are (or represent) Nash equilibria, but at other times people operate in cycles, repeating social structures that some of them rejected in the past.

Using the theory of evolutionary games, several authors have studied the setting up of conventions in a society, the making of its ethical concepts and its social institutions (see, for example, Binmore 1995, 1998; Young 1993, 1998; Vega-Redondo 1993). Other works study the evolution of cooperative behavior in the prisoner's dilemma (Glance and Huberman 1993, 1997). Those studies consider conflicts where large populations are involved in the long run. As explained above, environmentally cooperative behavior has been attained by some large populations able to overcome prisoner's dilemmas in the long run. An evolutionary approach offers the potential for interesting advances in understanding how environment awareness arises.

"Collective action" is the name applied in the literature to the process by which social organizations form and operate. Olson (1965), in an insightful heuristic study, pointed out the essential question related to the emergence of institutions, arguing that rational self-interested people should not participate in forming social organizations. Why, then, do social organizations emerge? He suggested that in forming such organizations people overcome a social dilemma similar to the prisoner's dilemma. That is, a social organization emerges to allow people to enjoy some common goods. But those goods are public goods, and nobody can be excluded from enjoying them. Why do people decide to expend some effort participating in those organizations? People could decide not to participate, but that inaction may lead to a situation where nobody can enjoy common goods. What, then, makes collective action possible? Olson thinks that rational, self-interested individuals will not act voluntarily participating in organizations in order to achieve their common interests. This chapter considers that the common goods dilemma (protecting them or not, facing expropriation and privatization in the latter case) is a prisoner's dilemma and may thus provoke a "tragedy" (of the commons), which people will try to overcome through forming an organization. However, hyperrationality and imposition are not invoked here in order to explain the emergence of voluntary organizations. People have limited rationality, but this can lead to the emergence of an organization through a limited learning process. The approach that is followed here is closer to that of Ostrom (1990, 1998).

There are some similarities between Ostrom's work and that presented here. She studied the ways in which communities involved in common pool resource dilemmas (of the prisoner's kind) overcame those dilemmas. She concluded that this was not possible in a one-shot game and considered a repetition process, where individuals have bounded rationality. Chance plays a role in her approach. Nonrational behavior by individuals appears at random. Repetition of the conflict might lead people to design norms in order to overcome the dilemma, when there is good communication and trust between people. Then, some level of cooperation is reached. Only a proportion of the community would cooperate. There are also differences between Ostrom's approach and that used in this chapter. In Ostrom's approach (see Ostrom et al. 1994), the process of repetition of the common pool resource dilemma is a game similar to the repeated prisoner's dilemma that Axelrod analyzed (1984), where the players are always the same. Ostrom assumed that bounded rationality means that individuals have inherited an acute sensitivity for learning norms that increase their own long-term benefits when confronting social dilemmas with others who have learned and value similar norms. She found (in her analysis and laboratory experiences) that people do not act according to Nash equilibria, or to finite backward induction. In this chapter, on the other hand, the conflict that is repeated (in the organization game) is not a prisoner's dilemma. In a large population, encounters occur between small groups of the members of that population drawn at random. Also, the process of repetition is a perturbed Markov process; it is not a repeated game. The bounded or limited rationality implies that individuals are just able to gather small amounts of information in a stochastic way and to improve their payoffs in response, without necessarily achieving the optimum. They also make mistakes in a stochastic way. Finally, the equilibria of the process are stochastic, and they describe which social behavior patterns occur with highest probability. In the studied dilemma, there are two cases where stochastically stable equilibria are Nash equilibria. In another case, the stochastically stable equilibria are vertices of some cycles.

This chapter will study communities that face a prisoner's dilemma, where cooperation means collaboration in the protection or production of a public good that is nonexcludable. The members of each community consider whether or not they should voluntarily form an organization that forces them to cooperate, and the emergent organization dynamics are analyzed, with the conflict repeating in the long run following the evolutionary game approach. The approach follows Kandori et al. (1993), Kandori and Rob (1995), and Young (1993, 1998) in considering a large population that is involved, through different communities, in a large number of similar conflicts in the long run and in each period. People learn in a limited way from experience. They look for an improvement in their expected payoff. However, they make mistakes in each period, with small probability. Thus, the Markov process with a system of regular perturbations is obtained and their stochastically stable equilibria are studied. Each strategic structure of society (the vector of the strategies that people choose) is a Markov process state. The stochastically stable equilibria are the states where people form an organization with minimum size, such that each member's payoff is larger than the payoff it can obtain by itself. Just a part of the community participates, in general, in the organizations. But people who form the organization change. There are people who choose to free ride. The different aspects of the model will be illustrated with a real case, the experience of the Tepoztlán community (see Rosas 1997 for a chronology of events).

The model only allows a qualitative analysis. However, it can be useful as an instrument for environmental and ecological policy. In the first place, it can help to convince policymakers of the important roles that different communities may play in the solution of environmental problems. Second, it may be useful as an instrument to measure the effectiveness of policies that try to encourage the environmental activities of communities.

The main objective of the chapter is to study how an environment-protective social pattern emerges within a community that has dispersive and destructive tendencies. This cooperative social pattern should be the result of dynamics between people who behave realistically; that is, they are not very rational or informed. The key concept is the stochastically stable equilibrium that expresses the emergent patterns of social behavior. The stochastically stable equilibria that were found here mean that within a community of sufficient size and social productivity, a part of the population will surely become a nucleus able to bind members of the community and maintain a healthy relationship with the environment.

The chapter is organized as follows. In section 14.2, the experience of the Tepoztlán community is summarized. In section 14.3, a particular prisoner's dilemma is described and the organization game established (which is almost equivalent to the Okada and Sakakibara extensive game, 1991), and their strict Nash equilibria considered. Section 14.4 explains how the conflict is repeated in the long run by a large population, under a learning-mistaking dynamic. The Markov process is established and its stochastically stable equilibria studied. In section 14.5 other versions of the model are mentioned, and their possible applications to the study of some real situations are considered. Those new versions are concerned with the possibility of accumulation of the public good, as well as the possibility that the members of the community have different trends to their participation because they receive different benefits from the public good. Finally, section 14.6 provides some conclusions and policy implications of the presented model.

14.2 Tepoztlán community

Tepoztlán is a town in the state of Morelos, Mexico, close to Tepozteco Hill. In 1937 this hill was declared a national park and in 1988 it became part of a protected zone within the Ajusco-Chichinautzin Biological Corridor. At Tepozteco Hill two geographical regions come together: a forest of pines and holm oaks and a jungle. For this reason, it has a great diversity of flora and wildlife. The hill also accumulates the rainwater that subsequently irrigates a great deal of the state of Morelos. This privileged zone is the communal property of the inhabitants of Tepoztlán.

The Tepoztlán community mostly comprises peasants and artisans with ancient pre-Hispanic traditions. The town is a market and trading center with a balance of the traditional and the modern; artists, intellectuals, retired people, and other groups who have settled there enjoy a good relationship with the established community. Though the area attracts many tourists, the impact of tourism has not been obtrusive.

The strong links among the Tepoztecos (Tepoztlán villagers) are built and rebuilt in the neighborhood life and especially in the fiestas that embrace the entire district. The organization is loose, however, when there is not a serious conflict. The decisions about important communal problems are taken in assemblies in which most representatives are seniors who have considerable experience and who enjoy the trust of the community. Occasionally, other individuals will participate in the discussion of particular problems.

There have been several attempts to transform the Tepozteco into a major tourist destination with associated infrastructure, but the villagers' opposition has thus far prevented this. The strongest attempt came at the end of 1994 in the form of a proposed golf club at the center of a tourist, industrial, and residential project. On 240 hectares of the protected zone, the plan envisaged construction of a golf camp, a clubhouse, a high-tech corporation park, and a service zone including hotels, restaurants, expensive stores, and a complex of 700 luxury residences, each with a pool. Enormous environmental damage would have resulted from construction of the complex. Of greatest magnitude would have been the water problem, given the proposed club's requirement for a water flow of 90 liters per second. In comparison, the current water usage by the Tepoztlán community of 50 liters per second already creates water supply problems, especially during the dry season. This scarcity has worsened since the construction, in a neighboring community, of a residential complex of large luxurious houses with gardens and pools (a much smaller complex than the golf club would be).

The club was never built. The development company started construction but was later forced to suspend and, finally, abandon the project because of community opposition. The members of the community formed a strong organization, the Comité de Unidad de Tepoztlán (CUT; Unified Tepoztlán Committee), comprising a majority of community adults, whose main goal was to stop the execution of the club project. After several months of struggle they achieved their aim.

Conviction that construction of the club would be harmful to the community, and the subsequent formation of the CUT, was a gradual process that took place in the context of contradictory expectations about the outcomes that would result from building the club. There was concern, most strongly voiced by the seniors, about the negative consequences to the community and to individual livelihoods if the project had been constructed, including water scarcity and loss of communal land and rights, and evolution from a community to a multitude of people at the service of the club.

However, the club had another meaning for the Tepoztecos. Each one of them evaluated the odds of obtaining an income from the club construction that would improve their poor economic situation. They would imagine the club already built and they would make plans concerning the new opportunities available to them. Many people were sure that they would find jobs as maids, cooks, or gardeners. Some of them would sell handicrafts, flowers, or other agricultural products to the new hotels and restaurants and to the tourists attracted by the club. Others were interested in becoming tourist guides or taxi drivers. The teenagers were enthusiastic about the possibility of working as caddies. Those that were owners of restaurants, hotels, or greenhouses were certain to extend their businesses. And so on. From the beginning of construction of the complex there would be opportunities to obtain extra income; the development company had advertised that 13,000 people would be necessary to build the club, and masons, plumbers, painters, electricians, and gardeners would be hired.

The club project therefore brought a dilemma to the people of Tepoztlán. For all of them it was clear that the water scarcity problem would worsen, community life would weaken, and communal land was threatened. It was also clear to them that if everyone acted as they were tempted to act, the problem would be very grave, much worse than if the community acted cooperatively.

The process whereby the people of Tepoztlán organized concerted opposition to the club went through several stages, as outlined in the following subsections.

14.2.1 First three months of 1995

Announcement of the club project provoked an arousal of individual interests and a significant weakening of community bonds. Only the seniors and a few others were in clear opposition to the club and acted accordingly, organizing assemblies in order to study the laws that established the protected area in Tepoztlán, and communal rights over the land and resources. Club opponents began to encourage the Tepoztecos to protest against the project, and to promote and spread the idea that an organization with wide participation and with the capacity to engage the cooperation of the majority of people was much needed. On the other side were a few determined club supporters, mainly owners of various business. Many others, while recognizing that there would be both damages and benefits, were of the opinion that communal rights, in particular those related to water, could coexist with the club. They also thought that the most senior and socially conscious members of the community would act as a counterweight to club activities and safeguard communal interests. The club became an increasing topic of conversation, contradictory rumors abounded, and it was not clear whether opposition to the club among the Tepoztlán community would take an organizational form. Generally, however, the number of opponents to the club and their commitment increased.

14.2.2 From April 1, 1995 to August 23, 1995

In April 1995 the first large assembly met to discuss the situation. They decided to oppose the club and to campaign against it. "No to the club" graffiti started appearing and well-supported demonstrations took place, but the opposition still lacked strength and unity, and in August the mayor of Tepoztlán granted the development company permission to start building work. Large earthmovers arrived in Tepoztlán to "clean the hill".

14.2.3 From August 24, 1995 to April 10, 1996

On 24 August, a multitude of protesters took control of the town hall and denied access to the municipal council. An assembly was summoned, attended by a large percentage of the local population, and the CUT was formed. On September 3 the development company and the local government organized an alternative assembly intended to give the impression that construction of the club had community support. The police allowed entry to project supporters only, but again thousands of townspeople mobilized and the police and authorities were forced to vacate the premises. Tepoztlán held its own elections and chose a seven-member council, all opposed to the development. A September 1995 poll in the *Reforma* national newspaper showed that 86 percent of the Tepoztlán population opposed the club and supported the CUT. The Tepoztlán movement was given coverage by the national news media and started to attract the support of several influential groups, including students and indigenous populations.

Neighborhood units were organized to maintain discipline in the town while demonstrations continued, with representatives of the movement visiting Cuernavaca to press the governor or Mexico City to press the federal environmental authorities. The federal government arrested three members of the CUT and, on April 10, 1996, police used considerable force in halting a convoy of Tepoztecos traveling in buses to Chinameca to attend an anniversary commemoration of the death of Mexican revolutionary Emiliano Zapata; many were injured and an elderly man killed. National outrage ensued, and the following night the development company made a television announcement that they had decided to abandon the project to build a golf club in Tepoztlán.

14.2.4 From April 11, 1996 to November 1996

The government decided to negotiate with the CUT, offering to liberate imprisoned representatives of the movement and to officially prohibit the building of golf clubs in the protected area. In exchange they asked for a commitment of the CUT to new local elections, with the participation of the official political parties. The Tepoztecos rejected the offer, because in their opinion the local authorities were already elected and worked well. At this stage, however, support for the CUT had somewhat declined, with a number of people considering that the movement had achieved its goal and the organization was no longer required.

14.2.5 From December 1996 to March 1997

At the end of the year members of the CUT decided to organize the election of representatives following the same rules they had used in 1995 to elect the autonomous local government. These representatives were then presented as candidates for the official elections in alliance with the leftwing Democratic Revolutionary Party (PRD). Following a four-month election campaign that involved considerable mobilization of Tepoztecos, the CUT candidates achieved victory over the long-time ruling Institutional Revolutionary Party (PRI), with more than 70 percent of the vote. After the elections, the government officially prohibited construction of projects such as the club inside the protected zone.

14.2.6 March 1997 to the present day

After the plans to build the club were abandoned, community conditions gradually reverted to the preproject status quo. The traditional communal and neighborhood assemblies again became the main modes of organization and the foci of discussions about the issues affecting Tepoztlán. The CUT steadily diminished in importance, becoming an inconsequential organization before finally disappearing.

14.3 The models

14.3.1 Prisoner's dilemma

In this conflict every participant has the option of acting selfishly or cooperatively. Independently of what other actors decide, each individual actor considers that selfish action is the best option, and will thus choose that option. However, all these selfish individuals acting together will build the worst possible outcome, since each one of them would have been in a better situation if everyone had cooperated. These are the special characteristics of the prisoner's dilemma; characteristics that would have surprised Adam Smith and that did surprise game theorists when that famous example was designed. Nevertheless, for people who study modern environmental problems, the prisoner's dilemma is not a curiosity, but a simple model that captures an essential aspect of the behavior of numerous human groups.[1]

A very simple model of the prisoner's dilemma is used in this chapter, and will be illustrated with the experience of the Tepoztlán community. Section 14.2 described and explained the dangers the villagers believed they faced because of the club project, and the benefits they believed the project would bestow. It might help to clarify why this conflict should be categorized as a prisoner's dilemma.

In order to design the game all the villagers will be considered as players. That set of people will be denoted as N. Each player has two possible strategies, either cooperation (C), which means to oppose the club and act accordingly, or not to cooperate (NC), supporting the club construction and taking advantage of it. The payoff functions do not express real earnings but the players' beliefs about the outcomes of building or not building the club. The existence of the club would affect the payoffs in two directions. First, people might obtain extra payoffs if they worked for the club in some capacity. Second, people might lose some common goods, such as water. The possibility of maintaining those common goods might depend, at least partially, on the number of people who choose C (to oppose the club).

Constructing a simple example might help the discussion of payoffs. Let us assume that there are two players. If a player j chooses C, it obtains γ' units of utility for itself. On the other hand, if a player chooses NC (to obtain advantage from the club), it could obtain for itself a larger payoff γ. The cost of cooperation (choosing C) is the utility lost by players who cooperate, that is, $d = \gamma - \gamma'$. On the other hand, for each player who chooses C, each of two players would obtain β units of utility. The extra payoff that player j would obtain from the club depends on the strategy that j chose (C or NC). It would be zero if it chose C. The extra payoff that player j would obtain from the common goods depends on the number of players who chose C, but j might choose NC. In fact, that extra payoff arising from all the players who cooperate (social productivity) should be a monotonic increasing function of the number of players who are choosing C, and β would be the marginal social productivity.

The following 2×2 matrix game expresses the conflict:

$$\begin{array}{c} \\ C \\ NC \end{array} \begin{array}{cc} C & NC \\ \left(\begin{array}{c} (\gamma' + 2\beta, \gamma' + 2\beta) \\ (\gamma + \beta, \gamma' + \beta) \end{array} \right. & \left. \begin{array}{c} (\gamma' + \beta, \gamma + \beta) \\ (\gamma, \gamma) \end{array} \right) \end{array}$$

If $\beta < d = \gamma - \gamma'$, ($NC$, . . ., NC) is the unique Nash equilibrium of the game. However, if $n\beta > d$, it is a tragedy, because if the two players had chosen C, they would obtain a better payoff.

In the general conflict with n players, γ denotes the utility that each individual player believes it could obtain for itself if the club were built, and it took advantage from this (choosing NC). If a player decided not to benefit from the club, acting against it and trying to preserve communal rights, it would obtain γ', which is smaller than γ. Besides, choosing C implies dedicating some time to taking action against the club, which would involve loss of utility d'. The cost of cooperation, $\gamma - \gamma' - d'$, is denoted d. On the other hand, every member of the community will obtain β units of utility for each individual who chose C (acting for communal rights). It does not matter if j chose C or NC, j would obtain βr units of utility more if r people chose C. Why did people believe that the cooperation of every member of the community would allow each individual to earn β units of utility more? Because they thought that if opposition to the club were great enough, the club could not use the community groundwater (and other community resources) as it wanted. In this chapter it is assumed that people believed the club would be built, and, at the same time, the pressure of some community members could place a degree of restraint on the club's activities. The potential of the community rights over water would be a monotonic increasing function of the number of people who are choosing *C*. It is assumed the function is linear.

When all players have chosen a strategy, the notation $\sigma = (\sigma^1, \sigma^2, \ldots, \sigma^n)$ will be used, where σ^j is either *C* or *NC*, according to what *j* chose. σ is a profile of strategies.

For each profile of strategies σ and each individual player *j*, s_σ will denote the number of people different from *j* who chose *C* according to σ. The payoff that each player believes it will earn depends on its choice and that of all other players. If *j* chose *C* it would expect to earn $\gamma + \beta(s_\sigma + 1) - d$. On the other hand, if it chose *NC* it would expect to earn $\gamma + \beta s_\sigma$. Let f_j be the function that associates to each σ the payoff for *j*. Then the possible payoffs may be written in the following way:

$$f_j(\sigma) = \begin{cases} \gamma + \beta(s_\sigma + 1) - d & \text{if } \sigma^j = C, \\ \gamma + \beta\, s_\sigma & \text{if } \sigma^j = NC. \end{cases}$$

It is clear from the context (section 14.2) that $\beta < d$, but $\beta > \frac{d}{n}$, for the Tepoztlán community. That is, the cooperation cost *d*, which is $\gamma - \gamma' - d'$, is larger than the utility one player adds to itself because of its cooperation (β). However, the utility that everybody's cooperation adds to each player (βn) is larger than the cost *d*. Then, for each player, it is better to choose *NC* (to support the club construction), but this will provoke a tragedy. The game is an *n*-person prisoner's dilemma.

14.3.2 Strict Nash equilibrium in a game

Strict Nash equilibrium is a strategy profile such that if a player *j* changes its strategy and the others do not, the payoff of *j* is smaller than the previous payoff (Nash 1950). Strict Nash equilibria are pure strategy Nash equilibria, but each pure strategy Nash equilibrium might not be strict.

In the designed game the unique Nash equilibrium is (*NC*, . . ., *NC*) and it is strict, because $\beta < d$ implies $\gamma + \beta - d < \gamma$.

The conflict within the Tepoztlán community, when there is not a particular situation causing extra pressure, can be expressed through a similar prisoner's dilemma. In that case, *C* would mean participation in some tasks in order to guarantee the care of the communal resources and to prevent some people acting in an abusive way. On the other hand, *NC* would mean to obtain all of the advantages from the communal resources and to avoid all of the communal tasks.

There are games with a similar payoff function, but they are not prisoner's dilemmas. There is not a general reason why the two inequalities $\beta < d$ and $\beta > \frac{d}{n}$ should be satisfied. If $\beta > d$, the unique Nash equilibrium would be $(C, \ldots, C)$ and it would also be strict. If $\beta < d$ and $\beta < \frac{d}{n}$ were true, the unique Nash equilibrium would be $(NC, \ldots, NC)$, but it would not mean tragedy. Figure 14.1 shows how each player's Nash equilibrium payoff ($f_j(\sigma^*)$) changes when β changes. The size of the population n, individual productivity γ, and the cost of cooperation are constant.

For example, in a society where communal property—the land and other resources, traditions, beliefs, history, production techniques—leads to community links vital to every individual, it would be the case that $\beta > d$, as is the case in those communities that stay unified for centuries. In other cases, it is quite possible that β is much smaller than $\frac{d}{n}$, as in those Mexican communities whose members decide to migrate illegally to the United States, despite the risks involved. In that case, in the game that expresses the conflict, C would represent staying in the original community, and NC would represent migrating to the United States. The individual productivity that migrants believe they could obtain in the United States would be γ. However, people that choose NC (migration) experience some hardship; d_s represents the utility lost because of such hardship. Therefore, people believe that the utility derived from migration would be $\gamma - d_s$. That utility would be larger than γ', the individual productivity that they usually obtain in their community. If it were not larger nobody would choose NC. Besides, the unitary social productivity β, which is obtained from the common resources and from people who choose C, is very small, because those communities are very poor, lacking significant resources. Why do people who migrate obtain utility from people who choose C? Because they retain close links with and affection for the nonmigrating community, whose members have chosen to cooperate. On the other hand, d' would be zero, because no effort is involved in staying. Is $d = \gamma - d_s - \gamma'$ larger than β? Is it larger than βn? In fact, thousands of Mexican people migrating to the United States have stated that inequalities are

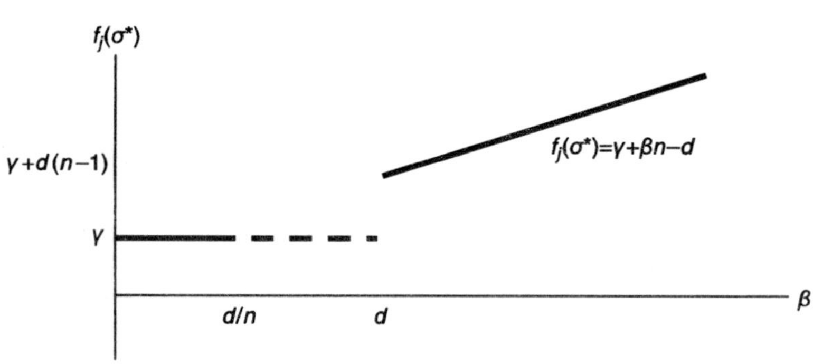

Figure 14.1 Changes in player's Nash equilibrium payoff as β changes.

dramatically true for them. Then, the Nash equilibria of that conflict is that everybody migrates (they choose *NC*), but it is not a tragedy for each member of the community. However, it is a social tragedy. The same could be said of the members of Moroccan communities that risk drowning when going across the Mediterranean Sea in a small crowded boat.

This chapter considers the case where $\beta < d$ but $\beta > \frac{d}{n}$ as in the Tepoztlán community. In such a case no one would choose an ecological stance without some factor forcing them to do so, although there is a natural number k^* larger than 1, but smaller than n, such that if k^* members of the community decided C, the utility of each person in N would be larger than γ. Therefore, within the community a new conflict (game) arises in which each of its members will try to obtain the best advantages from cooperation, even by using mandatory means. Some observations will be made before studying the organization game.

14.3.3 Observations on the model

Four observations will be made on the model.

First, though the model has been applied to a specific kind of conflict, its elements can be reinterpreted or adapted to the study of other cases, for example where C means the production of a physical good. More detailed payoff functions (corresponding to prisoner's dilemmas) may be constructed that fit more closely to particular studied cases. However, the payoff function used here has been chosen because it shows the main properties of the model in an explicit way. Those properties are individual productivity γ', unitary social productivity β, costs d and d', and community size n. Another advantage of the model is that the estimation of the parameters is simple.

Second, people from real communities may have very different individual productivity and cooperation costs, and the public good protected may give them different benefits. On the other hand, they may cause very different types and levels of environmental damage, with some industries or nations producing considerable damage and showing little interest in cooperation. In such circumstances, the game would not be symmetric.

Third, it may be more realistic to consider that the utility from cooperation is not lineal. Therefore, marginal productivity would not be a constant β, but a function that would depend on the number of people who choose C. That utility function has to be a monotonic increasing function. In a conflict where cooperation (C) provokes the production of a physical good, β might be concave, as it usually is in a production function. In the case where C provokes a social force, as in the studied example, β could be convex, because the social force grows in that way.

Fourth, it is assumed that people believe that the golf club will be built in any case, but the opposition of some (or many) people from Tepoztlán would maintain the community rights (maybe partially) over water. In fact, the opposition of a majority of the community stopped the club project. Another game could be designed considering a critical number of protesters (say s', which is larger than 1) in which the club would not be constructed. Then, if the number of players who chose *C* was greater than s', everybody would obtain all the benefits from the communal resources βn. A player who chose *C* (opposing the club) would only derive a small utility γ'. from its individual productivity, and it would lose d' because it had to use some of its time opposing the club. A player who chose *NC* would derive a larger utility γ because of its individual productivity (deriving advantages from the club). On the other hand, if the protesters were fewer than s', the club would be constructed, and all the rights over the communal resources would disappear. Then, a person who chose *NC* would just earn γ, while a person who chose *C* would earn $\gamma' - $ d′. If the assumptions about $\beta < d = \gamma - \gamma' - d'$ and $\beta > \frac{d}{n}$ were preserved, the game would not be a prisoner's dilemma. (*NC*, . . ., *NC*) is again strict Nash equilibrium, but there are other strict Nash equilibria. They are all the profiles where exactly s' people choose *C*, whoever they were. The critical number s' would be exogenous.

14.3.4 The organization game

The Tepoztlán experience tells us that people form organizations in order to overcome the effects of their prisoner's dilemma. While the communal links were strong, it can still be said that, prior to the conflict with the club, the organizational structure was less developed and looser than under the CUT regime. But in any case, only a small part of the community participated in organizations. The CUT was a voluntary organization; each individual could choose to be a member or not. Nobody was punished for not participating in it. As a consequence people outside the CUT could benefit from the club, while free riding on the efforts of the organized people who strove to retain their communal rights. On the other hand, if any members of the organization did not keep their word and supported the club in any significant way they would have been rejected by their fellow members, placing them in a worse situation than if they had respected their responsibilities as members of the CUT.

In order to design the organization game, the CUT experience will be considered. The members of the community will again be the players. The options available will be the strategies *P* and *NP*, respectively.

In order to define the payoff function of the organization game, the following assumptions corresponding to section 14.2 will be established. Players who choose *NP* would continue believing that they would obtain benefit from the club, and from the efforts of the CUT members. On the other hand, if a player chooses *P* but it supports the club in any way, there is an implicit punishment. This means that its payoff would be worse than if it had respected its responsibilities as a member of the CUT. A part of the effort of the members of the CUT (people who choose P) would not be used directly to defend the community against the club, but to guarantee that the organization works properly (organization cost). That cost will be modeled as though that part of the effort were done by some of the members of the CUT, say n_0, who did not obtain any income, but received γ' units of utility from the CUT. That is, the organization cost will be $\gamma' n_0$, and it will be denoted by d_0.

Then, if the organization were formed by s persons, the payoff of j, a member of the community, would be $\gamma + \beta(s - d_0) - d$ or $\gamma + \beta(s - d_0)$, respectively, depending on j's choice, P or NP. The CUT would have disappeared quickly if the size of the organization had not been large enough for each of the members to obtain a payoff greater than γ. That is, if s were large enough, $\gamma + \beta(s - d_0) - d$ should be larger than γ. Then, the size of the organization should be larger than $d_0 + \frac{d}{\beta}$. The least integer number larger than $d_0 + \frac{d}{\beta}$ will be denoted as s^*. It is the minimum size of an efficient organization, that is, an organization size such that it allows people to earn more than γ. s^* might be smaller or larger than n or equal to it, depending on the values of the community parameters (β, γ, γ', d', d_0, n). If s^* were larger than n, nobody would think about forming an organization, and the organization game would not mean anything.

As before the utility or payoff for each individual depends on everybody's decision. σ denotes $(\sigma^1, \sigma^2, \ldots, \sigma^n)$, where σ^i is either P or NP according to what i has chosen. Given σ and j, we will call s_σ the number of players other than j that chose P. The payoff function for each individual is defined by using the following table (analogous to the one used for the prisoner's dilemma):

$$h_j(\sigma) = \begin{cases} \gamma + (s_\sigma + 1)\beta\left(1 - \dfrac{d_0}{s_\sigma + 1}\right) - d \text{ } \textit{if } s_\sigma \geq s^* - 1 \text{ and } \sigma^j = P, \\[2ex] \gamma + s_\sigma\beta\left(1 - \dfrac{d_0}{s_\sigma}\right) \textit{ if } s_\sigma \geq s^* \text{ and } \sigma^j = NP, \\[1ex] \gamma \text{ otherwise.} \end{cases}$$

This function is similar to that found in Okada and Sakakibara 1991.

It is possible to interpret and modify functions h_j and the involved parameters in order to study other conflicts (see section 14.3.3).

It will be assumed that s^* is smaller than n or equal to it. It is not difficult to find the organization game's Nash equilibria in pure strategies. They are described in proposition 1, whose proof is in the Appendix at the end of this chapter.

Proposition 1: If s^* is smaller than n or equal to it, a profile σ of pure strategies is the Nash equilibrium of the organization game if and only if the number of people that choose P is equal to s^* or is smaller than $s^* - 1$. The equilibria where s^* people choose P are strict, the others are not.

Observe that ($NP, \ldots, NP$) is always Nash equilibrium, and ($P, \ldots, P$) is Nash equilibrium if and only if s^* is equal to n. Therefore, when the size of the community is equal to the least integer number larger than $d_0 + \frac{d}{\beta}$, the participation of all members of the community is Nash equilibrium. That is a very particular case.

As explained in section 14.3.2, when there is not a situation creating unusual pressure within Tepoztlán, there are conflicts that can be expressed through similar prisoner's dilemmas. Then the organization game makes complete sense, and proposition 1 has the same meaning in that context. The correspondent s^* would, of course, be smaller than that determined in the case when construction of the club was planned.

However, it is not enough to know what the Nash equilibria are. In many cases they mean acceptable situations or sound very intuitive. But, why do people have to choose Nash equilibria strategies? How do they reach those behavior patterns? Besides, the Nash equilibrium payoffs of several games, the organization game in particular, are very different for each player. Which equilibria would occur?

It is important to study what social patterns would appear when the conflict is repeated period after period by people with limited rationality, as in real conflicts such as that at Tepoztlán. Young's dynamics (1998) will be used to study the long-run emergent patterns in the conflict.

14.4 Repetition of the conflict by a large population

It is necessary to model the Tepoztlán organization process using adequate dynamics. As explained in section 14.2, the emergence of the CUT was a gradual process. Many conflicts similar to the organization game took place daily among different groups of Tepoztecos in which the decision to act individually or in an organized way was involved. Day by day such conflicts arose in one or another neighborhood, in the fields, at the market, in schools, or in other settings. The emergence of the CUT did not change the fact that relationships between people occurred mainly at the local level. Decisions taken in local assemblies involved members of the local community. Representatives of local assemblies met in higher-level assemblies, where decisions taken at local level were discussed. The same is true of more informal settings; in a large demonstration, for example, people from every neighborhood or sector would walk together with their own posters.

The proposed golf club development had many different effects on the community and people had to adopt a position on the issue, often changing their way of thinking and acting according to their own and others' experiences over time. Of course, each person did not have full information about other people's behavior or attitudes; rather, snippets of information were gathered from gossip in the neighborhood, market, or church, from attendance at local assemblies, from watching television or listening to the radio. Nobody gathered information systematically; instead, it was received at random. According to the information obtained people took the best decisions they could, although sometimes they made mistakes. Also, rules were loose; it was possible for a member of the group to free ride in a subtle way, leading to fluctuations in the real size of the CUT. The pattern of events in Tepoztlán is known; it would be interesting to consider a model that achieved similar long-run patterns to those that occurred there. Are the dynamics of Young and of Kandori et al. a useful model for that purpose?

Those authors model the repetition of the conflicts (which are expressed by a strategic m-person game) by a dynamic that depends on a learning process that happens in the long run. However, people who play that repeated game are not always the same m players. Instead it is assumed there is a large population N (the population of Tepoztlán, for example) that comprises many community conflicts that are expressed, in each period t, by an m-person game (it may be the organization game), and that game is repeated time after time. There are many conflicts of this kind in each period and in the long run, period after period. The population N is partitioned in subpopulations $N_1, N_2, \ldots, N_m$, one for each player in the studied game. For each j, N_j is large (the size is, n_j, $\sum_{i=1}^{m} n_i = n$). A group of m persons, one of each N_i, can face a community conflict (organization game). In a conflict such as that in Tepoztlán it is coincidental which groups of individuals will face one another in one period. But it is assumed that each of those conflicts happens among just m people drawn at random from each N_i. It is also assumed that people from N learn by experience, but they are myopic. The way people gather information in real conflicts such as that involving the Tepoztecos is stochastic, because whom would they get information from? What they hear from one day to another is coincidental. The model assumes that people can only access a small sample of information. They act according to that information, but they also make mistakes in each period, with a small probability, and choose a strategy that is wrong according to any possible sample of information. That is the approach of Kandori et al. (1993), Kandori and Rob (1995), and Young (1993, 1998). They expressed that process as a perturbed Markov process and studied its stochastically stable equilibria, which are the unique states of the process that happen with positive probability in the long run.

Applying that approach to the organization game, the process was as follows. In each period t, each j chooses one strategy, either P or NP, and uses its selected strategy to participate in several organizational conflicts during the period. Who are the participants in these encounters? It is incidental, depending on a range of factors including job, neighborhood, family, hobbies, political or religious ideas, and social concerns. It is assumed that each encounter happens with $m - 1$ players drawn at random, one of each N_l, $l \neq j$. Besides, people are not very rational. They are myopic and learn by their own and others' experiences. According to those experiences, which are only a sample of all the encounters that take place, each person j learns in a limited manner. That is, j obtains a sample of information about a small number of conflicts, and the strategies that people from each N_i chose in dealing with those conflicts, in period t, and decides to use, in $t + 1$, the strategy (P or NP) that is the best response to that sample. It is assumed that the size of those samples is k, and is the same for each person and in each period. k is small in relation to the size of each N_i. Besides, with a small probability, each person in N deviates from the strategies suggested by the learning dynamics, as happens in real life. The deviations of each j are independent of the deviations of others.

If everybody in N chose a strategy, in a period t, the strategy structure of society would be a vector $z = (z^1 = (z^1_1, \ldots, z^1_{n_1}), \ldots, z_m = (z^m_1, \ldots, z^m_{n_m}))$, where z^j_i means the strategy that was chosen by the member number i of the population N_j. z^j_i could be P or NP. Therefore, Z will denote the set of all those vectors.

Each $z = (z^1 = (z^1_1, \ldots, z^1_{n_1}), \ldots, z_m = (z^m_1, \ldots, z^m_{n_m}))$ could be associated to a profile of mixed strategies of the organization game in a natural way. That is, z is associated to the profile of frequencies $x = (x^1, x^2, \ldots, x^m)$, where x^j is equal to $\left(\frac{n^j_1}{n_j}, \frac{n_j - n^j_1}{n_j}\right)$, and n^j_1 is the number of people from N_j who chose P. It will be said that z represents the profile x.

A learning dynamic can be modeled as a perturbed finite Markov process, with Z as its set of possible states (see Kandori et al. 1993; Kandori and Rob 1995; Young 1993, 1998) in order to study the long-run behavior. The term Q_{zz}' of the Markov matrix is determined as follows:

> Given a state z, a sample of size k of z is a vector
>
> $z(k) = (z^{1(k)} = (z^1_{i_1}, \ldots, z^1_{i_k}), \ldots, z^{m(k)} = (z^m_{j_1}, \ldots, z^m_{j_k}))$,
>
> where $z^{r(k)}_{l_s}$ is one of the coordinates of z^r (without replacement).

Given a pair of states z and z', if there is not a combination $\{a_i\}$ of m samples, one for each player, allowing transition from z to z' because of the best replies of each player (according to the profiles of mixed strategies that correspond to the frequencies in the samples), then $Q_{zz'}$ is defined as zero. If there are combinations of samples with the mentioned properties then $Q_{zz'}$ is defined as the probability of any of those combinations being selected by the m players. The matrix, whose entries are $Q_{zz'}$, is denoted as Q.

On the other hand, it is taken into account that, with a very small probability η, each person i can choose a strategy that is not the best reply to any possible sample. The probability of each j making that mistake is independent of others' mistakes. There is, then, for each η, another Markov matrix that is a perturbation of Q:

$$Q'_{zz}(\eta) = (1-\eta)^n Q'_{zz} + \sum_{y \in d(z),\, y \neq z'} \sum_{J_{yz}' \neq \phi} q^{J_{yz'}}_{yz'} \eta^{|J_{yz}|} ('1-\eta)^{n-|J_{yz}'|} Q'_{yz'}$$

where $q^{J}_{xx'}$ is the probability of going from x to x', when only the members of J are making mistakes, x and x' are in Z, and $J \neq \theta$. $J_{xx'}$ means a subset of N such that $q^{J_{xx'}}_{xx'}$ is positive.

For each η, $Q(\eta)$ is an irreducible matrix very close to Q when η is small. Then, there exists a unique distribution vector $q(\eta)$ such that $q(\eta)Q(\eta) = q(\eta)$. The vanishing order of $Q_{zz'}(\eta)$ is equal to the minimum number of errors needed to go from z to z'. This number of errors will be called the cost of going from z to z'.

$q_z(\eta)$ is the probability of state z occurring when the rate of making mistakes is η. The surviving states when the probability of making mistakes tends to zero would be the behavior patterns that will prevail in the long run. A theorem of Young (1993), based on some results of Freidlin and Wentzell (1984), proves that the limit of $q(\eta)$ as η tends to zero exists and is a distribution vector q^* such that $q^* Q = q^*$.

> *Definition 1:* A state z in Z is a stochastically stable equilibrium of the process if and only if the q^* coordinate that corresponds to z is positive.

Vector q^* describes the probability of state z occurring in the long run. Only the stochastically stable equilibria have a positive probability of occurring in the long run. Then, the process cannot remain in the same state, but visits more frequently the stochastically stable equilibria. In fact, when the probability of making mistakes η is small enough, the process can be in any state, but the probability of the majority of those states occurring is very small, and those of the stochastically stable equilibria are high.

Assume that a finite strategic game is analyzed with dynamics similar to those described here. The stochastically stable equilibria of the game might represent the Nash equilibria of the game or not. It is possible that the stochastically stable equilibria are vertices of a cycle of the dynamics that do not represent any Nash equilibrium, but the repetition of situations that people visit time after time.

Figure 14.2 shows a simple scheme of part of the flux. Each of the vertices in the graphic is a possible state (a strategy structure of society). Each continual arrow between two vertices shows that it is possible to go from one state to another without mistakes. Each dotted arrow between two vertices represents a situation where it is necessary to make mistakes in order to go from one state to another. If there are no arrows between two vertices, then it is not possible to go from one state to another with or without mistakes.

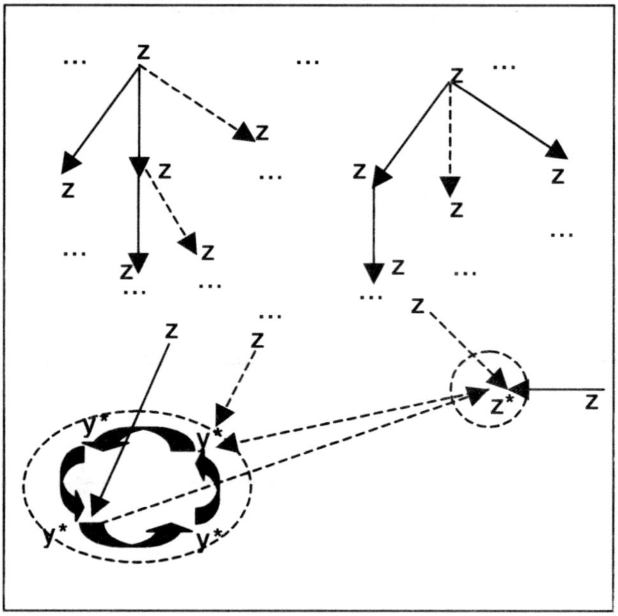

Figure 14.2 Possible changes in the strategy structure of society during the learning process.

Those vertices that have an asterisk are stochastically stable equilibria. They might be isolated, because it is not possible to get outside each one without mistakes, or they might belong to a cycle. There is a sequence of continual arrows that join each state that is not a stochastically stable equilibrium to one that is. Therefore, the process always reaches some stochastically stable equilibrium, but again leaves it.

Young (1993, 1998) proves that each stochastically stable equilibrium in a weakly acyclic game is associated to a strict Nash equilibrium, if the sample is complete enough, that is, if $\frac{k}{\min\{N_i\}}$ is small enough. It will be proved that the organization game is weakly acyclic, but first that concept will be defined.

> *Definition 2:* A game is weakly acyclic if, for each pure strategies profile $\sigma(1)$, there is a sequence of pure strategies profiles $\{\sigma(1) = (\sigma^1(1), \ldots, \sigma^m(1)), \ldots, \sigma(s) = (\sigma^1(s), \ldots, \sigma^m(s))\}$, such that $\sigma(s)$ is strict Nash equilibrium and, for each $i = 2, \ldots, s$ there is a player j_i who changed its strategy from the profile $\sigma(i-1)$ to the profile $\sigma(i)$, and $\sigma^{j_i}(i)$ is a j_i best reply to $\sigma(i-1)$. All of the rest of players did not change their strategies from $\sigma(i-1)$ to $\sigma(i)$.

> *Proposition 2:* The organization game is weakly acyclic.

The proof is not difficult, and it can be seen in the Appendix at the end of this chapter.

Therefore, the following theorem is a corollary from propositions 1 and 2 and the Young's theorem that establishes that each stochastically stable equilibrium in a weakly acyclic game with a complete sample represents a strict Nash equilibrium.

In this case, a complete sample means that $\dfrac{k}{\min\{N_i\}}$ is small enough that people can choose the combinations of samples a sufficient number of times to allow the stochastically stable equilibria to be reached.

> *Theorem:* If $\dfrac{k}{\min\{N_i\}}$ is small enough, the stochastically stable equilibria are those where all the people in exactly s^* of the $N_1, N_2, \ldots, N_m$ choose P (participation) and all the rest of the members of N choose NP.

This theorem deals with a large population where many communities are involved in the organization game. It means that if people are myopic enough (that is, their sample size is small enough) some people, exactly s^*, in each community, will form, almost surely, an organization in the long run. Of course, in each period the members of the emergent organization might be different. With small probability, there would be larger organizations or there would not be any organization at all. Recall that the size s^* is linked to the organization game parameters that in some sense characterize the communal conflict.

How are these stochastically stable equilibria related to the behavior of the Tepoztlán community? In a qualitative manner they reflect the long-run behavior of that community. They reflect more adequately the normal times in Tepoztlán, because the process has been repeated for many years, and in the long run the people of Tepoztlán often organize assemblies at local level (wider-level assemblies occur less often). Only some members of the community—neighbors, farmers, vendors, depending on the kind of assembly—participate in them. Different people participate in each assembly, although some people always participate and some others never do. The sizes of the assemblies are very similar. In the conflict concerning the golf club development, the process was similar, but that conflict happened during less than three years. At that time, when the CUT was founded, the assemblies were much larger and took place more often. There were also demonstrations and protests, which are forms of organization. The process was concentrated in a short but rich period. The number of participants from the community in each protest was similar after the CUT was formed, but that number was much larger than usual. According to the model this was because of the growth of parameters γ (the individual productivity people believed the constructed club allowed them to obtain) and d' (the time that people who participated in the CUT had to spend in assemblies, demonstrations, protests, and so on).

However, there are some facts that are not very well represented in the model. In the following section they will be explained, and some possibilities explored to improve the model to accommodate them.

14.5 Possible modifications to the model

14.5.1 Different tendencies to cooperate

In Tepoztlán there has been, for a long time, a community with a constant tendency to organize. According to the stochastically stable equilibria, each member of the community participates in an organization with the same probability. In the real community, there are people who are more interested in particular problems and these people change through time, but it is not true that every one participates with the same frequency. Moreover, there are people who always participate. They are more conscious or active or experienced. There are some people who never take part. Why do people have different tendencies to participate and cooperate? People from real communities, as in Tepoztlán, may have very different individual productivity and cooperation costs. Besides, common goods may give them different benefits. It is also obvious that some industries or nations produce great environmental damage and are not very interested in cooperating. Because of that, each individual could face the organization dilemma in a different way.

The simplest case would be another prisoner's dilemma where each player receives different benefits from the public good. For example in Tepoztlán, seniors obtain large benefits from the common goods. For them β_i should be larger than d. On the other hand, there are people whose β_i should be smaller than $\frac{d}{m}$, for example people who have business or political interests in Tepoztlán but do not live there. In a model with those differences, seniors would always participate in the organization against the club and the others would never participate.

In the case of the Tepoztlán community, this consideration would make the model more realistic. However, there are some interesting problems that might be studied as regards this improvement to the model, such as the migration problem and its environmental effects. For example, in a country A which is a destination for migrants, native inhabitants form an important group. For them β_i is large. Illegal immigrants who live in overcrowded conditions and just want to send money to their families would form another group. For them β_i is almost zero. Immigrants who have some sort of legal status might form another group. They may even have acquired the nationality of A, and perhaps have brought their family to A. For them β_i is close to the value for the native inhabitants of A. These groups have different behaviors that might be explained by this kind of model.

14.5.2 Accumulation of communal stock through cooperation

The community of Tepoztlán has, over the years, been faced with several attempts to build tourist projects: two different attempts (separated by 40 years) to build a golf course, and one attempt to build a scenic train. As each of those projects approached the implementation stage, the community conflict parameters γ and d' were altered, and it was necessary to build larger organizations to face the problem. After these projects were abandoned, the community recovered its usual dynamics. To study each case it is necessary to change the parameters, as is usual in comparative statistics. This process is not very elegant, and does not explain why such projects are presented from time to time.

Until now, it has been assumed that the organization game repeats under the same conditions year after year. But it could be considered that there is a sort of accumulation of environment-protecting stock through the establishment of laws and decrees or social recognition of the need for environmental protection. It is necessary to find a way to properly measure these pseudo-stocks.

A somewhat different approach from the one used in sections 14.3 and 14.4 could be established taking into account this accumulation, and might model the cycles observed in the Tepoztlán community. This sort of accumulation occurred, for example, when Tepozteco Hill was decreed a national park and became the property of the community, forbidding to some extent outside exploitation of the lands, forests, and water. Subsequently the decree making the hill a reserved zone solidified that prohibition. The community managed to prevent all three of the proposed tourist projects. Locally, at the level of neighborhoods, schools, and marketplaces, the community achieved similar partial measures. This meant that the communal rights over natural resources would acquire a more stable status and would provide larger individual productivity as the community accumulation increased. Nobody could be excluded from this sort of public good stock, even if someone did not participate in achieving it. It is also considered that the social stock depreciates, at a slow rate, period after period, in the sense that the laws and decrees may continually loosen or they may not be observed as before and, at the same time, the communal watchfulness also relaxes.

The accumulation-learning process will be briefly explained. It is assumed that the community is large ($m > s^*$), and m and β are constant in time. The repetition of an organization game similar to the one previously studied will be considered period after period, although in an altered form, because the individual productivity and the size of the minimum efficient group change period after period.

In the beginning of the process there is not any accumulated stock, that is, e_0 is equal to 0, so $s_0^* = s^*$. If, up to a certain period, e units of the public good have been accumulated, each individual productivity would have been γ'_e, which is equal to $\gamma' + \beta e$. As individual productivity grows, then the cost of organization, which is an increasing function of that productivity, also grows. The community would be immersed in an organization game, but with the parameter γ'_e instead of γ'. The smallest size group that allows people to obtain more than $\gamma' + \beta e$ is s_e^*, which is the least integer that is larger than $n_0(\gamma' + \beta e) + \frac{d}{\beta}$. Even if m were bigger than s^*, s_e^* could change its relation with m, in some period $t > 0$. It could continue to be smaller than the size of the population, or reach it. In one period s_e^* will surely overcome that size. For this reason the stochastically stable equilibria of the process would be vertices from some cycles. The set of possible states of this process is $Z \times E$, where E is the set of possible stocks. Then the stochastically stable equilibria are couples (z, e), where z is a strategy structure of society, and e is the accumulated stock of the common good. As before each player draws a sample of size k from z and chooses a best reply in the organization game when there is an accumulated stock e. Thus, people's behavior corresponds to a dynamic (Kandori et al. 1993; Kandori and Rob 1995; Young 1993, 1998) similar to the one explained in section 14.3, but now accumulation is another element affecting the strategy selection. People will achieve satisfactory stock e^*, stopping organization and cooperation as a result. Then depreciation will lead people into an organization–disorganization cycle as if they were trying to preserve the stock e^*. The cycles in the Tepoztlán people's process of organization might be explained in this way.

14.5.3 Effect of natural ties on the probability of conflict

In the model each member of the community could have an encounter with $m - 1$ members of the community drawn at random, but, in actuality, the probability of that encounter happening among, for example, neighbors or co-workers should be much higher, establishing an organization game with the characteristics of a spatial game similar to that described by Young (1998).

14.5.4 Some additional remarks

There are many considerations that could enrich the model in future work. It could be interesting, as it is more realistic, to study how organization dynamics are transformed when the parameters m and β are not constant. Of course, in real life β can grow, sometimes because of public policies. Both parameters might be functions that depend on time and the size of the organized group.

In this work, the qualitative behavior of a specific community (that of Tepoztlán) has been compared to the qualitative behavior predicted by the model. However, the study has not developed the important task of estimating the parameters that correspond to a real community in order to contrast, in a more precise way, the model predictions and observed community behavior. This could help to suggest, with better foundations, specific economic and social policies. The parameter estimations could be carried out using home surveys similar to those that are used to model consumer behavior.

The presented model was only intended to study the qualitative behavior of communities with environmental problems. These qualitative features will not be lost even if the model is enriched in the directions mentioned above or in many other possible directions, or if a precise parameter estimation is carried out. Nonetheless, in spite of its limitations, some essential aspects in the tendency of those communities to organize themselves can be studied, and in the relationship between those tendencies and the accumulation of community wealth. It is also stressed that the model dynamics reproduce qualitative properties that are seen in the real dynamics of emerging organizations within different social sectors. This may be useful to guide policymakers toward better environmental policies.

14.6 Conclusion and policy implications

In spite of its purely qualitative features, the approach that has been followed to study environmental problems is relevant to the consideration of real-world environmental disputes.

The qualitative conclusions obtained using this evolutionary game model may be reviewed as follows.

First, it is clear that large human conglomerations are among the main actors responsible for the modern environmental catastrophe. Other important actors are large corporations and national States, particularly the world powers, as they try to satisfy their huge economic and political interests. The model justifies the opinion that these human conglomerates are also main actors in facing the catastrophe. They may act both by forcing themselves to change their own environment-damaging activities, and by stopping those of the large corporations and State powers. If this is true, the evolutionary game approach is the proper one to discover what behavior patterns will emerge in the long run. Many policies are inspired by the idea that prohibitive laws, punishments, and several other projects can by themselves abolish negative behavior patterns. If these negative patterns are long-run equilibria of the process in which the communities are involved that idea may be fanciful and those laws and projects may not achieve their objectives. The model results suggest, for example, that the building of walls and fences and the establishment of punitive laws in order to stop the illegal migration from Mexico to the United States or from Morocco to Spain will not succeed by themselves. On the other hand, environmental protection laws, decrees, and punishments can be very useful if they strengthen the virtuous long-run equilibria of the process in which the community is involved. They may even transform the game conditions in order to obtain virtuous equilibria.

The qualitative analysis carried out in this work allows the following conclusions to be obtained: (a) if the direct payoff that members of the communities receive from the environment is very small in relation to the cost necessary to protect it, they are likely to engage in activities, either directly or indirectly, that will lead to the plundering of the environment; (b) if, on the other hand, the benefit is large enough (the larger the better), there will emerge from within the community a nucleus responsible for protecting the environment and other communal interests, either continuously or cyclically, watching over it and preventing the negative activities of other members of the community and of actors with greater capacity to damage, such as some corporations and States. This nucleus will also maintain the unity of the members of the community among themselves, as well as their link with their place of origin, helping reverse the trend toward emigration.

This approach can also be useful in studying other sorts of social failure, such as the increase of crime in a society, which have their origin in the disintegration of communal and social links and social capital.

How can a qualitative analysis help environmental policymakers (including national governments and international organizations) to design projects, protocols, and agreements? Of course, it can only do that in a qualitative way, and other models and quantitative tools must be used to develop specific projects. However, models such as this can be useful in distinguishing which kinds of policies are appropriate and which ones are counterproductive, and in evaluating policy effectiveness and efficiency. They can also help in studying how some policies can endanger the existence of many communities that have an important role to play in the protection of ecology and the environment. From the viewpoint of citizens of many developed countries it is clear that the environment and human rights should be protected. For them, this implies that not all economic and political activities are valid and that the relevant international treaties should be enforced. It is also clear that it is necessary to protect communities around the world against expropriation of their resources. That is, national governments and international organizations should consider the dangers produced when β diminishes to the point of almost vanishing. On the contrary, they should design policies that result in β increasing.

Note

1. The prisoner's dilemma is attributed to A.W. Tucker. For an early account of the game, see Luce and Raiffa 1957.

References

Axelrod, R. (1984) *The Evolution of Cooperation*, New York: Basic Books.

Binmore, K. (1995) *Game Theory and the Social Contract*, Volume 1, *Playing Fair*, Cambridge, MA, and London: MIT Press.

Binmore, K. (1998) *Game Theory and the Social Contract*, Volume 2, *Just Playing*, Cambridge, MA, and London: MIT Press.

Freidlin, M. and Wentzell, A. (1984) *Random Perturbations of Dynamical Systems*, New York: Springer-Verlag.

Glance, N.S. and Huberman, B.A. (1993) "The outbreak of cooperation", *Journal of Mathematical Sociology*, 17(4): 281–302.

Glance, N.S. and Huberman, B.A. (1997) "Training and turnover in the evolution of organizations", *Organization Science*, 8(1): 84–96.

Hardin, G. (1968) "The tragedy of the commons", *Science*, 162: 1243–8.

Kandori, M. and Rob, R. (1995) "Evolution of equilibria in the long run: a general theory and applications", *Journal of Economic Theory*, 65: 383–414.

Kandori, M., Mailath, G., and Rob, R. (1993) "Learning, mutation and long run equilibria in games", *Econometrica*, 61: 29–56.

Luce, R.D. and Raiffa, H. (1957) *Games and Decisions*, New York: John Willey and Sons.

Maruta, T. and Okada, A. (2001) "Stochastic stability of group formation in collective action games", KIER DP 536, Kyoto University.

Nash, J.F. (1950) "Equilibrium points in n-person games", *Proc. Nat. Acad. Sci.*, 36: 48–9.

Okada, A. and Sakakibara, K. (1991) "The emergence of the state: a game theoretic approach to the theory of social contract", *Economic Studies Quarterly*, 42(4): 315–33.

Olson, M. (1965) *The Logic of Collective Action*, Cambridge, MA: Harvard University Press.

Ostrom, E. (1990) *Governing the Commons: The Evolution of Institutions for Collective Action*, Cambridge: Cambridge University Press.

Ostrom, E. (1998) "A behavioral approach to the rational choice theory of collective action", *American Political Science Review*, 92(1): 1–22.

Ostrom, E., Gardner, R., and Walker, J. (1994) *Rules, Games, and Common-Pool Resources*, Ann Arbor, MI: University of Michigan Press.

Rosas, M. (1997) *Tepoztlán Crónica de Desacatos y Resistencia*, México: Ediciones ERA.

Vega-Redondo, F. (1993) "Competition and culture in an evolutionary process of equilibrium selection: a simple example", *Games and Economic Behavior*, 5: 618–31.

Weissing, F. and Ostrom, E. (1991) "Irrigation institutions and the games irrigators play: rule enforcement without guards", in R. Selten (ed.) *Game Equilibrium Models II: Methods, Morals, and Markets*, Berlin, Heidelberg: Springer-Verlag.

Young, H.P. (1993) "The evolution of conventions", *Econometrica*, 61: 57–84.

Young, H.P. (1998) *Individual Strategy and Social Structure: An Evolutionary Theory of Institutions*, Princeton, NJ: Princeton University Press.

Appendix

Proof of proposition 1: let σ be a profile such that the number of people who chose P is smaller than $s^* - 1$. No organization appears in that profile, and everybody earns γ. If a player j changed its strategy, and the others did not, it is still not possible to form an organization, and j would continue earning γ. These are Nash equilibria, whatever the values of the parameters are, and it does not matter if s^* is smaller than n, as is assumed, or if it is larger than n or equal to it. These equilibria are not strict.

Let σ be a profile such that exactly s^* people choose P, whoever they are. An organization is formed if $s^* \leq n$. Let j be a player who chose P. Then, j earned $\gamma + \beta(s^* - d_0) - d$, but if it changes its strategy, and the others do not, the organization cannot be formed, and j would earn γ, which is smaller than $\gamma + \beta(s^* - d_0) - d$. Let i be a player who chose NP in σ. i earned $\gamma + \beta(s^* - d_0)$; if it changes its strategy, and the others do not, it would earn $\gamma + \beta(s^* + 1 - d_0) - d$, which is smaller than $\gamma + \beta(s^* - d_0)$. Then σ is strict Nash equilibrium.

Let us examine the other profiles. Let σ be a profile of pure strategies $(\sigma^1, \sigma^2, \ldots, \sigma^n)$, such that the number of people who chose P is $s^* - 1$. No organization appears in that profile and each player earns γ. If j chose NP in σ, and it changes its strategy, and the others do not, there would be s^* players that would choose P, and an organization would appear. Then j would earn $\gamma + \beta\ (s^* - d_0) - d$, which is larger than γ. That is to say, σ is not Nash equilibrium. Let σ be a profile where the number of people who chose P is larger than s^*. Let j be a player who chose P. An organization, whose size is $s_\sigma + 1$, would appear, then j would have earned $\gamma + \beta\,(s_\sigma + 1 - d_0) - d$, but if it changes its strategy, and the others do not, an organization, whose size was s_σ, would appear. Then j would earn $\gamma + \beta\,(s_\sigma - d_0)$, which is larger than $\gamma + \beta\,(s_\sigma + 1 - d_0) - d$. Then σ is not Nash equilibrium. There is not another Nash equilibrium in pure strategies in the organization game.

Proof of proposition 2: let $\sigma(1)$ be a profile of pure strategies such that w, the number of people who chose P, is larger than s^*. If a player chose P in $\sigma(1)$, it will be called j_1. The payoff of j_1 in $\sigma(1)$ is $\gamma + \beta(w - d_0) - d$. A best reply of j_1 to $\sigma(1)$ is NP. Let $\sigma(2)$ be the profile where j_1 chooses NP and each one of the rest of the players chooses the same strategy that it chose in $\sigma(1)$. If the number of people who chose P in $\sigma(2)$ is s^*, $\sigma(2)$ is strict Nash equilibrium. If that number is larger than s^*, it could proceed by induction to build $\{\sigma(1), \sigma(2), \ldots, \sigma(s)\}$, such that $\sigma(s)$ is strict Nash equilibrium (there are s^* persons that choose P in $\sigma(s)$), and, for each $i = 2, \ldots, s$, there is a player j_i who changed its strategy from the profile $\sigma(i-1)$, where it chose P, to the profile $\sigma(i)$, and $\sigma^{j_i}(i) = NP$ is a j_i best reply to $\sigma(i-1)$. All of the rest of players did not change their strategies from $\sigma(i-1)$ to $\sigma(i)$.

Let $\sigma(1)$ be a profile of pure strategies such that w, the number of people who chose P, is smaller than s^*. If a player chose NP in $\sigma(1)$, it will be called j_1. The payoff of that player in $\sigma(1)$ is γ. If it changed its strategy, and $w + 1$ is smaller than s^*, it would again earn γ. If it changed its strategy, and $w + 1$ is equal to s^*, it would earn $\gamma + \beta\,(s^* + 1 - d_0) - d$. In both cases, P is a best reply of j_1 to $\sigma(1)$. Let $\sigma(2)$ be the profile where j_1 chooses P and each one of the rest of the players chooses the same strategy that it chose in $\sigma(1)$. If the number of people who chose P in $\sigma(2)$ is s^*, $\sigma(2)$ is strict Nash equilibrium. If that number is smaller than s^* the building of the search sequence could again proceed by induction. In a finite number of steps a strict Nash equilibrium would be reached.

Then the organization game is weakly acyclic.

15 Effects of alternative CDM baseline schemes under an imperfectly competitive market structure

Haruo Imai, Jiro Akita, and Hidenori Niizawa[1]

This chapter investigates how alternative clean development mechanism (CDM) baseline-setting methods perform when applied to a CDM project undertaken by a profit-maximizing firm in an imperfectly competitive industry. A Cournot oligopoly game model is used to analyze how CDM baselines affect output scale choice by the firm engaging in the CDM project as well as other firms in the industry. The theoretical analysis is applied and illustrated using a numerical example inspired by an actual CDM project currently under way.

15.1 Introduction

This chapter uses the standard Cournot oligopoly game analysis to show how alternative clean development mechanism (CDM) baseline-setting methods perform when applied to CDM emission reduction projects undertaken by profit-maximizing firms in an imperfectly competitive industry, analyzing particularly how CDM baselines affect the output scale of the firm engaging in the CDM project as well as other firms in the industry. The CDM was introduced in 1997 by the Kyoto Protocol to the United Nations Framework Convention on Climate Change (UNFCCC). Efforts have been made since to implement the CDM, and the number of CDM projects has continued to increase, though projections regarding the future of this mechanism are not entirely optimistic (Bohm and Carlen 2000).

There have been concerns about how to determine the amount of emission reduction rightfully attributable to a CDM project, thereby providing an appropriate incentive to each involved party. The determination requires not only measuring actual emissions accurately, but also imagining counterfactually what would have happened were it not for the CDM project. The latter question is known as a baseline-setting problem. When setting such baselines, it is tempting to take into consideration information that becomes available ex post facto, that is, after the CDM project is implemented, in addition to data available prior to the project. Incorporating ex post information, however, inadvertently runs the risk of distorting the incentives of project participants. Imai and Akita (2003) show how such a risk is associated with baseline-setting methods when CDM is applied to firms in isolation. This chapter extends the analysis to firms in an industry in which their output decisions are interrelated.

Baseline-setting methods may be broadly classified into ex ante methods and ex post methods. It was expected that ex ante methods would be primarily used, but most methods actually adopted fall into the ex post category.[2] Some methods exist that lie between these two extremes, namely conservative baseline and proxy baseline. Conservative baseline has been employed in real-life CDM methodologies, while proxy baseline is well suited to consideration of CDM firms operating in an oligopolistic industry.

The performance of these methods is compared here in an imperfectly competitive market, where the application of game theory has been found typically useful to examine the consequence of interdependence amongst firms in an industry. Specifically, the Cournot oligopoly model is employed, both as an analytically useful benchmark and as a realistic description of the mode of competition in some industries to which the CDM is applicable.

The results indicate that different baseline-setting methods indeed bring about different outputs, profits, emission credits of the firms, and consequently world emission levels. It is observed that in some cases, market interdependence could mitigate the above-mentioned adverse incentives generated inadvertently through baselines. The policy implications of these observations depend on the priorities of different policymakers with different positions and responsibilities. With this in mind, the analyses are applied and illustrated in a numerical example inspired by a real-life CDM project.

15.2 Background

15.2.1 Kyoto Protocol and the CDM

The Kyoto Protocol was negotiated in 1997 and came into effect in February 2005. More than 30 developed countries and transition economy countries have agreed to assume the treaty obligation to control their greenhouse gas emissions below their respective specified levels during the period from 2008 through 2012. Those countries are referred to as Annex I Parties (countries) because they are listed in Annex I of the United Nations Framework Convention on Climate Change. On the other hand, those countries that have ratified the Protocol but not assumed any obligation for greenhouse gas reduction are referred to as non-Annex I countries.

The Kyoto Protocol is notable for including "flexibility mechanisms", called Kyoto mechanisms, to help Annex I countries to attain their emission reduction targets. There are three such mechanisms: (a) emissions trading; (b) joint implementation; and (c) the clean development mechanism. With regard to emissions trading, environmental economists had discussed its working long before the development of the Kyoto Protocol.[3] The idea has already been put into practice to combat actual polluting activities; the acid rain program in the United States is an example.[4] More innovative are the CDM and joint implementation, both of which grant emission credits to investment projects reducing greenhouse gas emissions. In the case of the CDM, investment projects take place in non-Annex I countries. Issued credits are called certified emission reductions (CERs). In the case of joint implementation, investment projects take place in Annex I countries. Issued credits are called emission reduction units (ERUs). CERs and ERUs can be used to fulfill the emission reduction obligations of Annex I countries. While the emissions trading system under the Kyoto Protocol is a cap and trade system that imposes national caps on greenhouse gas emissions beforehand, the CDM and joint implementation are project based, and CERs and ERUs are issued after reducing emissions by means of project activities.

Since the Kyoto Protocol exempts non-Annex I countries from their obligations to control greenhouse gas emissions, the issue of how to contain emissions from non-Annex I countries is deemed important in the so-called post-Kyoto negotiation. The CDM is a way to promote emission reduction projects in developing countries, both directly through the projects themselves, and indirectly through raising awareness and transferring emission reduction technology by the project. In addition, the CDM also helps developed countries to economize their emission abatement costs by allowing part of their reduction to take place outside their own territories (where further reduction opportunities are presumably scarce and reduction costs are accordingly high). Such cost economization is desirable from the viewpoint of worldwide economic welfare, for the saved resource could be used elsewhere for the betterment of global equity and of humankind's ability to avoid further degradation of the natural environment.

The Kyoto Protocol provides merely a rough guideline for each mechanism, leaving the details to subsequent negotiations. The details of the mechanisms were not made clear before the Marrakech Accords in 2001 (UNFCCC 2002). In accordance with the decision of the Marrakech Accords to facilitate a prompt start for the CDM, the Executive Board of the CDM (CDM-EB) was set up in the same year. The CDM-EB[5] oversees the CDM process and makes decisions over practical issues, while more fundamental issues are handled by the so-called Conference of the Parties serving as the Meeting of the Parties to the Kyoto Protocol. However, the first meeting of the Joint Implementation Supervisory Committee was not held until February 2006.[6]

With regard to the current demands for CDM CERs, the European Union implemented emissions trading through its Emissions Trading Scheme (EU-ETS) in January 2005. On the one hand, CDM CERs are usable and thus demanded to achieve targets in the EU-ETS. On the other hand, national governments, including Japan, that foresee difficulty in achieving their national emission reduction targets have prepared schemes through which they can purchase CERs using public funds, which constitutes another major demand source of CDM CERs at present.

The Kyoto Protocol stipulates quantified emission limitation or reduction commitment only until 2012. Negotiations on emission targets beyond 2012, which have been taking place since the eleventh Conference of the Parties to the UNFCCC in 2005, are far from being complete. Should countries fail to reach an effective agreement, CDM CERs would be valueless beyond 2012. That is, whether or not the CDM is continued in the next commitment period (after 2013) is still uncertain. Conversely, the fate of the post-Kyoto regime depends on the role of non-Annex I Parties, which, in turn, is naturally affected by the benefits brought by the CDM.

That the future of the Kyoto Protocol CDM is not entirely clear does not make it sterile to study it in detail. The nature and working of this mechanism should be studied thoroughly and comprehensively, for analogous project-based mechanisms may well be needed to combat different but perhaps related environmental problems other than that of global warming in the future. By the same token, nor is it necessarily academic to examine options that have already been ruled out by the actual CDM-EB settlement.

15.2.2 Baseline issue of CDM

Prior to the Marrakech Accords, substantial discussions had taken place on how the mechanics of the CDM should be designed.[7] The topics covered in the discussions included (a) baselines (Ellis and Bosi 1999; Willems 2000); (b) which activities should be eligible as CDM projects; (c) how to ensure that issued credits are backed by actual extra emission reduction (the "additionality problem") (Chomitz 1998); (d) who can apply for CDM credits; and (e) whether credits can be freely traded. Some issues, including (b) the CDM eligibility question, are still unsettled and continue to be discussed at the Conference of the Parties, the subsidiary bodies of the Protocol, and the CDM-EB, while other issues have been more or less settled. For example, an additionality test[8] has been implemented and applied to approve several methodologies. While respecting the settlements and resolutions adopted so far by the Conference of the Parties, the subsidiary bodies, and the CDM-EB, their consequences should be carefully examined to see whether their design goals are likely to be achieved (if not already), and whether there is any room for improving their performances.

Unlike emissions trading, the CDM is a project-based scheme that by itself imposes no cap on emissions. Instead, it grants credits to projects based on their reduction of emissions from the "business as usual" (BAU) level that would have taken place were it not for the CDM. That is, the difference between the baseline and actual emissions achieved by a CDM project is the amount of CERs issued and awarded to the project. The credits issued are the CDM CERs, while the BAU reference level is the CDM baseline. Both in emissions trading and the CDM, it is imperative to ensure that emission levels are accurately measured. In emissions trading, once the measurement is accurately done, it only remains to see if it is within the tolerance of the given cap, and there is no residual ambiguity whatsoever. In the case of the CDM, however, it is not always a trivial matter to determine the baseline, as explained below. Thus, with the CDM, accurate measurement of emissions alone is not sufficient to ensure appropriate calculation of emission reduction. If the baseline is inappropriate, so will be the amount of issued CERs, which could potentially undermine the purpose of the Kyoto Protocol, even if emission levels are accurately monitored and measured.

Incidentally, joint implementation is as project based as the CDM. Joint implementation differs, however, in that both investing country and host country belong to the Annex I group, assuming an obligation to honor the assigned emission reduction levels. Hence, it is as though joint implementation is subject to a grand cap imposed on the participating countries, even though ERUs are calculated on a project basis using baselines analogous to those in the CDM. With the CDM, a host country is supposed to be a non-Annex I country, and there is no overarching cap imposed on the parties involved.

Thus, setting an inappropriate CDM baseline is tantamount to making the Kyoto Protocol emission reduction target effectively more stringent or lenient by way of an insufficient or excessive amount of issued CERs. And it is not always obvious to define what constitutes the appropriate baseline. The source of difficulty is an inherent one. That is, the baseline for a CDM project is meant to capture the emissions that would have prevailed were it not for the project, or furthermore, if there were no CDM to begin with. Either way, a CDM baseline almost by definition deals with hypothetical or counterfactual scenarios that exclude a CDM. Such scenarios are never realized once CDM projects are actually implemented. So, there is no way of verifying the appropriateness of the baseline in the light of actual emissions achieved by the projects, even after every relevant uncertainty is resolved.

The above-mentioned spirit of capturing what would have happened without the CDM, which underlies baseline setting, is also reflected in the additionality test. To pass this test, a candidate project must prove that it could never be put into place without the benefit and prospect of acquiring CDM CERs. One could reasonably argue that this additionality test constitutes an important part of the CDM baseline in a broad sense, though the latter is presumably broader in scope.

With regard to the baseline per se, the Marrakech Accords described a set of practical criteria for baseline setting, which state that a baseline emission level should represent such emissions that would have resulted in the absence of the proposed project. Such hypothetical emission levels can be estimated by means of (a) historical record; (b) industry average; or (c) optimal technology (in the language inspired by Fischer (2005)). Most of the time, (a) a historical record of some kind should be available. When historical data are nonexistent or inappropriate, (b) industry average is called for. If neither recourse is available, as in the case of CDM projects that involve innovative activities, it is necessary to resort to (c) optimal technology.

For a project to be eligible for CDM credits, it must first pass the additionality test. That is, additionality is a necessary condition for eligibility. This being the case, CDM methodologies submitted to the EB for approval mostly consist of proof of additionality and description of the baseline.

The CDM-EB has adopted a bottom-up approach in handling methodologies. Rather than spelling out comprehensive prescriptions for every possible methodology from the outset, the EB makes rules inductively based upon judgments made in the course of actually approving or rejecting concrete methodologies submitted to it. Thus, it is unlikely that already registered methodologies have exhausted every possibility. In particular, it is worth exploring alternative baseline-setting methodologies from various viewpoints.

To apply for CDM credits, one must first compose a methodology suitable for the project. Then the methodology must be sent to the EB for approval. Furthermore, this application must be conducted under the auspices of an "operational entity" accredited by the EB.

The EB approval process for new methodologies can be very time consuming. Proposed methodologies can be eventually turned down after a long period of deliberation. After several years of trial and error, however, a good number of methodologies and operational entities have by now been approved. Projects have already been approved and CERs actually granted. (The Kyoto Protocol permitted the issuance of CDM credits as early as in 2000, and possibly retrospectively.) So it seems that the CDM has been more or less successfully launched. In fact a concern might be the possibility that too many projects are approved, generating too many CERs, which may render the Protocol target too lenient, given that the United States and Australia have chosen to withdraw from the Protocol.[9] In fact, the secretariat of the UNFCCC announced on June 9, 2006 that their estimate of CDM CERs expected to be generated by the end of 2012 was as large as 1 billion tonnes of CO_2, which equals the sum of annual emissions of Spain and the United Kingdom combined.

15.2.3 Baseline methodology and interdependence of output

The remainder of this chapter investigates the functioning of alternative CDM baseline-setting methodologies when a CDM project is approved for a firm that belongs to an industry where firms strategically interact against one another. In particular, attention is focused on the possibility of interdependent output levels and the scales of operation of the firms within the industry.

In order to motivate this situation, consideration will first be given to the following hypothetical CDM project applicable to a firm whose emission level is proportional to its output level. Let x be the output. Then the emission level is $e \cdot x$ where $e > 0$ is a constant emission factor that is determined by the technology employed by the firm.[10] Now let this firm introduce a new technology (with investment costs I), so that its emission factor decreases from

e to e^{CDM} where $0 < e^{CDM} < e$. If the output x remains unchanged, then the emissions should decrease from $e \cdot x$ to $e^{CDM} \cdot x$. In general, however, output will be affected by the change in technology. Real-life examples of such projects can be found on the CDM website of the UNFCCC. In particular, registered projects using methodology AM0008 or ACM0005 are typical.[11] Under AM0008,[12] firms plan to switch from a fuel that is inexpensive but emits more greenhouse gases to a fuel that is more expensive but emits less greenhouse gases. Under ACM0005,[13] a firm in the cement industry switches its raw material to reduce emissions. There are in fact many other CDM projects that share this generic feature of reducing the emission factor. The above-mentioned projects are singled out primarily because their associated industrial characteristics are nearest to those envisaged by the following analysis.

The CERs (credits) generated from this project equal $B - e^{CDM} \cdot x^{CDM}$, where B is the baseline emission level defined by the methodology adopted by the project, and x^{CDM} is the output level after the CDM project started. (To be more precise, the CER amount is $\max\{B - e^{CDM} \cdot x^{CDM}, 0\}$ because the CER amount must be nonnegative.) Now, how is the baseline B set? The above-mentioned methodologies essentially set B equal to $e \cdot x^{CDM}$, that is, the pre-CDM emission factor times post-CDM output level. (ACM0005 permits ex post updating of e by examining current emission factors of the other brands of cement in the same industry. For simplicity, it is assumed that the emission factor remains at e.) This method utilizes observed ex post output level to yield the baseline. Therefore, depending on the scale of operation, the baseline varies ex post facto. This type of baseline is here referred to as an ex post baseline. (Laurikka (2002) called it a "relative baseline" while Fischer (2005) called it a "rate-based baseline". Here the naming convention used in Imai and Akita 2003 is retained). This baseline also resembles the baseline defined using the average emissions methodology described in the Marrakech Accords.

Some methodology (such as AM0017[14]) imposes an upper limit y to ex post output x^{CDM}, so that the baseline becomes $B = e \cdot \min\{x^{CDM}, y\}$. We may call this a conservative baseline because it rules out the possibility that an excessive number of CERs get issued when the output turns out to be extremely high. If the output upper limit y is set equal to the pre-CDM output level x, the conservative baseline never exceeds the pre-CDM emissions level $e \cdot x$. This observation motivates another variation where the baseline is set equal to the pre-CDM emissions level, $B = e \cdot x$. This will be referred to as an ex ante baseline. There are other baseline methods available when considering a project for a firm operating in an industry (and in fact this situation is analogous to the one for the industry average baseline given in the Marrakech Accords, in that the method exploits the fact that there are firms engaging in the same activity without applying for CDM credits). First, a baseline may be set using output levels of other firms in the industry that are not engaged in a CDM project. This may be referred to as an intra-industry proxy baseline.

Second, a baseline may be set using the output levels of firms in an industry similar to but not the same as the industry the CDM firm in question belongs to. This may be referred to as an inter-industry proxy baseline. Based on the assumption that the other industry remains unaffected by the CDM in one industry, then the inter-industry proxy baseline can be identified with the ex ante proxy baseline, whereas the intra-industry proxy baseline can be identified with the ex post proxy baseline, for within the industry, other firms' behavior may be affected by the CDM project. These terms will be used for the analysis below.

In a situation where a profit-maximizing private firm operates in an industry where firms strategically interact against one another, different baseline methodologies can lead to different output, profit, credit, total emission, and welfare levels. There are many instances where CDM projects are applied in firms that interact with other firms within industries. The presence of other firms in the same industry has been explicitly mentioned in some approved methodologies and project design documents. In economics, it is customary to suspect interdependent strategic behaviors among firms in an oligopolistic environment. When it comes to CDM methodology-related discussions, however, it is intriguing that there has been little emphasis put on such interdependence of actions among firms.

The interdependence of particular interest to this study is that among firms' output levels. In order to exemplify such output interdependence, a game-theoretic model of oligopolistic industrial organization, the Cournot competition model, is used. It is well known that the interdependence among firms' output levels in an oligopolistic industry can be quite complex and diverse. In terms of rudimentary textbook taxonomy, industries are classified according to the mode of regulation (for example regulated or unregulated industries), the mode of competition (for example perfectly competitive, imperfectly competitive, or monopolistic), or the pattern of organization (for example collusive or antagonistic). Precise data on industrial structures in the non-Annex I countries are unfortunately not easy to come by, though available documentation and reports suggest that significant changes in the industrial structure are taking place.

Since the purpose of this study is to identify the impact of the CDM and its methodologies on the interdependence of outputs, situations where firms' outputs are somehow fixed regardless of the presence of CDM are of little interest. Regulated industry where output levels are subject to regulation, or collusive industry where firms' outputs are fixed by collusive agreement, may fall into this category. For this reason the electricity generation industry, which tends to be heavily regulated, has not been considered here. Even if outputs are interdependent, the details of the interdependence could be quite diverse, as numerous theoretical models indicate.[15] Thus, the presence of empirical evidence is desirable to select an appropriate model while maintaining simplicity for the sake of the demonstration. In fact, an abundance of empirical results have been recorded, especially with the increasing interest being shown in technologies that fall within the scope of "new empirical industrial organization" (NEIO). One typical NEIO method employing a conjectural variation model is very useful for identifying the Cournot model in real-life market situations. That is, whenever the coefficient of conjectural variation is shown to be 0, then the situation is identified with the Cournot model. However, there has been much criticism regarding the use of a static conjectural variation model as a general description of the mode of competition in an oligopolistic industry. To address this problem, numerous different methods have been brought into the discipline to test the mode of competition.

The present analysis draws on the view expressed in the survey by Reimer and Stiegert (2006) that the most persuasive test may be the menu approach, in which several industrial organization models are tested against each other and the model that fits best is considered as the relevant model. Studies of some food industries, including that by Dong et al. (2006), support a quantity-setting oligopoly model, of which the Cournot model is a special case. Deodhar and Sheldon (1996), using other methods, provide evidence supporting application of the Cournot competition model in particular food sectors. In addition, Reimer and Stiegert (2006) list several empirical studies that do not necessarily attempt to identify or assume any particular model from the menu but support the authors' observations concerning the magnitude of the profit margin. Their results indicate that competition in the industries under study is imperfect but market performance is not much different from that of a competitive industry, suggesting that the effects of market imperfection need not be overemphasized. Profit margins were found to be small, implying that collusive behavior may not be prevalent, pointing to the relevance of the noncooperative Cournot model. Based on these studies for food and agriculture industries, it is tentatively concluded that the empirical results tend to support the application of the Cournot competition model, especially when industries are faced with international competition.

Now, project 0455[16] (registered in September 2006) is an umbrella fuel-switching project involving both food and drink industries, and is used here as a good example of a real-life project motivating the use of the Cournot competition model in this analysis. The Cournot competition model has been used as a benchmark in the NEIO and other approaches.[17] So an analysis based on the Cournot approach has its own merit independent of the empirical support.

This is not to claim any particular robustness for the comparative statics results derived from the Cournot model used here. It is acknowledged that the Cournot model is not universally applicable, and that the Cournot assumption by itself does not yield unequivocal comparative statics results—the results can be so diverse that the directions of change depend on a number of other parameters. Such being the case, the central message of this chapter is simply that interdependence among firms can be a factor when considering alternative baseline-setting methods, thus it has to be properly taken into account when contriving methodology-related policies.

Most of the 321 projects officially registered as of October 2006 are either small-scale or non-CO_2-related projects; most of the rest are related to electricity generation or other forms of energy. For these projects, it is harder to judge the applicability of the Cournot model, and thus of this analysis. Small-scale operations are difficult to investigate in detail other than through their project design documents. Non-CO_2-related projects are usually greenfield projects requiring a different dimension of methodology. The electricity and energy sector tends to be subject to regulation although the exact situation may vary. Among the remaining projects, it is reported that the Indian cement industry is quite collusive and that the industry environment is rapidly changing. It appears difficult to capture the situation using a model as simple as the one used in the present analysis.

15.3 Market equilibrium output and emissions in response to CDM

The previous section provided an overview of the consequences of alternative CDM baseline-setting methodologies when firms' outputs are interdependent. This section further elaborates on and exemplifies how exactly the output interdependence is affected by alternative baseline methodologies, while focusing attention on the Cournot duopoly model. First, the notion of Cournot–Nash equilibrium outputs is explained, together with the pre-CDM market equilibrium situation. Second, the linear compensations scheme is explained. Third, an ex ante baseline (inter-industry proxy baseline) that does not depend on the firm's own output is discussed. Fourth, the ex post baseline is considered. It is argued that an ex post baseline exhibits an in-built bias when output is endogenously adjusted in response to the CDM. Fifth, the intra-industry proxy model is discussed as a way to remedy this bias while retaining the virtues of the ex post baseline. Finally, the findings are summarized.

15.3.1 Cournot oligopoly and pre-CDM market situation

Consider a Cournot oligopoly market model where a number of firms producing a homogeneous product compete for customers in an industry. The market demand condition is expressed by an inverse demand schedule that indicates market price for total quantity demanded. In equilibrium, the total quantity demanded equals the total quantity produced, which is the sum of the outputs by each individual firm. Thus the total quantity produced in the industry determines the market price by way of the inverse demand function. Then, the profit of one firm depends not only on its own output, but also on all other firms' outputs by way of the industry output and the market price.

In Cournot competition, each firm maximizes its profits taking other firms' decisions into consideration. Their choices are said to be in Cournot–Nash equilibrium if each firm chooses the output level that maximizes its own profit given other firms' output choices.

For simplicity, it is assumed that demand is linear, and that the marginal production cost of each firm is independent of its output. It is also assumed that the greenhouse gas emission level is proportional to output in each firm. That is, the emission level of each firm equals the product of its own constant emission factor and its own output level.

> *Lemma 1:* Cournot–Nash equilibrium output level x_i^* of firm i in an oligopolistic market consisting of n firms with linear inverse market demand $P(q) = 1 - q$ is given by
>
> $$x_i^* = \frac{1}{n+1} - c_i + \frac{1}{n+1}\sum_j c_j,$$
>
> where c_j is the constant marginal cost of the j_{th} firm. Hence, x_i^* is decreasing in c_i and increasing in c_j $(j \neq i)$.
> The equilibrium emission level is given by $Ex_i^* = e_i x_i^*$, where e_i is a constant emissions factor of firm i.

15.3.2 CDM project and a linear compensation scheme

Suppose only firm 1 is given an opportunity to undertake a CDM investment project, which increases production marginal cost and decreases the emission factor. Firm 1's emission level then equals the reduced emission factor times its output. Firm 1 gets compensated for its emission reduction in the form of the share of CDM CERs it receives. The emission reduction that generates CDM CERs is measured by the amount by which the emission level falls short of the emission baseline level prescribed to the firm. Thus, firm 1 receives compensation net of fixed initial investment cost depending on the emission reduction achieved. For simplicity, it is assumed that the net compensation depends linearly and positively on the reduction.

As in the pre-CDM situation, firm 1's profit depends on its own output and all other firms' outputs. With CDM, however, the outputs affect firm 1's profit also by way of the compensation deriving from the CDM CERs generated. How exactly its own and other firms' output affects the measured emission reduction by firm 1 depends, as shall shortly be expounded, on what baseline scheme is adopted.

15.3.3 Ex ante baseline (inter-industry proxy baseline)

The ex ante baseline is first considered. This baseline is supposed to define the business as usual situation; that is, the CDM baseline should capture what would have happened were it not for the CDM. The ex ante baseline answers this question by setting the emission baseline equal to the product of the pre-CDM emission factor and the pre-CDM equilibrium output level. The ex ante baseline prescribed for firm 1 is therefore independent of the post-CDM output level choice made by firm 1. The larger the output becomes, the larger the resultant emission becomes, and the smaller the measured emission reduction becomes. With this ex ante baseline, firm 1 behaves as if it were faced with an increased marginal production cost compared to the pre-CDM situation, for two reasons. First, the CDM project, while reducing the emission factor of the firm, is assumed to increase its output production marginal cost. Second, because the measured emission reduction is a decreasing function of output, producing one more marginal unit of output now incurs an opportunity cost of foregoing achieving extra emission reduction and its associated compensation income.

The next proposition summarizes the consequences of this ex ante baseline scheme.

Proposition 2: recall the setting of lemma 1. Suppose firm 1 voluntarily undertakes a CDM investment project, which increases production marginal cost c_1 to c_1^{CDM}, and decreases the emission coefficient e_1 to e_1^{CDM} ($< e_1$). Furthermore, suppose an ex ante emission baseline $b^{ex\ ante} = e_1 x_1^*$ (where x_1^* is the equilibrium pre-CDM output level) is adopted. Then, the Cournot–Nash equilibrium output level $x_i^{ex\ ante}$ is given by

$$x_1^{ex\ ante} = x_1^* - \frac{n}{n+1}\Delta c_1^{ex\ ante},$$

$$x_i^{ex\ ante} = x_i^* + \frac{1}{n+1}\Delta c_1^{ex\ ante} \text{ for any } i \neq 1, \text{ where}$$

$$\Delta c_1^{ex\ ante} \equiv c_1^{ex\ ante} - c_1 = (c_1^{CDM} - c_1) + \beta e_1^{CDM} > 0.$$

The total emission reduction of firm 1 is

$$b^{ex\ ante} - e_1^{CDM} x_1^{ex\ ante} = (e_1 - e_1^{CDM})x_1^* + e_1^{CDM}\frac{n}{n+1}\Delta c_1^{ex\ ante}.$$

The intuition behind the proposition above is as follows. In the absence of uncertainties or any predictable change in the environment, the ex ante baseline may be deemed as the true baseline in the sense that it captures the emission level that would have taken place if firm 1 undertook no CDM project. The ex ante baseline is completely independent of ex post output. Then, producing an extra unit of output neither loosens nor tightens the baseline, so that the CDM project only increases the production marginal cost and adds the marginal opportunity cost of foregoing additional units of emission reduction. Because of this effective increase in marginal cost, firm 1's output is reduced.

15.3.4 Ex post baseline

In the previous section it was noted that the CDM methodologies mentioned there essentially resort to what is called the ex post baseline. It was also noted that the ex post baseline level varies ex post facto depending on the scale of operation. That is, unlike the ex ante baseline, the ex post baseline prescribed for firm 1 is not independent of the post-CDM output level choice made by firm 1. To be precise, the ex post baseline is set equal to the product of the pre-CDM emission factor and the post-CDM output of the firm. With regard to this ex post baseline, it is seen that the emission reduction of firm 1 is now an increasing function of firm 1's output. This is in striking contrast to the ex ante baseline case, where a firm's emission reduction is found to be a decreasing function of its own output. On the one hand, the marginal production cost becomes greater because of the CDM by assumption. On the other hand, producing one more marginal unit of output leads now to increased emission reduction. The latter effect may be large enough to offset and even reverse the former effect, leaving the net effect generally ambiguous.

The next proposition summarizes the consequences of this ex post baseline scheme.

Proposition 3: recall the setting of lemma 1. Suppose firm 1 voluntarily undertakes a CDM investment project, which alters its production marginal cost from c_1 to c_1^{CDM}, and its emission coefficient from e_1 to e_1^{CDM} ($< e_1$). Furthermore, suppose an ex post emission baseline $b^{ex\,post} = e_1 x_1^{ex\,post}$ is adopted. Then, the Cournot–Nash equilibrium output level $x_i^{ex\,post}$ is given by

$$x_1^{ex\,post} = x_1^* - \frac{n}{n+1}\Delta c_1^{ex\,post} = x_1^{ex\,ante} + \frac{n}{n+1}\beta e_1,$$

$$x_i^{ex\,post} = x_i^* + \frac{1}{n+1}\Delta c_1^{ex\,post} = x_1^{ex\,ante} - \frac{1}{n+1}\beta e_1 \text{ for any } i \neq 1, \text{ where}$$

$$\Delta c_1^{ex\,post} \equiv c_1^{ex\,post} - c_1 = (c_1^{CDM} - c_1) + \beta(e_1^{CDM} - e_1)$$
$$= \Delta c_1^{ex\,ante} - \beta e_1.$$

Lemma 4: the ex post baseline $b^{ex\,post}$ overstates the emissions that would have taken place without the CDM project, that is,

$$b^{ex\,post} = e_1 x_1^{ex\,post} > e_1 x_i^* = b^{ex\,ante},$$

if $\Delta c_1^{ex\,post} \equiv c_1^{ex\,post} - c_1 = (c_1^{CDM} - c_1) + \beta(e_1^{CDM} - e_1) < 0$. The ex post baseline therefore tends to baseline overestimation when the effect of the production marginal cost increase dominates, and to baseline underestimation when the effect of the emission coefficient decrease dominates.

The total emission reduction amount of firm 1 is

$$b^{ex\,post} - e_1^{CDM}\, x_1^{ex\,post} = e_1 x_1^{ex\,post} - e_1^{CDM}\, x_1^{ex\,post}$$

$$= (e_1 - e_1^{CDM})\, x_1^* - (e_1 - e_1^{CDM})\frac{n}{n+1}\Delta c_1^{ex\,post}.$$

The intuition behind the proposition above is as follows. The ex post baseline depends on the ex post output of firm 1, which is chosen by the firm itself. Then, producing an extra unit of output loosens the baseline by units of emission equal to the pre-CDM emission factor, while increasing actual emissions by units of emission equal to the post-CDM emission factor. Since a CDM project reduces the emission factor, the pre-CDM factor is strictly greater than the post-CDM factor. Thus, the CDM project increases marginal production cost on the one hand, while introducing additional marginal revenue because of the emission factor reduction on the other hand. If the latter emission-saving effect of the CDM is dominant, firm 1's ex post output will become larger than its pre-CDM equilibrium level because of the effective decrease in marginal cost, leading to a baseline overestimation.

15.3.5 Intra-industry proxy baseline

The previous subsection confirms that the ex post baseline is not independent of the CDM project, failing to capture properly what would have happened without the project. On the other hand, the ex post baseline permits dynamic updating of the information on which the baseline is calculated. That is, the ex post baseline is inappropriate because the ex post output is under the direct influence of the CDM project. The ex ante baseline, however, is completely free of CDM influence, but throws away all potentially useful ex post information.

A way to combine the best of the two approaches is to search for a proxy variable that is both reasonably independent and informative. Such a proxy variable may be sought either from within the same industry as firm 1 (intra-industry proxy baseline), or from a similar industry (inter-industry proxy baseline). With regard to the inter-industry proxy baseline, there is much less reason to worry that it may be affected by firm 1's output or the CDM. Therefore its working is analogous to that of the ex ante baseline, which is fixed. With regard to the intra-industry proxy baseline, it should be expected that firms' output levels are interrelated. The chief concern is the extent to which the intra-industry proxy is independent of firm 1's output, and thus of the CDM. The following considers a firm $k(\neq 1)$ that is comparable to firm 1 but is not subject to CDM, and its output is used as a proxy for that of firm 1. So the intra-industry proxy baseline is set equal to the product of the pre-CDM emission factor of firm 1 and the post-CDM output of firm k.

The next proposition summarizes the consequences of this ex post proxy baseline scheme.

Proposition 5: recall the setting of lemma 1. Suppose firm 1 voluntarily undertakes a CDM investment project, which alters its production marginal cost from c_1 to c_1^{CDM}, and its emission coefficient from e_1 to e_1^{CDM} ($< e_1$). Furthermore, suppose an intra-industry proxy emission baseline $b^{proxy} = e_1 x_k^{proxy}$ is adopted, where $c_1 = c_k$. Then, the Cournot–Nash equilibrium output level x_i^{proxy} is given by

$$x_1^{proxy} = x_1^* - \frac{n}{n+1} \Delta c_1^{proxy} = x_1^{ex\ ante},$$

$$x_k^{proxy} = x_1^* + \frac{1}{n+1} \Delta c_1^{proxy} = x_k^{ex\ ante} \text{ for any } k \neq 1, \text{ where}$$

$$\Delta c_1^{proxy} \equiv c_1^{proxy} - c_1 = (c_1^{CDM} - c_1) + \beta e_1^{CDM} = \Delta c_1^{ex\ ante} > 0.$$

Lemma 6: the intra-industry proxy baseline b^{proxy} overstates the emissions that would have taken place without the CDM project, that is,

$$b^{proxy} = e_1 x_k^{proxy} > e_1 x_k^* = e_1 x_1^* = b^{ex\ ante}.$$

The intra-industry baseline therefore tends to baseline overestimation no matter how weak the effect of the production marginal cost increase. This bias, however, vanishes as the number of firms n approaches infinity. Total emission reduction amount of firm 1 is

$$b^{proxy} - e_1^{CDM} x_1^{proxy} = e_1 x_k^{proxy} - e_1^{CDM} x_1^{proxy}$$

$$= (b^{ex\ ante} - e_1^{CDM} x_1^{ex\ ante}) + e_1 \frac{1}{n+1} \Delta c_1^{ex\ ante}.$$

The intuition behind the proposition above is as follows. First, note that the intra-industry proxy baseline has a mixed feature. It does incorporate ex post information, namely the ex post output level of firm $k (\neq 1)$. At the same time, it is nearer to the ex ante baseline in that it is fixed from the viewpoint of firm 1 and is affected by the CDM project only to the extent that firm k's proxy output is indirectly affected by firm 1's output. That explains why firm 1's output is unambiguously smaller than its pre-CDM equilibrium level, and firm k's output is greater than its pre-CDM equilibrium level, leading to the baseline overestimation result.

15.3.6 Discussion

This section briefly summarizes the results obtained above and compares the alternative baseline methodologies.

With regard to the ex ante baseline, it was noted that the CDM baseline is supposed to capture the emissions that would have taken place if firm 1 undertook no CDM project. In this sense, the ex ante baseline may be deemed nearest to what the true baseline should be. Thus, the ex ante baseline becomes completely independent of ex post output. Then, the CDM project increases production marginal cost and adds the marginal opportunity cost of foregoing additional emission reduction. Hence followed the reduction in output.

With regard to the ex post baseline, it was noted that the ex post baseline depends on a post-CDM firm 1 output that the firm itself chooses. Thus, producing more output loosens the baseline. This baseline-loosening effect more than offsets the increase in actual emissions because the CDM reduces the emission factor. Thus, the CDM project increases marginal production cost on the one hand, while introducing additional marginal revenue on the other hand. If the latter emission-saving effect of the CDM dominates, firm 1's ex post output becomes larger than the pre-CDM level, leading to a baseline overestimation.

With regard to the intra-industry proxy baseline, it was noted that the baseline has a mixed feature, incorporating ex post information while being nearer to the ex ante baseline in that it is perceived as fixed from the viewpoint of firm 1, and is affected by the CDM only through the output interdependence between firm 1 and firm k.

Although the ex ante baseline and the intra-industry proxy baseline are not identical to each other, they are both perceived as given from the viewpoint of firm 1. Hence, the result is obtained that the post-CDM firm 1 output under the intra-industry proxy baseline coincides with that under the ex ante baseline, and is smaller than the pre-CDM level. In either case, undertaking a CDM project tends to output reduction because of the production marginal cost increase and the emission credit opportunity cost. The output decrease for firm 1 in turn tends to an output increase for any other firms, including firm k. The intra-industry proxy baseline therefore exceeds the ex ante baseline, and the intra-industry proxy baseline grants a larger amount of emission reduction credit than the ex ante baseline.

On the other hand, the ex post baseline is perceived to depend on the output choice by firm 1. That is, when firm 1 decides its output level, it takes into account that increasing output practically loosens the emission baseline that it is faced with. Thus, the output level unambiguously exceeds the ex ante or intra-industry proxy output level. However, firm 1's post-CDM output under the ex post baseline may possibly fall short of the pre-CDM level if the production marginal cost increase because of the CDM dominates. Therefore, the ex post baseline may actually understate the appropriate (ex ante) baseline.

15.4 A numerical illustration

This section illustrates the analyses conducted in the previous section using a numerical example inspired by project 0455, mentioned in section 15.2.3. The project concerns the food sector, for which there exists some evidence in support of the Cournot competition model, at least in the international market. Furthermore, the project employs baseline methodology AM0008, which makes reference to both ex ante and ex post baseline emission levels.

The project involves conversion from oil to natural gas as a source of fuel in several factories, including those of two firms, say firm A and firm B, operating in the food sector. The data provided by the project design document include the ex ante baseline and the estimated emission reduction. The ex ante baseline employed is called a dynamic baseline because it describes the hypothetical future course of emissions without the CDM, as expected before the project period. The project started in 2004. For simplicity, only two points in time are considered, namely the years 2004 and 2006. Table 15.1 summarizes key data for firm A and firm B, adopted from the project design document. The lower panel indicates that the ex ante baseline oil use without the CDM in 2004 equals 148,386 GJ (gigajoules) for firm A, and 34,173 GJ for firm B, totaling 182,559 GJ. Expected consumption of natural gas with the CDM in 2004 is 147,143 GJ for firm A and 31,447 GJ for firm B, totaling 178,590 GJ. Table 15.1 also indicates that switching fuel from oil to natural gas requires investment of 283 million COP (Colombian pesos) for firm A and 51 million COP for firm B, totaling 334 million COP.[18] The project design document establishes that the project is not commercially viable if no CDM credits are granted, as indicated by the negative net present value figures shown in the table. The above-mentioned fuel consumption data can be converted and expressed in terms of tonnes of CO_2 equivalent (tCO_2e) using emission coefficients publicized by the Intergovernmental Panel on Climate Change and the conversion rate stipulated in the Kyoto Protocol. Then the 2004 baseline emission level becomes 11,389 tCO_2e for firm A and 2,623 tCO_2e for firm B, totaling 14,012 tCO_2e, whereas the 2004 forecast project emission level becomes 9,250 tCO_2e for firm A and 1,977 tCO_2e for firm B, totaling 11,227 tCO_2e. The difference—14,012 tCO_2e minus 11,227 tCO_2e = 2,785 tCO_2e—is the predicted amount of CDM CERs generated.

The data summarized in Table 15.1 by themselves do not suffice to make the theory derived in the previous section applicable to the present case. The following set of auxiliary assumptions and interpretations are therefore introduced:

- The firms behave in accordance with the Cournot competition model with *n* firms. The current scale of operation of each firm is in equilibrium.
- The demand function permits linear approximation in the neighborhood of equilibrium with respect to quantity changes induced by the CDM project.

- The ex ante estimates given in the project design document are calculated under the assumption that the firms maintain their pre-CDM scales of operation throughout the duration of the project.
- Fuel consumption is proportional to the scale of operation.
- Production marginal cost mostly consists of energy marginal cost.
- The capacity constraint does not bind the firms' scale of operation either in the short run or in the long run.
- The CDM project contract makes the CDM firms the residual maximizers. That is, the contract specifies that a fixed amount of benefit should accrue to investing parties other than the CDM firms, while the CDM firms should receive all the residual benefits from the CDM CERs generated.

Table 15.1 Summary data for two food companies in project 0455 for years 2004 and 2006

		Firm A	*Firm B*
Average fuel consumption (GJ/year)[a]		148,386	34,173
Investment (million COP)[b]		283	51
Net present value (without credits: million COP)[b]		−3,802	−1,535
	Firm	*2004*	*2006*
Projected fuel price (COP/MBtu: high scenario):[c]			
Residual fuel oil		9,129	10,228
Natural gas		14,694	17,328
Baseline residual fuel oil consumption (GJ)[a]	A	148,386	177,994
	B	34,173	34,173
Natural gas consumption in the project (GJ)[a]	A	147,143	176,503
	B	31,447	31,447
Project emissions (tCO_2e)[d]	A	8,323	9,983
	B	1,779	1,779
Leakage[e]	A	927	1,112
	B	198	198
Ex ante project emissions (cum leakage)[e]	A	9,250	11,095
	B	1,977	1,977
Ex ante baseline emissions	A	11,389	13,661
	B	2,623	2,623

Sources: Project design document of project 0455: Umbrella Fuel-Switching Project in Bogotá and Cundinamarca (http://cdm.unfccc.int/UserManagement/FileStorage/QLUPBJXGIPMY5WVXK2KIQ28VX3KU3C). The project involves eight firms in different industries and the project is for 10-year periods. Estimated data from 2004 through 2013 are provided in the project design document.

a. GJ = gigajoules.

b. COP = Colombian pesos.

c. The fuel price was adjusted to COP for consistency of the data, although the document lists it as US\$. MBtu = million British thermal units.

d. tCO_2e = tonnes CO_2 equivalent.

e. Leakage means indirect emission increase arising from the emission reduction project (such as transportation demand for fuel for shipping natural gas).

These auxiliary assumptions are admittedly heroic. Consequently, the analyses below should not be construed as judgmental in any way regarding the appropriateness of project 0455 or methodology AM0008. The following discussion is tentative and expository in nature.

Next, an attempt will be made to pin down the effect of alternative CDM baseline methods, namely the ex ante baseline and the ex post baseline, on the equilibrium outputs and emissions. In addition to the data summarized in Table 15.1, in 2004 the price of oil was about US$3.00 per million British thermal units (MBtu, approximately equal to 1 GJ), the price of natural gas was about $5.60 per MBtu, and the price of CDM CERs was about $5.00 per tCO_2e. Recall that Table 15.1 indicates baseline fuel consumption totals for 2004 of about 180,000 GJ for both oil and natural gas. Given the energy prices above, the energy consumptions translate respectively to about $540,000 for oil and $1,008,000 for natural gas. Because of the assumption above that the project design document baseline oil consumption calculation presumes an unchanged scale of operation, the ratio of the total energy costs equals the ratio of average energy cost. Then average energy cost equals marginal energy cost since fuel consumption is assumed to be proportional to the scale of operation. Furthermore, energy marginal cost is assumed to be the primary component of overall marginal cost. Hence, energy conversion from oil to natural gas without the benefit of CDM credit effectively makes the production marginal cost 1.87 (= $1,008,000/$540,000) times larger. This yields a reasonable measure of the ratio of postproject marginal cost to preproject marginal cost under the ex ante baseline in that it presumes that the scale of operation remains unchanged.

Meanwhile, recall that in Table 15.1 the expected total amount of CDM CERs generation in 2004 equaled 2,785 tCO_2e. This translates to $13,925 (= 5.00[$/tCO_2e] × 2,785[tCO_2e]) given the CER price of $5.00 per tCO_2e. Thus, the ratio of postproject marginal cost cum marginal CER revenue to preproject marginal cost is 1.84 (= {$1,008,000–$13,925}/$540,000), which yields a reasonable estimate of marginal cost ratio under the ex post baseline in that it incorporates the effect of postproject change in output on CER income. The ratio of postproject marginal cost to preproject marginal cost is therefore estimated to be 1.87 under the ex ante baseline, and 1.84 under the ex post baseline. The difference between these two numbers is not large, reflecting the low emission price and the low emission reduction amount and high energy costs.

In the light of proposition 2 in section 15.3.3, the ex ante baseline case marginal cost ratio yields the conclusion that output will be smaller than the pre-CDM level by $1.87 \times c \times \frac{n}{n+1}$, where n is the number of firms in the industry and c is the preproject marginal cost (adjusted to make the measurement commensurable to the demand level normalization). Likewise, in the light of proposition 3 in section 15.3.4, output will be smaller than the pre-CDM level by $1.84 \times c \times \frac{n}{n+1}$ under the ex post baseline. Note that these effects on output change from the pre-CDM level become larger as the number of firms in the industry grows larger. At this point, recall that proposition 3 says that the postproject output level could exceed the preproject level if the CER marginal revenue contribution outweighs production marginal cost increase. In this particular instance, however, the CER marginal revenue effect is too weak to reverse the effect of the increased energy marginal cost. As the output of the CDM firm decreases, other firms' outputs increase from their pre-CDM levels by $1.87 \times c \times \frac{1}{n+1}$ under the ex ante baseline, and by $1.84 \times c \times \frac{1}{n+1}$ under the ex post baseline. Thus, total output will decrease by $1.87 \times c \times \frac{n+1}{n+1}$ under the ex ante baseline, and by $1.84 \times c \times \frac{n+1}{n+1}$ under the ex post baseline.

Finally, the effect of baseline choice on the world emission will be considered, commencing with the following accounting relationship:

Total emission = emission by Annex I (A)
+ emission from the CDM firm (EC)
+ emission from other firms in the industry (EO)
+ emission from the rest of the system (R).

(This assumes that firms' activities take place within the country under consideration. If a competing firm's production activity is located in an Annex I country ratifying the Kyoto Protocol, the induced emission reduction arising from a decline in the output of the foreign firm as a repercussion of an increase in the CDM firm's output may be used up by other parties in some Annex I country, so that the effect pointed out could be cancelled out.)

Component (R) presumably remains unchanged by the CDM, and it is likely that the total CER credits will be used up by the Annex I countries to achieve their assignment level (A). Thus, the net effect of a CDM project on the world emission equals:

(Change in emissions by other firms in the industry: dEO)
+ (gap between the baseline emission level and the emission level that would have taken place if the firm did not adopt the project: dB).

For the ex ante baseline, $dB = 0$ because the total emission is evaluated at the old level. For the ex post baseline, dB is generally nonzero because the firm's incentive is shifted to make the output scale larger (or smaller as in the present example).

Furthermore, the world total emission is affected also by changes in outputs produced by other firms in the industry. In response to the operation scale change in the CDM firms, other firms in the same industry (possibly using older technology) would typically adjust their operation scales and emissions, which in turn affects the world total emission. But it seems that this effect is not usually taken into account in baseline calculation. Our tentative numerical calculation indicates that the fuel conversion project implemented by the CDM firms reduced their own output and emissions under either the ex ante baseline or the ex post baseline, but the effect was partly offset by increasing the output and emissions of other firms in the same industry.

Analogous computations can be carried out using the year 2006 data summarized in Table 15.1 in conjunction with year 2006 price data. In the interim, however, the price of oil has risen sharply, and there is a paradoxical situation where oil is more expensive than natural gas. At the same time, the price of CERs has soared to around \$12.00. Given these change in the price data, almost every result obtained for 2004 is reversed for 2006. The post-project marginal cost becomes smaller (as opposed to larger) under both the ex ante baseline and ex post baseline, so the ratio of postproject marginal cost to preproject marginal cost equals 0.71 for the ex post baseline and 0.76 for the ex ante baseline. Output scales of the CDM firms increase (as opposed to decrease) while those of other firms decrease (as opposed to increase). Thus, in this case, the effect of the CDM on the world emission level arising from the reaction of other firms in the same industry works favorably (as opposed to unfavorably).

15.5 Conclusion and policy implications

This chapter has studied how alternative CDM baseline-setting methods perform when applied to a CDM project undertaken by a profit-maximizing firm operating in an imperfectly competitive industry. First, the general background situations pertaining to the Kyoto Protocol and the CDM were explained. The CDM baseline-setting problem was then discussed, showing how different baseline-setting methods, particularly ex ante baseline and ex post baseline, could affect the endogenous choice of output scale and emissions made by the CDM firms as well as other firms in the same industry. This situation was formally analyzed using a standard Cournot oligopoly market model and the intuitions underlying the main results were discussed. Finally, a tentative numerical illustration was conducted applying the analytical results in a numerical example inspired by a real-life CDM project.

Policymakers with different positions and responsibilities may draw different policy implications from the analysis offered in this chapter. National policymakers may be naturally more interested in the CDM CER incomes and overall profits that accrue to firms in their own countries. Policymakers in a CDM host country would also care about the impact on their consumers' welfare. Policymakers in international organizations committed to the preservation of global environmental integrity may regard the world total emission level as most important. With this in mind, the policy implications of the analysis in this chapter will be briefly reviewed and discussed.

First, the analysis starts with a simple observation that the marginal cost (net of CDM CER marginal income) of the firm with the CDM project is lower under the ex post baseline than under the ex ante baseline. CER income is increasing in output scale under the ex post baseline, while it is decreasing in output scale under the ex ante baseline.

Second, it is supposed that interfirm competition in the relevant industry is reasonably well characterized by the Cournot oligopoly model. Then the observation immediately yields the result that the ex post baseline leads to more credits and higher total output than under the ex ante baseline. Furthermore it is straightforward to show that the market price is lower, and thus consumers are better off under the ex post baseline than under the ex ante baseline. It can also be shown that the total profit of the firms in the industry is higher under the ex post baseline than under the ex ante baseline in relatively weak conditions.

Third, with regard to the world greenhouse gas emission level, the net increase in emission is larger under the ex post baseline than under the ex ante baseline because of the larger total output. However, it should also be noted that there is a sense in which this effect on the world emission under the ex post baseline is less pronounced under an oligopoly than under a complete monopoly, because in an oligopoly the output expansion of the CDM firm is counterbalanced by the output contraction of other firms in the same industry. Put another way, the ex post baseline does not punish the CDM firm for its own output expansion, but it does not compensate either for the spillover effect that discourages other firms, which incidentally may well be operating with less energy-efficient technologies.

The above-mentioned policy implications, however, mostly hinge on the validity of the Cournot competition assumption. For that matter, this analysis is not meant to claim any general applicability and should be deemed at best a first-order approximation to far more complex CDM realities. The details of the comparative statics results are by no means robust, and may sensitively depend on model specifications and assumptions. (For instance, within the framework of Cournot competition, perturbing the linearity of the demand schedule at a small scale would leave the analysis more or less intact; but once one allows an arbitrary nonlinear demand schedule, then there could be examples in which some comparative statics results are quite different.) The situation is somewhat analogous to that of so-called strategic trade policy literature (see Brander 1995). When an industry is collusive, the effect of the CDM and the baseline becomes even more difficult to estimate. This analysis has, however, highlighted one important aspect of CDM implementation that has strangely been left unscrutinized, namely the incentive consequences of CDM baseline-setting methods that operate through the interrelationship between firms in an imperfectly competitive market.

As more real-life experiences of actual implementation of CDM are accumulated, further insights will be gained into which specific model is more applicable when and where. But for now, it seems premature to draw any sweeping conclusions, and it is better to remain open minded with regard to the effects of market interactions, even if their full manifestations have not yet been witnessed.

The effect of the CDM on emissions and output in the presence of market interactions could be so complex that no one-size-fits-all general theory could possibly provide a general prescription for all conceivable real-life CDM projects that are to appear in the future. Nevertheless, it is to be hoped that whenever policymakers evaluate individual CDM projects or CDM methodologies, they carefully examine and take into account the relevant market structure and potential market interactions. It is important to keep in mind the market interaction issues that were emphasized in this chapter in the discussion of the future of the CDM within the post-Kyoto regime, and perhaps in the discussion of some future similarly project-based schemes to combat other global or regional environmental problems.

Notes

1. The authors are grateful for the financial aid from the Monbu Kagakusho Grants-in-Aid for Scientific Research, and from the Nikkei Shorei Fund. They also appreciate the precious comments provided by the anonymous referee and the editors.
2. "The ex post calculation of baseline emission rates may only be used if proper justification is provided. Notwithstanding, the baseline emission rates shall also be calculated ex ante and reported in the draft CDM-PDD in order to satisfy the requirements for identification of the elements of a baseline methodology agreed by the Executive Board at its eighth meeting." (EB10 Report Annex 1, 2003, http://cdm.unfccc.int/EB/010/eb10repan1.pdf (accessed January 4, 2007)).
3. Tietenberg 2001 is a convenient volume with a collection of papers on emissions trading.
4. www.epa.gov/airmarkets/arp/index.html (accessed October 24, 2006).
5. http://cdm.unfccc.int/EB.
6. http://ji.unfccc.int/Sup_Committee.
7. The negotiation text of the Marrakech Accords included a number of options. Some issues are covered by Stewart (2000).
8. Tool for the demonstration and assessment of additionality (version 02), November 28, 2005 (http://cdm.unfccc.int/methodologies/PAmethodologies/AdditionalityTools/Additionality_tool.pdf (accessed October 9, 2006)).
9. For the current overview of the CDM and the international carbon market, see Capoor and Ambrosi 2006 and Michaelowa 2005.
10. An emission factor is the estimated average emission rate of a given pollutant for a given source, relative to units of activity.
11. http://cdm.unfccc.int/EB. "AM" means a methodology approved by the CDM-EB. "ACM" means a consolidated methodology that has been approved.

12. http://cdm.unfccc.int/UserManagement/FileStorage/CDMWF_AM_446454474 (accessed October 9, 2006).
13. http://cdm.unfccc.int/UserManagement/FileStorage/CDMWF_AM_2R-F4A6B593SVHDHULWXASWD1OWQ7VB (accessed October 9, 2006).
14. http://cdm.unfccc.int/UserManagement/FileStorage/AM0017_version_2.pdf (accessed October 9, 2006).
15. For instance, in their menu study, Carter and MacLaren (1997) test Cournot (static quantity setting game), Bertrand (static price setting game), and Stackelberg (games with various leader firms for each strategic variable, where a leader firm chooses its strategy first and the follower firms choose their strategies after observing this). Also among methodologies utilized by NEIO, static as well as dynamic versions are considered (see Bresnahan 1989).
16. http://cdm.unfccc.int/Projects/DB/DNV-CUK1150715630.86/view.html.
17. Usually, the Cournot model is regarded as the typical model describing the situation between monopoly and perfect competition and so, in theory, investigation using the Cournot model is applicable to most real-life situations. This is also the case in empirical works, although the Cournot mode of competition is not necessarily the mode that is most frequently confirmed (its prediction is too simple to accommodate the variety of results occurring in reality, as pointed out by Aiginger (1996)). In the NEIO method, the Cournot model corresponds to coefficient 0 in the conjectural variation studies. With approaches using other methods, the Cournot model is still worthy of primary consideration. For instance, most empirical studies surveyed by Richardson (1990) included Cournot as an element of the menu, while Harrison (1994) only considers the Cournot model for her empirical research.
18. 1,000 Colombian pesos = US$0.5 (January 2007).

References

Aiginger, K. (1996) "Confronting the implications of the Cournot model with industry and firm data", *Small Business Economics*, 8: 365–78.

Bohm, P. and Carlen, B. (2000) "Cost-effective approaches to attracting low-income countries to international emissions trading: theory and experiments", Research Papers in Economics 2000–12, Department of Economics, Stockholm University.

Brander, J. (1995) "Strategic trade policy", in G. Grossman and K. Rogoff (eds) *Handbook of International Economics, Volume 3*, Amsterdam: Elsevier Science Publishers.

Bresnahan, T. (1989) "Empirical studies of industries with market power", in R. Schmalensee and R. Willig (eds) *Handbook of Industrial Organization, Volume 2*, Amsterdam: Elsevier Science Publishers.

Capoor, K. and Ambrosi, P. (2006) *State and Trends of the Carbon Market 2006*, World Bank and International Emissions Trading Association.

Carter, C. and MacLaren, D. (1997) "Price or quantity competition? Oligopolistic structures in international commodity markets", *Review of International Economics*, 5: 373–85.

Chomitz, K.M. (1998) *Baselines for Greenhouse Gas Reductions: Problems, Precedents, Solutions*, prepared for the Carbon Offsets Unit, World Bank.

Deodhar, S. and Sheldon, I. (1996) "Estimation of imperfect competition in food marketing: a dynamic analysis of the German banana market", *Journal of Food Distribution and Research*, 27: 1–10.

Dong, F., Marsh, T., and Stiegert, K. (2006) "State trading enterprises in a differentiated environment: the case of global malting barley markets", *American Journal of Agricultural Economics*, 88: 90–103.

Ellis, J. and Bosi, M. (1999) "Options for project emission baselines", OECD and IEA Information Paper.

Fischer, C. (2005) "Project based mechanisms for emission reductions: balancing trade-offs with baselines", *Energy Policy*, 33: 1807–23.

Harrison, A. (1994) "Productivity, imperfect competition and trade reform", *Journal of International Economics*, 36: 53–73.

Imai, H. and Akita, J. (2003) "On the incentive consequences of alternative CDM baseline schemes", in T. Sawa (ed.) *International Frameworks and Technological Strategies to Prevent Climate Change*, Tokyo: Springer Verlag.

Laurikka, H. (2002) "Absolute or relative baselines for JI/CDM projects in the energy sector?", *Climate Policy*, 2: 19–33.

Michaelowa, A. (2005) "CDM: current status and possibilities for reform", HWWI Research Paper 3, Hamburg Institute of International Economics.

Reimer, J. and Stiegert, K. (2006) "Imperfect competition and strategic trade theory: evidence for international foods and agricultural markets", *Journal of Agricultural and Food Industrial Organization*, 4: 1–25.

Richardson, J. (1990) "Empirical research on trade liberalization with imperfect competition: a survey", NBER Working Paper W2883.

Stewart, R. (2000) *The Clean Development Mechanism, Building International Public-Private Partnerships under the Kyoto Protocol: Technical, Financial and Institutional Issues*, United Nations, UNCTAD/GDS/GFSB/Misc.7.

Tietenberg, T. (ed.) (2001) *Emissions Trading Programs Volume I and II*, International Library of Environmental Economics and Policy, Aldershot: Ashgate.

UNFCCC (United Nations Framework Convention on Climate Change) (2002) *Report of the Conference of the Parties on Its Seventh Session, Held at Marrakesh from 29 October to 10 November 2001. Addendum Part Two: Action Taken by the Conference of the Parties*, vols. I and II. Available online: http://unfccc.int/resource/docs/cop7/13a02.pdf (accessed January 21, 2007).

Willems, S. (2000) "Framework for baseline guidelines", OECD Information Paper.

Appendix

Proof of lemma 1

Proof: given the linear demand assumption, the profit of firm i is $\pi_i = (1 - \Sigma_j x_j)x_i - c_i x_i = (1 - c_i - x_i - \Sigma_{j \neq i} x_j)x_i$. This is maximized when $x_i^* = \frac{1 - c_i - \Sigma_{j \neq i} x_j^*}{2}$, given $x_{-i} = x_{-1}^* \equiv (x_1^*, x_2^*, \ldots x_{i-1}^*, x_{i-1}^*, \ldots, x_n^*)$. Thus, in equilibrium, for all i, we have $x_i^* = 1 - c_i - \Sigma_{j \neq i}\ x_j^* - x_i^* = 1 - c_i - \Sigma_j\ x_j^*$, which yields $\Sigma_i\ x_i^* = n - \Sigma_i c_i - n\Sigma_j x_j^*$, that is, $\Sigma_j x_j^* = \frac{n}{1+n} + \frac{1}{1+n}\ \Sigma_i c_i$. Hence, $x_i^* = 1 - c_i - \Sigma_j$ $x_j^* = 1 - c_i - \frac{n}{1+n} - \frac{1}{1+n} \Sigma_i\, c_i = \frac{1}{1+n} - c_i + \frac{1}{1+n} \Sigma_j\, c_j$.

Proof of proposition 2

Proof: since $c_1^{ex\,ante} = c_1 + \Delta c_1^{ex\,ante}$, we find

$$x_1^{ex\,ante} = \frac{1}{n+1} - c_1^{ex\,ante} + \frac{1}{n+1}\left(\sum_{j\neq 1} c_j + c_1^{ex\,ante}\right)$$

$$= \left(\frac{1}{n+1} - c_1 + \frac{1}{n+1}\sum_j c_j\right) - \frac{n}{n+1}\Delta c_1^{ex\,ante},$$

$$x_i^{ex\,ante} = \frac{1}{n+1} - c_i + \frac{1}{n+1}\left(\sum_{j\neq 1} c_j + c_1^{ex\,ante}\right)$$

$$= \left(\frac{1}{n+1} - c_i + \frac{1}{n+1}\sum_j c_j\right) + \frac{n}{n+1}\Delta c_1^{ex\,ante} \text{ for } i \neq 1.$$

Total emissions reduction amount is

$$b^{ex\,ante} - e_1^{CDM} x_1^{ex\,ante} = e_1 x_1^* - e_1^{CDM}\left(x_1^* - \frac{n}{n+1}\Delta c_1^{ex\,ante}\right)$$

$$= (e_1 - e_1^{CDM})x_1^* + e_1^{CDM}\frac{n}{n+1}\Delta c_1^{ex\,ante}.$$

Proof of proposition 3 and lemma 4

Proof: since $c_1^{ex\,post} = c_1 + \Delta c_1^{ex\,post}$, we find

$$x_1^{ex\,post} = \frac{1}{n+1} - c_1^{ex\,post} + \frac{1}{n+1}\left(\sum_{j\neq 1} c_j + c_1^{ex\,post}\right)$$

$$= \left(\frac{1}{n+1} - c_1 + \frac{1}{n+1}\sum_j c_j\right) - \frac{n}{n+1}\Delta c_1^{ex\,post},$$

$$x_i^{ex\,post} = \frac{1}{n+1} - c_i + \frac{1}{n+1}\left(\sum_{j\neq 1} c_j + c_1^{ex\,post}\right)$$

$$= \left(\frac{1}{n+1} - c_i + \frac{1}{n+1}\sum_j c_j\right) + \frac{n}{n+1}\Delta c_1^{ex\,post} \text{ for } i \neq 1.$$

Total emissions reduction amount is

$$b^{ex\ post} - e_1^{CDM} x_1^{ex\ post} = (e_1 - e_1^{CDM}) x_1^{ex\ post}$$

$$= (e_1 - e_1^{CDM}) x_1^* - (e_1 - e_1^{CDM}) \frac{n}{n+1} \Delta c_1^{ex\ post}$$

$$b^{ex\ ante} - e_1^{CDM} x_1^{ex\ ante} = (e_1 - e_1^{CDM}) x_1^* + e_1^{CDM} \frac{n}{n+1} \Delta c_1^{ex\ ante}$$

Proof of proposition 5 and lemma 6

Proof: since $c_1^{proxy} = c_1 + \Delta c_1^{proxy}$, we find

$$x_1^{proxy} = \frac{1}{n+1} - c_1^{proxy} + \frac{1}{n+1}\left(\sum_{j \neq 1} c_j + c_1^{proxy}\right)$$

$$= \left(\frac{1}{n+1} - c_1 + \frac{1}{n+1}\sum_j c_j\right) - \frac{n}{n+1}\Delta c_1^{proxy},$$

$$x_k^{proxy} = \frac{1}{n+1} - c_k + \frac{1}{n+1}\left(\sum_{j \neq 1} c_j + c_1^{proxy}\right)$$

$$= \left(\frac{1}{n+1} - c_k + \frac{1}{n+1}\sum_j c_j\right) + \frac{n}{n+1}\Delta c_1^{ex\ post} \text{ for } k \neq 1.$$

Given $c_1 = c_k$, we find $x_k^* = x_1^*$. Hence $b^{proxy} = e_1 x_k^{proxy} > e_1 x_k^* = e_1 x_1^* = b^{ex\ ante}$. Total emissions reduction amount is

$$b^{proxy} - e_1^{CDM} x_1^{proxy} = e_1 x_k^{proxy} - e_1^{CDM} x_1^{proxy}$$

$$= e_1\left(x_k^* + \frac{1}{n+1}\Delta c_1^{proxy}\right) - e_1^{CDM}\left(x_1^* - \frac{n}{n+1}\Delta c_1^{proxy}\right)$$

$$= (e_1 - e_1^{CDM}) x_1^* + e_1^{CDM} \frac{n}{n+1} \Delta c_1^{proxy} + e_1 \frac{1}{n+1} \Delta c_1^{proxy}$$

$$= (b^{ex\ ante} - e_1^{CDM} x_1^{ex\ ante}) + e_1 \frac{1}{n+1} \Delta c_1^{ex\ ante}$$

Index

Note: *italic* page numbers denote references to Figures/Tables.